A DICTIONARY
OF ZOOLOGY

A DICTIONARY

OF

ZOOLOGY

by

A. W. LEFTWICH, B.SC., F.Z.S.
Senior Biology Master, Wellingborough Grammar School

CONSTABLE AND COMPANY LIMITED
LONDON

D. VAN NOSTRAND COMPANY, INC.
Princeton, New Jersey
TORONTO   NEW YORK   LONDON

First published 1963
Second Edn. (with appendix) 1967
Reprinted 1968

Set in Times Roman type
for Constable and Company Limited, London, W.C.2,
and printed in Great Britain by
Cox & Wyman Ltd., London, Reading, and Fakenham

AUTHOR'S PREFACE

IN compiling the present work the author has had in mind primarily the needs of university students and senior grammar school pupils, but it is hoped that many naturalists and others interested in zoology will also find a use for it. It will be realized that no work of this kind can ever be as complete as could be desired. New terms are being added every day to the scientific vocabulary and, even if it were possible to include them all, there would be an inevitable tendency to give short or ambiguous definitions. This, in the author's opinion, is the principal defect common to most dictionaries of science and, as far as scientific terms are concerned, to the ordinary English dictionary. Bearing this in mind the author has endeavoured to be as complete and unambiguous as possible within the space of a single volume. Of necessity he has had to assume that the reader will be familiar with the basic biological terms to be found in any school textbook.

Most dictionaries of science confine themselves to defining in general terms the words in common use; medical dictionaries only include anatomical terms relating to the human body. Other dictionaries of biology may define the same terms in a broader sense relating them to comparative anatomy and to evolution but the number of organs and structures so defined is usually severely limited. Names of groups and families of animals are, with few exceptions, entirely omitted.

The author has, he believes for the first time, included definitions of all the principal phyla and classes of animals as well as a large number of orders, suborders and families. He has not included any particular genera or species unless these have features of very special interest. A student wishing to obtain an accurate definition of a particular group of animals, for example *Aphetohyoidea*, *Ornithischia* or *Pentatomidae*, would normally have to refer to a detailed textbook of zoology, palaeontology or entomology. The author has attempted to remedy this defect and to include most such terms in the present work.

With changing ideas and differences of opinion on the subject of classification, many names of groups of animals have become or are becoming obsolete. In the course of his reading, however, the student of zoology will meet many such words. The author has therefore included a large number but has mentioned newer or alternative names where these exist.

Most of the words used in zoology are derived from Greek and

Latin; many more from the former than from the latter. The author has found in the course of many years of teaching experience that students with no knowledge of these languages will spell words incorrectly; will fail to perceive obvious relationships between words and will regard each name as an individual entity to be learnt parrot-fashion. A study of the classics is the obvious way of overcoming these difficulties but for better or for worse such studies are dying a lingering death in the grammar schools. The author has adopted a method which may be helpful, namely that of including a number of root-words and prefixes with their meanings in their natural position in the dictionary. For example the meaning of *Odonto-* will be found before Odontoblast, Odontoceti, Odontoid, etc.; that of *Cten-* before Ctenidia, Ctenophora and Ctenopoda. Further information which may be helpful will also be found in the Appendix on Classification and Nomenclature. The difficult matter of the endings of words is also discussed in this appendix.

Finally the author would like to acknowledge his gratitude to the many authors whose works he has consulted in the compilation of this book. There are some fifty of these, both English and American, and it would be invidious to pick out any for particular mention. Most of the standard textbooks of zoology and many more specialized works have been consulted. The facilities offered by the library of the Zoological Society have been found most helpful. The author would like to express his thanks also to his colleague and friend Mr. R. E. Knight who has read the manuscript through and has offered many helpful criticisms and suggestions. Inevitably some mistakes will have crept in and there will no doubt be some important omissions. The author would be most grateful if readers would draw his attention to these.

A.W.L.

PREFACE TO SECOND EDITION

THE present edition of *A Student's Dictionary of Zoology* has been renamed *A Dictionary of Zoology* firstly because the publishers felt that this latter title gave a better indication of the scope of the book, and secondly to be a companion volume to George Usher's *Dictionary of Botany*. The book is substantially unaltered but about 450 new definitions and a few corrections have been added in the form of a supplement. A number of critics have suggested that the book would be improved by the inclusion of more terms relating to physiology and ecology. We have therefore included about 50 of the former and 40 of the latter in the supplement. From the outset, however, we decided to limit the scope of the present work by excluding elementary and well known words as well as those commonly found in medical and chemical dictionaries. In view of recent important discoveries, however, we have included definitions of DNA, RNA and the Genetic Code. Many biochemical terms and the chemical formulae of physiological substances omitted by us may be found in Usher's *Dictionary of Botany* mentioned above.

We should like to thank the following who have been kind enough to draw our attention to a few errors and have made many helpful suggestions: *The Times Educational Supplement*, Mr. W. B. Yapp of Birmingham University (in J. Inst. Biol.), Dr. Ivan Goodbody of the University of the West Indies, Dr. E. Elkan of the British Herpetological Society, Mr. Marcus Chambers, Mr. Bernard Longden and last but not least, my colleague Mr. M. J. Parkin who has given invaluable help in sections dealing with physiology and genetics. Most of the additional matter suggested by these biologists has been incorporated in the supplement. We have also included a number of new words most of which have been coined and have appeared in the literature within the last fifteen years. Such words are being collected by Dr. N. W. Pirie, F.R.S. on behalf of the Biological Council and have been published in the *Journal of the Institute of Biology*.

Finally may we reiterate that we should welcome any suggestions by readers for improving or enlarging the present work. All such suggestions will be carefully considered when preparing subsequent editions.

A. W. LEFTWICH.

Wellingborough.
August, 1966.

A

A-, AB-. Latin prefix: off, from, away, apart.

A-, AN-. Greek prefix: not, without.

ABACTINAL (ABORAL). The surface of the body opposite to the mouth in an animal showing radial symmetry.

ABAMBULACRAL SURFACE. The upper or 'dorsal' surface of a star-fish or similar Echinoderm, *i.e.* the surface opposite to the ambulacral grooves (see *Ambulacral*).

ABDOMEN. (1) The part of the body of a vertebrate between the thorax and the pelvic girdle.
(2) The posterior part of the body of an insect or other Arthropod.

ABDOMINAL PORES. Small openings connecting the coelom with the cavity of the cloaca in fish, some amphibia and reptiles.

ABDUCENS NERVE. The sixth cranial nerve of a vertebrate, innervating the posterior rectus muscle of the eye.

ABDUCTOR. A muscle which pulls a limb away from the main axis of the body (*cf. Adductor*).

ABOMASUM. The fourth part of the stomach of a ruminant, sometimes called the 'rennet stomach'.

ABORAL. The side of an animal away from the mouth in those animals which have no definite dorsal and ventral surfaces, or whose dorsal and ventral surfaces change position in the course of development (*e.g.* in a star-fish).

ABRAMIDINA. Bream and similar fresh-water fish having the anal fin elongated and the abdomen compressed. They belong to the group Ostariophysi which have the air-bladder connected to the ear by a chain of small bones.

ACANTH-, ACANTHO-. Prefix from Greek *Akanthos*: a thorn, spine.

ACANTHARIA (ACTIPYLAEA). A suborder of Radiolarians whose skeleton is said to be composed of spicules of Strontium Sulphate. The group is recognizable by the fact that the pores in the central capsule, through which the pseudopodia pass, are gathered together in clusters.

ACANTHOBDELLIDAE. Minute barnacles whose females live in hollows excavated in the shells of molluscs; and whose males are degenerate (complemental males) living within the mantle cavity of the female (see *Complemental Male*).

ACANTHOCEPHALA. Spiny-headed worms. Parasitic worms similar to Nematodes in structure but having an eversible proboscis provided with hooks for attachment to the intestines of the host. The larvae usually inhabit small crustaceans and only develop into adults when the first host is eaten by a vertebrate. A common species, for example, is found as a larva in fresh-water shrimps and as an adult in ducks.

ACANTHOPTERYGII. Spiny-rayed fish. A large class comprising the majority of salt-water bony fish; varying considerably in size and shape but all having

1

the dorsal and ventral fins supported by rigid spines. The pectoral girdle is attached to the skull by the post-temporal bones and the pelvic fins are usually displaced far forwards.

ACARINA. Ticks and mites. Arachnids, mostly minute, having a rounded body in one piece and having four pairs of legs without gnathobases. Most are either scavengers or ectoparasites. The blood-sucking ticks usually have an elongated proboscis but free-living mites often have mouth-parts which are leg-like, clawed or sensory.

ACCIPITRINAE. Hawks. Carnivorous birds with curved, hooked beaks, powerful claws and a feathered head and neck. The latter distinguishes them from vultures which have these parts naked.

ACCOMMODATION. Changing the focus of the eye by altering the curvature or the position of the lens.

ACEPHALOUS. Without a head. The term is also used to describe creatures whose heads are much reduced or are invaginated into the thorax, as in certain insect larvae.

ACETABULUM. The socket for the articulation of the femur in the pelvic girdle of a tetrapod vertebrate. It is made up of parts of the ilium, ischium and pubis.

ACETYLCHOLINE. A chemical substance produced at the synapses and nerve endings of the parasympathetic nerves. Its effects are the exact opposite to those produced by *sympathin*, a substance similar to adrenalin, which is produced by the sympathetic nerves. Acetylcholine, for example, slows down the heart and sympathin speeds it up.

ACICULUM. An internal needle-like chaeta or bristle giving support to the external chaetae on certain Polychaete worms. Free-swimming polychaetes such as *Nereis* have on each segment a pair of flap-like projections known as parapodia on which are a number of external chaetae. Each parapodium contains one aciculum thicker than the ordinary chaetae, having a skeletal function and acting as a point of origin for the parapodial muscles.

ACIDOPHIL. A cell whose cytoplasm contains granules readily stainable with acidic stains such as eosin. Examples are certain types of white blood corpuscles.

ACIPENSEROIDEI (ACIPENSERIDAE). Sturgeons. Large bony fish inhabiting fresh or salt water in the Northern Hemisphere: having a heterocercal or unsymmetrical tail, an elongated snout without teeth and having five rows of bony plates running the length of the body in the skin.

ACOELA. An order of marine Turbellaria (free-swimming flatworms) having no functional pharynx and no gut-cavity. The inability to feed in the normal way is overcome by having large numbers of symbiotic algae embedded in the body. These carry out photosynthesis and so provide a source of nourishment for the animal.

ACOELOMATE. Without a coelom, as for example the Platyhelminthes, Nematodes, Rotifers and a few other simple invertebrates.

ACONTIA. Stinging threads attached to the mesenteries within the body-cavities of certain sea-anemones. When the animal is disturbed they may be shot out of the mouth or out of special pores in the body-wall. Stinging is brought about by numerous nematoblasts on these threads.

ACOUSTICO-LATERALIS SYSTEM. The somatic afferent components of the seventh, eighth, ninth and tenth cranial nerves of vertebrates; bringing sensory impulses from the special receptors of the ear, skin and lateral line (if present) to the hind-brain.

ACR-, ACRI-, ACRO-. Prefixes from Greek *Akros*: top, point, tip.

ACRANIA. Chordate animals without a true brain or skull: comprising the Hemichordata or Acorn-worms, the Urochordata or Sea-squirts and the Cephalochordata or Lancelets. Sometimes the name Acrania is restricted to the latter group.

ACRASIALES. Masses of amoeboid cells clustered together.

ACRIDOIDEA. Grasshoppers and crickets: one of the four superfamilies of Orthoptera, distinguished from other insects of this order by having large hind femora for jumping. Stridulation is a characteristic feature of the males and there are organs of hearing in both sexes.

ACRIDIDAE. Short-horned grasshoppers and locusts: one of the families of Acridoidea distinguished by having antennae shorter than the body. Most of them have the habit of migrating in large numbers.

ACROCEPHALY. Having a dome-shaped head. A term used in anthropology (*cf. Platycephaly*).

ACRODONT. A condition present in certain reptiles in which the teeth are fused to the edge of the jaw-bone.

ACROMION PROCESS. A continuation of the spine of the scapula forming a projection for articulation with the clavicle in many mammals and birds.

ACROSOME. The pointed anterior projection of a spermatozoon by which it is able to penetrate the ovum.

ACROTROPHIC OVARIOLES. A type of ovariole found in certain beetles and plant-bugs in which the nutritive cells are only at the apex of each ovariole. They are connected to the developing eggs by strands of protoplasm which become gradually longer as the eggs progress towards the oviduct.

ACTIN-, ACTINO-. Prefix from Greek *Aktis, aktinos*: a ray.

ACTINISTIA. An order of fish represented by the Coelacanthidae (*q.v.*) which range from the Upper Devonian Period to the present day. The body and head are covered with bony plates and the air-bladder is also ossified. The paired fins consist of short blunt scaly lobes around which the fin-rays are attached like a fan (see *Coelacanthidae*).

ACTINOPTERYGII. A large class comprising all the commonest and most familiar fishes: cod, herring, eel, salmon etc., and in addition some, such as the sturgeons and gar-pikes, which used to be termed 'ganoids'. The name is given from the fact that the paired fins are only supported by horny fin-rays and have no bony or cartilaginous skeleton.

ACTINOTRICHIA. Small unjointed horny fin-rays at the edges of the fins in many bony fish. They correspond to the *ceratotrichia* of cartilaginous fish and are distinct from the *lepidotrichia* or bony supports which form the main skeletal framework of the fin.

ACTINOZOA (ANTHOZOA). Solitary or colonial Coelenterates without any free-swimming medusa stage and with the coelenteron divided by mesenteries. The group includes most of the common sea-anemones and corals.

ACTINULA. A type of larva produced by certain colonial Coelenterates such as *Tubularia*. It is like a small polyp, complete with tentacles, developing within a cup of the gonophore.

ACTIPYLAEA. See under *Acantharia*.

ACULEATA. The stinging forms of Hymenoptera including bees, ants and wasps. They belong to the group Apocrita which have a constricted waist and an ovipositor modified to form a sting.

AD.   Latin preposition used as a prefix: to, at, towards, into.

ADAMANTOBLAST (AMELOBLAST).   An enamel-forming cell in a tooth.

ADAMBULACRAL OSSICLES.   Small skeletal plates on either side of the ambulacral grooves of a star-fish near the tube-feet.

ADAMBULACRAL SPINES.   Spines standing on the adambulacral ossicles of a star-fish.

ADDUCTOR.   (1) A muscle which pulls a limb towards the main axis of the body.
    (2) A muscle which pulls two parts together, as for example those which close a bivalve shell.

ADECIDUATA.   Mammals in which the mucous membrane of the uterine wall does not come away with the placenta after the birth of a foetus.

ADELGINAE.   Parasitic plant-lice of minute size producing galls at the bases of young conifer shoots.

ADELOCODONIC.   A term applied to coelenterate zooids which remain attached to the colony, as in the corals.

ADEPHAGA.   One of the two suborders of Coleoptera; the other being the Polyphaga. Adephaga usually have filiform antennae and five-jointed clawed tarsi. The larvae are of the *campodeiform* type (*q.v.*). The group includes the large water beetle *Dytiscus*, the ground beetles *Carabus* and *Calosoma*, the Tiger Beetle *Cicindula* and the aquatic Whirligig Beetle *Gyrinus*.

ADIPOSE FIN.   The posterior dorsal fin of salmon and similar fish; containing much fatty tissue and having delicate unjointed horny rays instead of the usual jointed bony fin-rays.

ADIPOSE TISSUE.   Connective tissue containing large cells in which are stored droplets of fat.

ADORAL.   The side of an organism on which the mouth is situated, *e.g.* the under side of a star-fish.

ADRADIAL.   The positions on the disc of a jelly-fish halfway between the per-radial and the inter-radial positions (see *Radial Canal* and *Inter-radial*).

ADRENAL GLANDS.   Endocrine glands situated near the kidneys in all higher vertebrates; secreting a number of hormones, the chief of which is *adrenalin*. In lower vertebrates such as fish there are two sets of glands: *inter-renals* between the kidneys and *supra-renals* above the kidneys. The adrenals of higher animals are formed by a fusion of these two.

AEGITHOGNATHOUS.   A term used by Huxley in the classification of birds to indicate that the maxillo-palatine bones do not unite with one another or with the vomer; the latter being truncated in front. The group includes more than half the total number of living birds, the chief being the *Passeriformes*.

AELUROIDEA.   A group of carnivorous mammals comprising three families, *viz.* Felidae (cats, lions, tigers etc.), Viverridae (civets), and Hyaenidae (hyaenas). The name *Feloidea* is tending to supersede Aeluroidea (see Appendix on Classification).

AEPYORNIS (AEPYORNITHIDAE).   Large running birds from Madagascar which have probably become extinct in geologically recent times. They were considerably larger than the ostrich and differed from it in having four-toed feet. The eggs, which are frequently found in the sand and gravel of lakes, had a capacity of more than two gallons.

AEROBE (AEROBIC).   An organism which can only live in the presence of free oxygen.

AEROBIC RESPIRATION. See *Respiration.*

AESTHETASCS. Sensory or olfactory hairs on the antennae or antennules of many arthropods.

AESTHETES. Sense organs having the structure of simple eyes; found scattered over the dorsal surface of the shell plates in certain molluscs of the class Amphineura.

AFFERENT. Conducting towards any part of the body:
(1) Nerves which conduct impulses towards the spinal cord or brain.
(2) Vessels which convey blood towards a particular organ such as the gills of a fish.

AFTERBIRTH. See *Placenta.*

AFTERSHAFT (HYPORACHIS). A small duplicate vane arising from the base of the rachis in certain feathers. It is normally found on *contour feathers,* the small feathers arranged in rows over the surface of a bird's body.

AGAMETES. An alternative name for *Schizozoites* or *Merozoites;* spore-like cells produced asexually by certain protozoa.

AGAMIDAE. Lizards of the Old World: without bony scales, with teeth consisting of incisors, canines and molars, and whose eyes have lids.

AGAMOGONY (SCHIZOGONY, MEROGONY). Asexual reproduction or cell-division of a protozoon to form schizozoites or agametes.

AGAMONT. A 'vegetative' stage in the life-cycle of many protozoa in which they reproduce asexually an indefinite number of times before giving rise to *gamonts* which reproduce sexually.

AGGLUTINATION OF BLOOD. A clumping together of the red corpuscles when blood from two different animals is mixed or when blood of different groups within the same species is mixed. It is due to the interaction of two substances: an *agglutinin* in the plasma and an *agglutinogen* in the red corpuscles.

AGLOSSA (AGLOSSINAE). Frogs and toads which have no tongue, have complex lungs and are entirely aquatic with broadly webbed feet. The tadpoles have two spiracles but have no true internal gills. The gill-slits are apparently used for filtering micro-organisms which form the food. External gills, present when the tadpole hatches, disappear at a very early stage to be replaced by lungs. The African Clawed Toad *Xenopus* and the Surinam Toad *Pipa* are in this group.

AGLYPHA. Non-poisonous snakes belonging to the family Colubridae; distinguished from others of the group by having all the teeth solid and not grooved.

AGNATHA. Fish-like creatures with no true jaws. They include the Cyclostomes (lampreys) and the fossil Ostracoderms (*q.v.*).

AGROMYZIDAE. Cambium Borers. Minute flies whose small white larvae bore into the wood of willow trees. They belong to the order Diptera and the suborder Cyclorrhapha.

AIR BLADDER (OF FISH). A sac-like structure, varying much in shape and size, lying dorsal to the alimentary canal in most bony fish. It is sometimes connected to the oesophagus by a duct which may, however, close up before the fish reaches maturity. Its main function is to give buoyancy to the fish but in the Dipnoi or Lung-fish it is highly vascular and can be used for respiration when the fish is out of water.

AIR SACS. (1) Alveoli of the lungs: millions of small sacs forming the main part of the lungs of mammals and birds. By dissolving in a thin film of water, oxygen and carbon dioxide are able to diffuse into and out of the blood through the walls of these alveoli.

(2) Air sacs of birds. These are large bags acting as air-reservoirs connected to the lungs and to the cavities of the bones. During respiration the compression and expansion of these sacs causes air to be swept rapidly through the alveoli of the lungs. There is thus a very efficient respiratory exchange.

(3) Air sacs of insects. Thin-walled dilatations of the tracheae which, by their elasticity, give increased ventilation to the respiratory system in many swiftly flying insects.

AISTOPODA. Extinct snake-like amphibians of the Carboniferous Period without limbs or pectoral girdle.

ALAR. Pertaining to wings.

ALARY MUSCLES (OF INSECTS). A series of small triangular wing-like muscles inserted in the pericardial walls of an insect. The contraction of these muscles causes blood to flow from the perivisceral cavity into that of the pericardium and thence through the small openings or *ostia* into the long dorsal tubular heart.

ALA TEMPORALIS. The ascending process by which the upper jaw is attached to the neurocranium in the mammalian embryo (see *Autostylic*). It later becomes ossified to form the alisphenoid bone.

ALAUDIDAE. Larks: small brown passerine birds, nesting on the ground; having short tail, broad wings and beak of moderate length.

ALBINISM. Absence of pigment in skin, hair, eyes, feathers etc.

ALBUMEN. The chief protein contained in white of egg. The name is also sometimes given to any similar protein; as for instance one of those contained in blood.

ALCEDINIDAE. Kingfishers. Birds with large head, short wings and tail and a long beak: the group includes the green kingfisher of Europe, the black and white kingfisher of Africa and the Kookaburra of Australia.

ALCIDAE. Auks, razorbills, guillemots, puffins and similar marine fish-eating birds with short wings, heavy body and webbed feet. The Great Auk, a flightless bird, has been extinct since about 1860.

ALCYONARIA. Polyps which are either solitary or in colonies and are distinguished from others by having eight internal mesenteries and eight feathery tentacles. There is usually an internal skeleton of calcareous spicules embedded in the mesogloea. 'Sea-fans' and 'sea-pens' belong to this group and also the corals known as 'Dead Men's Fingers', 'Organ Pipe' and 'Precious Coral'. The majority of large corals, however, belong to another group, the Zoantharia.

ALEPOCEPHALIDAE. Deep-sea salmon-like fish without adipose fin or air bladder and sometimes having small phosphorescent spots.

ALEYRODIDAE. Minute white flies of the suborder Homoptera, having wings and body covered with fine powdery wax. The best known British species are the Cabbage White-fly and the Greenhouse White-fly. The latter, which is very destructive to cucumbers and tomatoes, has been a pest in England for about forty years but may have been introduced originally from Brazil.

ALIENICOLAE. Parthenogenetic viviparous female aphids, generally wingless, which reproduce in enormous numbers on herbaceous plants throughout the spring and summer. In the autumn some winged forms known as *sexuparae* are produced and these usually fly to an evergreen shrub where they give rise to normal males and females. Mating takes place and eggs are laid from which new generations of alienicolae emerge in the following spring.

ALIMENTARY CANAL. The gut or passage through which food passes and in which it is digested and absorbed. It may consist of a number of regions: oesophagus, stomach, intestines etc., but usually the actual digestion takes place only in the central part known as the mid-gut or mesenteron which is lined with cells of endodermal origin. At the anterior and posterior ends there is frequently an invagination or folding in of epidermal cells forming the *stomodaeum* and the *proctodaeum* respectively.

ALISPHENOID. The principal bone forming the side of the middle region of the mammalian cranium; probably homologous with the *Epipterygoid* of a reptile. It is bounded by the *Basisphenoid* below and by the *Parietal* above.

ALISPHENOID CANAL. A passage by which the External Carotid Artery passes through the alisphenoid bone.

ALLANTOIC BLADDER. A urinary bladder formed from an outgrowth of the cloaca; homologous with the allantois of higher vertebrates. Such a bladder is found, for example, in the frog.

ALLANTOIC PLACENTA. A placenta which is largely formed from the allantois (*q.v.*), as in most mammals.

ALLANTOIS. A sac or bladder which grows out from the hind-gut in the embryos of reptiles, birds and mammals. As it enlarges it becomes very vascular, extending outside the embryo, and is the chief organ by which dissolved food and oxygen are absorbed. In birds and reptiles it partly surrounds the yolk-sac and draws food from it, but in mammals it fuses with the *chorion* (*q.v.*) and helps to form part of the placenta.

ALLELOMORPHS. Pairs of contrasted characters which are inherited according to the principles of Mendel, one character usually being dominant and the other recessive. They are carried by genes on homologous chromosomes.

ALLOIOGENESIS. Alternate sexual and asexual reproduction.

ALLOTHERIA. Primitive mammals of the Jurassic Period. The fossil remains are very fragmentary and have been classed with the Monotremes and with the Marsupials.

ALTRICES. Birds such as doves and pigeons which are hatched naked or nearly so and are for some time incapable of looking after themselves.

ALULA. (1) Of Birds: see Bastard Wing.
(2) Of Insects: a part of an insect's wing forming a small separate lobe close to the body.

ALVEOLI. (1) Minute air-sacs which make up the lungs of a vertebrate.
(2) Acini or secretory cavities of a gland.
(3) Any tiny pit or cavity.

AMBERGRIS. A grey wax-like secretion found in the alimentary canal of the sperm-whale: used in the manufacture of perfume.

AMBLY-. Prefix from Greek *Amblus*: dull, dim.

AMBLYCEPHALIDAE. Snakes of Central and South America and S.E. Asia; resembling poisonous snakes in the features of the head and neck, but in fact quite harmless.

AMBLYOPSIDAE. Small pike-like fish, frequently blind and often without skin-pigment; inhabiting swamps and subterranean streams of North America.

7

AMBLYPODA. Large extinct mammals of the Eocene period having five toes with hoofs; probably they were ancestral Ungulates.

AMBLYSTOMATINAE. Newt-like amphibians some of which show the phenomenon of *neoteny* or delayed metamorphosis. The well known *Axolotl* is an example of this. It can retain its larval form with external gills, median fin and aquatic habits, while becoming sexually mature and able to reproduce itself. The phenomenon is caused by insufficient thyroid secretion; when thyroxin is administered a rapid loss of gills and growth of lungs and other features of a land animal occur.

AMBULACRUM. One of the five radial bands on which the tube-feet are borne on the lower surface of a starfish or similar Echinoderm.

AMBULACRAL GROOVE. A groove down the centre of each ambulacrum of a star-fish: containing radial nerves, water-vessels and haemal strands which run to the tip of each of the five arms. The tube-feet are borne in rows on each side of this groove.

AMBULACRAL OSSICLES. Transverse dermal ossicles situated in the roof of each ambulacral groove of a star-fish.

AMBULACRAL SURFACE. The lower or 'ventral' surface which bears the ambulacra and tube-feet of a star-fish. The terms *ambulacral* and *abambulacral* are preferable to dorsal and ventral because they do not correspond to the original dorsal and ventral surfaces of the larva.

AMELOBLAST. See *Adamantoblast*.

AMETABOLA. Insects which undergo little or no metamorphosis. When the young emerge from the eggs they closely resemble the parent and there is little change of form at each instar. Insects of this type, being primitively wingless, are also known as Apterygota: they include the Thysanura or Bristle-tails and the Collembola or Spring-tails.

AMIOIDEA (AMIIDAE). An order of fish comprising many extinct forms and represented at the present day by a single genus *Amia*, the Bow-fin from the fresh waters of the United States. In many ways this fish shows a mixture of primitive and advanced characters: it has a dorsal 'lung' by which a certain amount of air-breathing is possible; a vestigial spiral valve in the intestine, and a small valvular *conus* in the heart. The skin is covered with thin cycloid scales and the tail is practically homocercal.

AMITOSIS. Cell-division by simple constriction of the nucleus without the formation of chromosomes.

AMMOCOETE. The larva of a Cyclostome such as the lamprey which shows many primitive features and in some respects, notably the ciliary method of feeding, resembles *Amphioxus*.

AMMODYTIDAE. Sand-eels: small elongated carnivorous fish which swim in shoals near the shore and bury themselves in the sand.

AMMONITES (AMMONOIDEA). A large group of fossil molluscs of the class Cephalopoda; having a spiral shell divided by complex suture lines to form a number of chambers. Many species occurred in the later Palaeozoic and throughout the Mesozoic epochs, but they apparently became extinct in the late Cretaceous period. The nearest living relative is the *Nautilus*.

AMNION. The inner of the two foetal membranes in reptiles, birds and mammals; the outer being the chorion. The amnion and the chorion, which are at first continuous with one another, develop as folds of blastoderm rising up on each side of the developing embryo; meeting above it and enclosing it. The space between the embryo and the amnion is filled with fluid and is known as the amniotic cavity.

AMNIOTA. Reptiles, birds and mammals, *i.e.* land-living vertebrates whose embryos have an amnion, chorion and allantois.

AMNIOTIC CAVITY. See *Amnion.*

AMNIOTIC FLUID. The fluid filling the amniotic cavity (see *Amnion*).

AMOEBINA. An order of Protozoa consisting of a naked mass of protoplasm without shell or other skeletal structure; moving and feeding by the action of simple lobular pseudopodia, and reproducing by binary or multiple fission.

AMOEBOID. Having the qualities of an *Amoeba*: *i.e.* moving and feeding by the action of pseudopodia.

AMOEBOCYTES. Cells with amoeboid properties, *e.g.* certain types of white blood corpuscles.

AMOEBULAE. Small amoebae which emerge after multiple fission within a cyst.

AMPHI-. Greek prefix: on both sides.

AMPHIASTER. The twin asters which give rise to the spindle in cell division.

AMPHIBIA. Cold-blooded, smooth skinned vertebrates which start life as an aquatic larva breathing by means of gills and later undergo a somewhat rapid metamorphosis into the adult lung-breathing form. Frogs, toads and newts are well known examples.

AMPHIBIOTIC. Amphibian.

AMPHIBLASTIC. An alternative word for *telolecithal*: a term used to describe eggs which have most of the yolk on one side (the vegetative pole), and the nucleus on the other side (animal pole).

AMPHICOELOUS. Biconcave: a term used to describe vertebrae in which the centrum has a conical depression on both its anterior and its posterior face. Such vertebrae are found in most fish.

AMPHIGONY. Sexual reproduction.

AMPHINEURA. A primitive group of Mollusca probably very near to the hypothetical ancestral type; fossils similar to present day species have been found in Ordovician strata. The slug-like body is symmetrical and without a visceral hump. The head is minute and without eyes or tentacles. The loose dorsal skin forms a mantle which, in some types, secretes a shell of transverse plates; in others it merely contains a number of spicules. The name Amphineura is given on account of the simple nervous system which consists of a pharyngeal collar like those of Annelids and Arthropods, from which proceed two pairs of longitudinal cords with ladder-like connectives.

AMPHINUCLEOLUS. A nucleolus containing both *plastin* (which takes acid stains) and *chromatin* (which takes basic stains).

AMPHINUCLEUS. A nucleus containing two haploid sets of chromosomes.

AMPHIPNEUSTIC. (1) Of vertebrates: breathing with gills and with lungs at some stage in the life-history.
(2) Of insects: having only two pairs of functional spiracles; one anterior and the other posterior. Such a state is found in the larva of the common house-fly.

AMPHIPODA. Crustacea which include many fresh-water and marine shrimps and sand-hoppers. They belong to the class Malacostraca, subclass Peracarida and have the following features: a laterally compressed body with no carapace over the thorax; limbs of a variety of types, prehensile, hooked, flattened etc.; a brood-pouch for carrying eggs and young, formed from flattened coxae or *oostegites* on some of the thoracic limbs.

AMPHISTYLIC SKULL. A type of skull found in a few cartilaginous fish in which the upper jaw is attached to the neurocranium by both the Otic Process and the Hyomandibula (*cf. Hyostylic* and *Autostylic*).

AMPHITHERIIDAE. Primitive mammals of the Jurassic and Cretaceous Periods; apparently Marsupials of the Opossum type.

AMPHITOKY. Production of males and females parthenogenetically.

AMPHIUMIDAE. Large newt-like amphibia with small limbs; with teeth in both jaws and without eyelids. Some are entirely aquatic and some live in swamps. The American 'Hellbender' and the Japanese 'Giant Salamander' are included in this group. The latter, the largest known living amphibian, can be up to five feet in length.

AMPHORIDEA. A primitive type of fossil Echinoderm dating from the Ordovician Period. They were sac-like and sessile resembling a Crinoid without a stalk.

AMPULLAE OF EAR. Dilatations of the three semicircular canals of the vertebrate ear at the point where each enters the sacculus.

AMPULLAE OF LORENZINI. Small pits in the skin situated close to the lateral line canals or in patterns on the heads of many fish. At the base of each pit is a sense organ connected with a branch of the seventh cranial (facial) nerve. It is thought that these organs enable the fish to detect temperature changes.

AMPULLARY ORGANS. AMPULLARY CANALS. See *Ampullae of Lorenzini* above.

AN-, A-. Greek prefix: not, without.

ANA-. Greek prefix: upon, up, backwards.

ANABANTIDAE. Fresh-water fish of India, Africa and S.E. Asia including the Climbing Perch which can move on land and ascend trees by means of its fin-spines.

ANABOLISM. Chemical action in living organisms whereby complex substances are built up from simpler ones and energy is absorbed.

ANACANTHINI (GADIFORMES). Bony fish in which the median and pelvic fins are without spinous rays and the pelvic are usually displaced forwards to the jugular or thoracic position. The group includes cod, haddock, whiting, hake and many other marine fish as well as some fresh water forms.

ANADROMOUS. Migrating up rivers to spawn in shallow water (*cf. Catadromous*).

ANAEROBE. ANAEROBIC. Organisms which can live in the absence of free oxygen.

ANAL. Appertaining to or in the neighbourhood of the anus.

ANALOGOUS ORGANS. Organs which, although they may have quite different origins, perform the same functions in different animals, *e.g.* the kidneys of an amphibian and those of a mammal (*cf. Homologous Organs*).

ANAL PLATE. A thin sheet of skin formed by fusion of the wall of the hind-gut with the ectoderm in the vertebrate embryo. It marks the position where the future anus will be formed by perforation.

ANAMNIA (ANAMNIOTA). Vertebrates whose embryos do not have an amnion, *i.e.* Agnatha, Fish and Amphibia (see *Amniota*).

ANAMORPHOSIS. An addition of extra abdominal segments to an insect after it has hatched from the egg. In the Protura, a primitive group of wingless insects, eight segments are present in the abdomen at the time of hatching and three more develop later.

ANAPHASE. The phase in cell division when the chromosomes, having arranged themselves round the equator of the cell, move away from one another towards opposite poles.

ANAPOPHYSES. A pair of small prominences below the post-zygapophyses on certain vertebrae, *e.g.* on the lumbar vertebrae of mammals.

ANAPSID. A type of skull in which there is a complete covering of dermal bones with no temporal fossae; such a skull is found in the primitive reptile *Seymouria* and also in the Chelonia (tortoises and turtles).

ANASTOMOSIS. An intercommunication or network of nerves, blood-vessels etc.

ANATIDAE. Swans, geese and ducks: aquatic web-footed birds with long necks and having a broad flat beak which is edged with horny lamellae and covered by a sensitive membrane.

ANCYLOPODA. Primitive mammals of the Cretaceous and Eocene Periods forming a link between the Edentates and the Ungulates. The feet had three or five claws and locomotion was plantigrade (on the whole of the foot).

ANDRO-. Greek prefix from *Aner, andros*: man, male.

ANDROCONIA. Wing-scales and connected scent glands of male butterflies.

ANELYTROPIDAE. Viviparous burrowing lizards of the family Scincidae, living in S.W. Africa and in Mexico. They differ from others of the family in the following ways: the body is covered with bony scales and has abdominal ribs; there are no limbs; the eyes are concealed and there is no ear-opening.

ANEUPLOID. Having a chromosome number which is not a multiple of the basic haploid number.

ANGUILLIFORMES (APODES). Eels: elongated bony fish with few or no scales; with no pelvic fins and usually having spineless median fins confluent with each other and with the symmetrical pointed tail-fin.

ANGULAR BONE. A membrane-bone of the lower jaw in bony fish, reptiles and birds. During the process of evolution to mammals this bone has become separated from the jaw and modified to form the tympanic bone.

ANGULO-SPLENIAL. A bone of the lower jaw of some amphibia and reptiles, formed by fusion of the angular and the splenial bones.

ANIMAL POLE. That region of a telolecithal (unsymmetrical) egg in which the nucleus is situated and in which there is very little yolk. It is opposite to the *vegetative pole* where the yolk is aggregated.

ANISO-. Prefix: unlike, unequal; from Greek *an*: not, *isos*: equal.

ANISOCERCAL. Heterocercal: having the lobes of the tail-fin unequal.

ANISOGAMETES. ANISOGAMY. See *Heterogamete*.

ANISOPTERA. Stoutly built dragonflies which fold their wings horizontally when at rest and whose hind-wings are broader than the fore-wings. Their nymphs have an elaborate system of tracheal gills inside the rectum, and breathing is by pumping water in and out through the anus.

11

ANISOSPORES. HETEROSPORES. Spores which are of two kinds, *viz.* microspores and megaspores, or male and female spores.

ANNELIDA. Segmented worms which include the common earthworm, the leeches and many aquatic worms, either free-swimming, burrowing or tube-dwelling. The chief features distinguishing this phylum are: a body-wall made of circular and longitudinal muscles; chaetae or bristles arranged segmentally; a nervous system consisting of a pair of anterior ganglia, a collar round the pharynx, and a pair of ventral cords with ganglia in each segment; excretion is by segmentally arranged nephridia (ectodermal tubules) or coelomoducts (mesodermal tubules) or nephromixia (combinations of the two).

ANNIELLIDAE. Worm-like lizards of California, having no limbs, a snake-like head and large fangs.

ANNULUS. Any ring-like structure:
e.g. (1) A circular segment of an annelid worm.
(2) The externally visible rings, several to each body-segment, in a leech.
(3) A transverse circular groove, in which the flagellum lies when at rest, in protozoa of the group Dinoflagellata.
(4) A ring-like segment of the limb of a crustacean.

ANOBIIDAE. A family of small beetles with wood-boring larvae. The adults are usually reddish brown and are distinguished by the fact that the pronotum covers the head like a monk's cowl. The Death Watch Beetle is in this group.

ANOM-, ANOMO-. Greek prefix: irregular; from *A-*, without; *nomos*, a rule.

ANOMALURIDAE. African arboreal rodents somewhat resembling squirrels.

ANOMODONTIA (THEROMORPHA). Reptiles of the Permian and Triassic Periods which had certain mammalian features, *viz.* the body lifted high off the ground; the teeth in sockets and consisting of incisors, canines and molars. The reptilian characters included a compound mandible, a quadrate bone and a pineal foramen.

ANOMOPODA. Fresh-water crustacea including the well known *Daphnia* or water-flea; belonging to the Cladocera, a suborder of the Branchiopoda. Their most obvious feature is the pair of large bristly antennae used for swimming. The other limbs are small and are more or less covered by the two halves of the laterally compressed carapace. Eggs and young are carried in a brood-pouch formed by two dorsal plates extending upwards on either side of the abdomen.

ANOMURA. A suborder of Decapod crustaceans intermediate between lobsters and crabs. They include the well known 'Hermit-crab'.

ANOPLURA. Biting and sucking lice: small wingless insects parasitic on birds and mammals. The body is flattened and the legs have strong claws for clinging to the host. Antennae, eyes and cerci are reduced or absent. The loss of these and of the wings is considered to be a secondary feature connected with the parasitic habit.

ANOSTRACA. 'Fairy-shrimps': small semi-transparent crustaceans of the class Branchiopoda; without carapace and with numerous leaf-like appendages fringed with hairs. They habitually swim on the back.

ANSERIFORMES. Ducks, geese, swans etc., aquatic birds with a desmognathous skull (palatine bones united) and with a flattened beak having a horny rim round the edge.

ANT-, ANTE-. Latin prefix: before.

ANT-, ANTI-. Greek prefix: opposed to, against.

ANTAMBULACRAL (ABAMBULACRAL). The surface of an Echinoderm opposite to that on which the mouth and ambulacral grooves are situated.

ANTENNAE. Elongated sensory appendages on the heads of most arthropods. In typical crustacea there are two pairs, of which the first or smaller pair are usually called antennules. In insects, centipedes and millepedes the functional antennae are homologous with the antennules of crustacea, whilst the second pair tend to disappear. The antennae of insects are sometimes extremely specialized and very varied in form.

ANTENNAL GLANDS. Excretory organs opening at the bases of the antennae in some crustacea.

ANTENNULE. See *Antennae.*

ANTERGISM. Simultaneous contraction of opposing muscles.

ANTERIOR ABDOMINAL VEIN. A vein receiving blood from the Pelvic veins in Amphibians; running forwards in the mid-ventral line and so to the liver.

ANTERIOR CARDINAL SINUS. See *Cardinal Veins.*

ANTERIOR RECTUS. See *Eye muscles.*

ANTHO-. Prefix from Greek *Anthos*: a bud.

ANTHOMEDUSAE. See *Gymnoblastea.*

ANTHOZOA. See *Actinozoa.*

ANTHRACOTHERIIDAE. Extinct mammals of the Eocene and Miocene Periods having some resemblance to the hippopotamus.

ANTHROPO-. Root and prefix from Greek *Anthropos*: a man.

ANTHROPOIDEA. The group of Primates which includes all the true monkeys, apes and humans. They are divided into two main groups: the *Platyrrhini* or New World monkeys most of which have prehensile tails, and the *Catarrhini* or Old World apes and monkeys which are either tailless or have a non-prehensile tail.

ANTHROPOMORPHIDAE (SIMIIDAE). Erect or semi-erect, tailless apes and humans: the group includes gorillas, chimpanzees, orang-utans and gibbons. In newer systems of classification, however, humans are put into a separate family, Hominidae.

ANTIARCHA. Extinct fish of the Devonian Period having the head, body and pectoral fins covered by an armour of bony plates and having the eyes close together on the dorsal surface.

ANTILOPINAE. Antelopes, springboks, gazelles etc.: a subfamily of the Bovidae comprising fast-running animals usually having horns on both sexes.

ANTIPROTHROMBIN (HEPARIN). A substance present in blood and having the function of preventing clots from forming by keeping in check the change of *prothrombin* into *thrombin*. When an injury takes place, according to Howell's theory, a substance known as *thrombokinase* is released by damaged tissues and platelets. This neutralizes the *antiprothrombin* and allows a chain of reactions to take place bringing about the formation of a blood-clot.

ANTLERS. Branched horns formed of a bony substance, on deer and similar animals. Unlike the horns of cattle they are normally confined to the males and are shed every year.

ANURA (BATRACHIA). Frogs and toads: tailless amphibia with four limbs and with a smooth skin; without gills in the adult but developing from a larva

13

or tadpole which has external gills at first and internal gills later. The hind-limbs are usually enlarged for jumping and have webbed feet for swimming.

ANUS. The posterior opening of the alimentary canal.

AORTA. (1) In mammals: the large artery leaving the left ventricle of the heart and conducting blood to all parts except the lungs.

(2) Ventral Aorta: the main artery in fish and lower chordates through which blood is pumped forwards from the heart to the gills. A homologous vessel which becomes shortened and modified is also present in the embryos of land animals.

(3) Dorsal Aorta: the main artery through which blood flows to most of the middle and hind-parts of the body. In fish it receives blood from the gills but in land vertebrates it is a continuation of the systemic arch or arches.

AORTIC ARCHES. Paired arteries connecting the ventral with the dorsal aorta in vertebrates or their embryos. In fish they develop into the afferent and efferent branchial vessels. In land animals there are six embryonic pairs of arches of which the first two and the fifth usually become atrophied in the adult; the third, fourth and sixth become the Carotid, Systemic and Pulmonary arches respectively.

APHAN-. Prefix from Greek *Aphanes*: obscure.

APHANIPTERA (SIPHONAPTERA). Fleas: parasitic blood-sucking insects without wings; with a laterally compressed body and having the hind legs very much enlarged for jumping. The larvae are without legs and have biting mouth-parts. Pupae are enclosed in silken cocoons.

APHETOHYOIDEA. Extinct fish of varied form but all having a complete pair of hyoid gill-slits and a primitive type of jaw-suspension by ligaments only.

APHIDOIDEA (APHIDIDAE, APHIDES). Plant-lice which are extremely prolific and do much damage by sucking the sap of leaves and by exuding a sticky or waxy secretion from a pair of short abdominal tubes. They belong to the group Homoptera and usually have a very complex life-history. Wingless parthenogenetic females breed viviparously in very large numbers throughout the summer, but in the autumn winged insects are produced and these fly to another host plant where they reproduce sexually and lay eggs in the normal way. Winter is usually passed in the egg stage.

The following terms are sometimes used in connection with the life-history:
Fundatrix: the first wingless female of a season.
Fundatrigeniae: parthenogenetic female offspring of the fundatrix.
Migrantes: migratory winged females produced parthenogenetically.
Sexuparae: winged females which produce sexual males and females.
Sexuales: normal sexual males and females.

AP-, APO-, APH-. Greek prefix: away from, apart.

APICAL ORGAN (OF ANNELIDS). A ciliated plate at the anterior apex of the trochosphere larva of an annelid. In most forms it later develops into the prostomium of the adult and may bear tentacles or sense organs.

APIDAE. Honey-bees: aculeate Hymenoptera with social habits and specialization of types. In the case of the common hive-bee *Apis mellifica*, the males or drones are formed from unfertilized eggs by parthenogenesis, whereas workers and queen develop from fertilized eggs. A normal hive contains one queen, a few hundred drones and many thousands of workers. The queen is the only complete female capable of laying eggs; the workers are incomplete females.

Bees differ from wasps in the following ways: (1) they build brood cells and honey cells of wax instead of wood-pulp; (2) they feed entirely on pollen and nectar whereas wasps are partly carnivorous.

APIOIDEA. A superfamily of Hymenoptera including all the social and solitary bees but excluding the wasps which belong to the Vespoidea. Like all Hymenoptera they have two pairs of membranous wings which become hooked together in flight.

APLACENTALIA. Monotremes and Marsupials: primitive mammals which have no placenta.

APLACOPHORA. Worm-like molluscs of the order Amphineura; having no transverse shell-plates but only spicules embedded in the mantle (see *Amphineura* and *Polyplacophora*).

APOCRITA. Hymenopterous insects with a deep constriction between the thorax and abdomen. They include bees, ants and wasps.

APODA. See *Gymnophiona*.

APODEMES. Ingrowths of the cuticle of insects, crustaceans and other arthropods serving as points of attachment for the muscles. They may unite to form an internal framework known as the Endophragmal skeleton.

APODES. See *Anguilliformes*.

APODOUS. Without legs.

APODOUS LARVA. A degenerate type of insect-larva with no legs, a small head and few sense-organs; usually living where food is plentiful. The housefly, the bee and the weevil have larvae of this type.

APOPHYSIS. (1) A bony process to which muscles, tendons or ligaments are attached.

(2) A sternal apodeme of an insect or other arthropod, *i.e.* an ingrowth of the exoskeleton on the ventral side, serving for the attachment of muscles or the support of organs.

APOPLASTID PROTOZOA. Colourless individuals among protozoa which are generally coloured (Phytomastigina). They may be formed when the cell repeatedly divides so quickly that the coloured plastids cannot divide fast enough to keep pace.

APOPYLES. Pores leading from the flagellated chambers into the central cavity (paragaster) of a sponge.

APOSEMATIC. Having a warning coloration or a repellent smell (*e.g.* the skunk).

APPENDICULARIA LARVA. See *Ascidian Tadpole*.

APPENDICULAR SKELETON. That part of the skeleton contained in the fins or limbs as distinct from the axial skeleton which consists of the vertebral column and skull.

APPENDIX. See *Vermiform Appendix*.

APTERIA. Spaces on the surface of a bird's body between the rows of contour feathers, where the skin is naked or covered only with down.

APTERYGIDAE. Kiwis: primitive birds from New Zealand having very small functionless wings and simple hair-like feathers. They are about the size of a large hen, are insectivorous and nocturnal, hiding in holes in the daytime.

APTERYGOTA. Insects which are primitively wingless, undergo no metamorphosis and show many similarities to the crustaceans from which they have probably descended. They include, springtails, bristle-tails and silver-fish but not fleas, bugs and lice which are secondarily wingless as a result of their parasitic habit.

AQUEDUCT OF SYLVIUS. The canal linking the third and fourth ventricles in the vertebrate brain: also known as the *Iter*.

AQUEDUCTUS VESTIBULI. A name formerly used for the *Endolymphatic Duct* (*q.v.*).

AQUEOUS HUMOUR. The watery fluid in the anterior chamber of the vertebrate eye, between the lens and the cornea.

AQUILINAE. Eagles: carnivorous birds with curved hooked beaks and long talons and with the head and neck feathered. They fly powerfully and usually nest on mountain crags.

AQUINTOCUBITALISM (DIASTATAXY). A peculiarity of certain birds (*e.g.* ducks and pigeons) in which the fifth quill feather on the ulna (fifth cubital feather) is missing; there being a gap between the fourth and the sixth.

ARACHNIDA. A subphylum of Arthropoda, mostly terrestrial, including spiders (Araneida), scorpions (Scorpionoidea), king-crabs (Xiphosura), ticks and mites (Acarina) and several smaller groups. They are distinguished from other arthropoda by having the body divided into two regions: an anterior *Prosoma* or cephalothorax and a posterior *Opisthosoma* or abdomen. The former bears two pairs of prehensile and sensory appendages called *Chelicerae* and *Pedipalps* behind which are four pairs of walking legs; the latter bears no limbs.

ARACHNOID MEMBRANE. A thin membrane surrounding the vertebrate brain, between the vascular *Pia Mater* and the thicker *Dura Mater*.

ARANEIDA. Spiders: Arachnida in which the Opisthosoma or abdomen is separated from the Prosoma by a waist; is soft and unsegmented and bears several pairs of spinnerets. Silk produced by these is of various kinds and is used for web-making, for egg-cocoons or for trussing up victims etc. The prosoma, which is a combination of head and thorax, bears two *chelicerae* or poison fangs and two *pedipalps* or feelers with jaw-like bases known as *gnathobases*. There are also four pairs of walking legs. Breathing is by means of both tracheae and lung-books; the latter having a number of leaves or thin plates between which air can pass. Most spiders are carnivorous, using gnathobases to press their victims against the mouth and suck them dry.

ARBOR VITAE. A series of deep folds in the mammalian cerebellum appearing in vertical section like the branches of a tree.

ARCH-, ARCHE-, ARCHI-, ARCHAEO-. Prefixes from Greek *Arche*: beginning, *Archaios*: ancient, primitive.

ARCHAEOCETI. Primitive whales (see *Zeuglodonta*).

ARCHAEOPTERYX. The earliest known fossil bird of the Jurassic Period. Although possessing feathers it differed from all modern birds in having teeth, claws on the wings and a long jointed tail. It was about the size of a rook.

ARCHAEORNITHES. An order of extinct reptile-like birds containing only the one genus *Archaeopteryx* (*q.v.*).

ARCHENTERON. The primitive gut-cavity of an embryo in the gastrula stage; communicating with the exterior by the blastopore.

ARCHIANNELIDA. A class of small marine segmented worms usually without parapodia or chaetae. The loss of these and other features is now considered to be secondary, *i.e.* to have evolved by degeneration from more complex forms.

ARCHICEREBRUM (PROCEREBRUM). That part of the brain of an Arthropod formed by fusion of the paired ganglia of the first somite together with a pair of presegmental ganglia. Sometimes the term archicerebrum is restricted to the presegmental ganglia only.

ARCHIDICTYON (ARCHEDICTYON). An irregular network of small veins like those of a leaf in the wings of the primitive fossil insects *Palaeodictyoptera* of the Carboniferous Period. Most present-day insects show considerable simplification of wing-venation, but the wings of Mayflies (*Ephemeroptera*) closely resemble the primitive type.

ARCHINEPHRIC DUCT. See *Archinephros* below.

ARCHINEPHROS. An excretory organ consisting of a continuous series of tubules arranged segmentally throughout the length of the body and leading into a longitudinal duct, the Archinephric Duct, which passes back to the cloaca. Such an excretory system is the hypothetical ancestral type giving rise to the more specialized type found in present-day vertebrates in which tubules first form at the head-end (*pronephros*); later in the middle (*mesonephros*) and finally at the hind-end (*metanephros*). The nearest existing arrangement to the archinephros is in the Cyclostome *Bdellostoma*.

ARCHIPALLIUM. A dorsal thickening of the cerebral cortex, well developed in reptiles and connected primarily with the olfactory sense (*cf. Neopallium*).

ARCHIPTERYGIUM. A fin having one main cartilaginous axis with branches on each side resembling the veins of a pinnate leaf. Such fins are found in the Lung-fish *Ceratodus* and the fossil *Pleuracanthus*.

ARCI-. Prefix from Latin *Arcus*: a bow.

ARCICENTROUS (ARCOCENTROUS) VERTEBRAE. Vertebrae in which the centrum is almost completely enclosed by the neural and haemal arches spreading round it on each side, as in some fish (*cf. Chordocentrous*).

ARCIFERA. Frogs and toads having the following features: a well formed tongue; separate Eustachian Tubes; only one spiracle in the tadpole; a pectoral girdle in which the epicoracoid bones of the two sides overlap.

ARCIFORM MUSCLES. Arch-shaped muscles attached to the epidermis of earthworms beneath the seminal groove. By their successive contractions they form a series of troughs which help to pass the seminal fluid along the groove during copulation (see *Seminal Groove*).

ARCTIIDAE. Tiger-moths: large brightly coloured moths whose larvae are usually extremely hairy.

ARCTOIDEA. A group of carnivorous animals which includes bears, weasels, otters and badgers; formerly ranking as a superfamily but now generally included, with dogs, in the Canoidea.

ARCUALIA (DORSALIA). Rudimentary vertebrae consisting of small cartilaginous pegs rising from the notochordal sheath on each side of the nerve cord in Cyclostomata and in the embryos of other vertebrates.

ARDEIDAE. Herons and bitterns: wading birds with long legs and neck; belonging to the Tribe *Ciconiiformes* which also includes storks, flamingoes and ibis.

AREA OPACA (EMBRYOLOGY). See under *Area Pellucida* below.

AREA PELLUCIDA (EMBRYOLOGY). A circular area of transparent blastoderm overlying the subgerminal cavity and surrounded by the denser *Area Opaca* in the early embryonic development of birds etc.

AREA VASCULOSA (EMBRYOLOGY). An area of 'blood islands' surrounding the embryo of an amniote in its early stages. These blood islands later coalesce to form the beginning of the vascular system which spreads out round the embryo as far as the *Sinus Terminalis.*

AREOLAR TISSUE. The commonest form of connective tissue in the animal body; filling spaces, surrounding most of the organs and acting as packing material between the muscles and under the skin. It consists largely of fine white collagen fibres and thicker yellow elastic fibres in a jelly-like matrix.

ARGENTEA. A layer of silvery tissue between the choroid and the sclerotic in the eyes of fish and certain other vertebrates.

ARGENTEUM. A silvery reflecting layer in the skin of fishes; formed from *iridocytes* containing crystals of guanin.

ARISTA. A bristle-like appendage on the antenna of an insect.

ARISTATE. Having aristae on the antennae.

ARISTOTLE'S LANTERN. A complicated five-sided skeletal structure forming a framework for the mouth and jaws of a sea urchin. It has five movable teeth on its inner side and an elaborate system of arches and muscles for moving these.

AROLIUM. A median pad between the claws of an insect's foot.

ARRECTOR PILI MUSCLES (ARRECTORES PILORUM). Small muscles connecting the walls of the hair-follicles with the lower layer of the epidermis in mammals. They cause the hairs to become erect by a reflex action when the animal is cold or frightened.

ARTERIOLES. Small arteries branching from larger ones.

ARTERY. An elastic tube conveying blood away from the heart.

ARTHRO-. Prefix from Greek *Arthron:* a joint.

ARTHROBRANCHIAE. Plumose gills attached to the joints at the bases of the limbs of crabs, lobsters etc.

ARTHRODIRA. Fish of the Devonian Period having the head and anterior covered with large bony plates, but the hind-part either without armour or with large dorsal and ventral plates only. The tail was heterocercal and without scales.

ARTHROPLEONA. A suborder of Collembola or Spring-tails (primitively wingless insects) having a distinctly jointed body. This distinguishes them from the other suborder, the Symphypleona, in which the segments are fused together. Both groups have a powerful springing organ or furcula at the hind-end.

ARTHROPODA. A large phylum of invertebrates including Crustacea, Arachnida, Insects, Myriopoda and the fossil Trilobites. They have a segmented body, a thick exoskeleton and a large number of jointed appendages acting as jaws, legs, gills or sense organs. The nervous system is like that of the Annelida, consisting of a double ventral cord, two ganglia in each segment and a collar round the pharynx. A primitive brain is formed by fusion of some of the ganglia in the head segments.

ARTICUL-, ARTICULO-. Prefix and root from Latin *Articulus:* a joint. Diminutive of *Artus.*

ARTICULAMENTUM. The lower, flexible layer of the shell in the Polyplacophora, a group of molluscs whose shell consists of dorsal plates articulating with one another and enabling the animal to roll up.

ARTICULAR BONE. The hindmost of the lower jaw-bones of a reptile or bird articulating with the quadrate bone of the skull. In mammals it is reduced and modified to form the *malleus*, the outermost of the three auditory ossicles.

ARTIO-. Prefix from Greek *Artios*: of even number.

ARTIODACTYLA. Even-toed ungulates: sheep, cattle, pigs, hippopotami etc. having two or four hoofed toes.

ARYTENOID CARTILAGES. A pair of small cartilages strengthening the wall of the larynx in mammals.

ASCIDIACEA. Sea-squirts: sedentary Tunicates, either solitary or in colonies, having a degenerate nervous system, an atrium opening dorsally near the mouth and numerous gill-slits forming a basket-work. The whole organism is enclosed in a *test* or outer skin made of material resembling cellulose (see *Tunicata*).

ASCIDIAN TADPOLE. The fish-like larva of an ascidian having a tail, a notochord and a tubular nerve cord. These larval features indicate that ascidians, though outwardly very different from chordates, should nevertheless be included with them. They are therefore classed as lower chordates or *Protochordata*.

ASCIDIANS. See the two entries above.

ASCO-. Prefix from Greek *Askos*: a wine skin, leather bottle.

ASCON GRADE. The simplest type of sponge consisting of a hollow sac perforated by many pores and having at one end a single large opening or *osculum*. Water passes in through the pores and out through the osculum and is kept moving by the flagella of the collared cells or *choanocytes* which line the central cavity.

ASCOTHORACICA. An order of parasitic crustacea of the class Cirripedia like primitive barnacles with a large sac-like carapace or mantle.

ASEXUAL REPRODUCTION. Reproduction from only one parent; as, for example, the budding of a hydra or the simple fission of Amoeba.

ASILIDAE. Robber-flies. A family of Diptera having large hairy bodies, often conspicuously coloured: with long narrow wings, prominent eyes, strong legs for grasping their prey while flying, and a well developed proboscis for sucking the juices of the victim. They will often attack bees or other insects larger than themselves.

ASPID-, ASPIDO-. Prefix from Greek *Aspis, aspidos*: a shield.

ASPIDOBRANCHIATA. An alternative name for molluscs of the group usually known as Diotocardia (*q.v.*). The name indicates that the gills have two rows of 'leaflets' resembling those of the Shield Fern *Aspidium*.

ASPIDOCHIROTAE. The largest order of the Holothuroidea (sea-cucumbers): bilaterally symmetrical Echinoderms approximately cylindrical and able to move horizontally. The mouth is at one end and is surrounded by shield-shaped tentacles.

ASPIDORHYNCHIDAE. Extinct fish of the Oolitic and Cretaceous Periods having a laterally compressed high oval body covered by rhomboidal scales; a homocercal tail and a long pointed snout.

ASPIRIGERA. See *Holotricha*.

ASTAC-. Prefix from Greek *Astakos*: a lobster.

ASTACURA. Lobsters, crayfish etc.: decapod crustaceans with a long abdomen and a tail-fan. Most of them are bottom-dwellers, walking more than they swim.

ASTER-, ASTERO-. Prefix and root from Greek *Aster*: a star.

ASTEROIDEA. Star-fish: star-shaped or pentagonal echinoderms with the mouth underneath and having also on the under surface five open radial grooves (ambulacral grooves) along which are numerous tube-feet. The skin is covered with calcareous plates, spines and minute pincers called *pedicellariae* which probably protect the star-fish from any ectoparasites attempting to lodge on the surface.

ASTEROSPONDYLOUS. A term used to describe vertebrae of certain sharks in which the cartilaginous centrum contains a star-shaped ring of harder calcareous material to strengthen it.

ASTOMATA. A suborder of Protozoa belonging to the Holotricha: having mega- and micronuclei, uniform ciliation over the surface and no mouth. The absence of the latter is probably a form of degeneration due to their parasitic mode of life.

ASTRAGALUS. An alternative name for the *tibiale*, the tarsal bone adjacent to the tibia in the limb of a tetrapod vertebrate.

ATHECAE. Marine turtles of tropical seas having paddle-shaped clawless limbs and a leathery carapace without horny plates.

ATHERINIDAE. Silversides: mullet-like fish having an indistinct lateral line but having a silvery band along each side. They live in shoals off European shores.

ATLAS. The first cervical vertebra with which the skull articulates in all higher vertebrates.

ATOPOSAURIDAE. Small extinct lizard-like crocodiles of the Upper Oolite having short rounded snouts.

ATRICHORNITHIDAE. Scrub-birds: small sparrow-like birds of the grass lands of Australia; good mimics.

ATRIOPORE. An opening from the atrial cavity to the exterior in such creatures as *Amphioxus*.

ATRIUM. Latin *Atrium*: a court, vestibule.
(1) An auricle of the heart.
(2) A chamber surrounding the pharynx of *Amphioxus* and Tunicates. Water flows from the pharynx through the gill-slits into the atrium and hence out through the atriopore.

AUCHENORHYNCHA. A group of Homopterous insects including cicadas, frog-hoppers and leaf-hoppers. Like all Homoptera they have a head-flexure which causes the mouth-parts to be beneath the thorax and to point backwards when at rest. In the Auchenorhyncha this flexure is not so marked as in the other group, the Sternorhyncha, whose mouth-parts appear to arise between the fore-limbs.

AUDITORY CAPSULE. A bony or cartilaginous capsule enclosing the middle and inner parts of the ear. In Teleost fish this capsule is made up of five bones but during the course of evolution to higher vertebrates some are lost and some fused together. In mammals the capsule is further modified by the addition of certain reduced bones which in reptiles form part of the lower jaw.

AUDITORY NERVE. The eighth cranial nerve of a vertebrate, conveying sensory impulses from the inner ear.

AUDITORY OSSICLES. Small bones connecting the tympanic membrane with the *Fenestra ovalis* or inner ear-drum in terrestrial vertebrates. Their function is to convey sound vibrations across the cavity of the middle ear and to magnify the force of these vibrations by a lever action. In mammals there are three such ossicles: the *Malleus* or hammer, the *Incus* or anvil and the *Stapes* or stirrup. Birds, reptiles and amphibians, however, only have one, the *Columella auris* corresponding to the stapes.

AUERBACH'S PLEXUS. A network of small nerve fibres and ganglia between the two muscular coats of the intestine; able to cause local reflex arcs by which pressure of food in the intestine can bring about peristaltic movement.

AURICLE (AURICULA). (1) Any ear-like structure (Latin dim. of *Auris*: an ear).

(2) The first compartment (or pair of compartments) of the heart, into which blood flows from the veins.

(3) Five calcareous arches situated round the circumference of the jaw-skeleton known as *Aristotle's Lantern* in sea-urchins (Echinoidea). Muscles from these arches help to pump water to the gills (see *Aristotle's Lantern*).

AURICULAR FEATHERS. A tuft of large feathers surrounding the ear of an owl or similar bird and giving the appearance of a pinna.

AUTO-. Prefix from Greek *Autos*: self.

AUTOCOID. An alternative name for an internal secretion or hormone.

AUTOGAMY. Conjugation of gametes which have both been formed from the same original cell, as in some protozoa.

AUTOSAURI. An alternative name for the Lacertilia (lizards).

AUTOSOME. A typical or normal chromosome, not a sex chromosome.

AUTOSTYLIC. An arrangement in lung-fish and terrestrial vertebrates whereby the upper jaw is attached to the neurocranium by three processes: the *otic*, the *basal* and the *ascending* process.

AUTONOMIC SYSTEM. That part of the vertebrate nervous system which controls involuntary muscles, glands etc. consisting of the *sympathetic* and the *parasympathetic* systems having opposite effects. Nerves of the former leave the spinal cord ventrally and go to chains of ganglia one on each side of the vertebral column, or to median ganglia in the mesenteries and hence to the viscera. The chief nerve of the parasympathetic system is the *vagus* or tenth cranial nerve which also has branches leading to most of the viscera.

AUTOTOMY. The voluntary casting off of a part of the body when an animal is attacked, *e.g.* the tail of a lizard.

AUTOZOOID. An independent polyp capable of feeding itself.

AVES. Birds: warm-blooded, oviparous, bipedal, lung-breathing animals covered with feathers. The heart has four chambers and the aortic arch is on the right. The wing is formed from the anterior limb with three digits. In their general anatomy birds closely resemble reptiles from which they are considered to have evolved.

AVICULARIUM. A specially modified polyp in a colony of Polyzoa having the appearance of a bird's head and beak with 'jaws' which can snap shut.

AXIAL ORGAN. A spongy brown organ adjoining the axial sinus in star-fish and most Echinoderms. It consists of many spaces and contractile vessels which are believed to help in the circulation of the watery blood.

AXIAL SINUS. A part of the coelom of a star-fish or other echinoderm forming a sac immediately beneath the *madreporite* (the porous filter through which water enters). From this sinus water flows through the *stone canal* to the water-vascular ring and to the tube-feet (see *Stone Canal*).

AXIS CYLINDER. The central protoplasmic thread of a nerve fibre.

AXIS VERTEBRA. The second cervical vertebra in birds and mammals; having a forward projecting *odontoid process* which acts as a pivot on which the first vertebra, the *atlas*, can rotate when the head turns.

AXOLOTL. See *Amblystomatinae* and *Neoteny*.

AXON. A nerve fibre or protoplasmic process which transmits impulses away from the cell-body of a neurone.

AXONOST. The basal cartilaginous or bony support for the fin-rays of a fish.

AXOPODIUM. A pseudopodium with a stiff internal supporting axis, *e.g.* in the *Heliozoa*.

AXOSTYLE. A stiff rod of protoplasm acting as an internal skeletal support in certain protozoa such as *Trichomonas*.

AZYGOS VEINS. Veins receiving blood from the wall of the chest in reptiles, birds and mammals; believed to be homologous with parts of the posterior cardinal sinuses of fish.

B

BALAENOIDEA (MYSTACOCETI). Toothless whales with large heads and with baleen or 'whalebone' hanging from the palate acting as a strainer for the small creatures which form its food. The group includes the *Right-whales* and the *Rorquals* (see also *Whalebone*).

BALANCERS. Reduced and functionless hind-wings in Dipterous insects (see *Halteres*).

BALEEN. See *Whalebone*.

BALISTIDAE. Tropical fish of coral reefs having minute rough scales or movable scutes, incisor-like teeth and sharp spines on the fins. Many have poisonous flesh.

BARBELS. Tactile spines or bristles on the jaws of some fish.

BARBICELS. Small hooked processes by which the barbules of a feather interlock (see *Barbule*).

BARBS (OF FEATHERS). Lateral lamellae forming the vane of a feather.

BARBULES (OF FEATHERS). Small processes on each side of the barbs of a feather, interlocking by means of hooks known as *hamuli* or *barbicels*.

BARTHOLIN'S GLANDS (GLANDS OF DUVERNOY). Glands near the entrance to the mammalian vagina secreting a mucous fluid which facilitates the entry of the penis during copulation.

BASAL GRANULE. A granule at the base of each flagellum in flagellated protozoa; usually connected by protoplasmic threads or rhizoplasts to the nucleus and to the parabasal body.

BASALIA. Basal skeletal elements in the paired fins of a fish.

BASAL PROCESS. A bony or cartilaginous process by which the upper jaw is attached to the floor of the neurocranium during the development of the skull in terrestrial vertebrates (see *Autostylic*).

BASEMENT MEMBRANE. A structureless or fibrillar membrane to which epithelial cells are attached and on which they rest.

BASIBRANCHIAL. A median ventral cartilage or bone running along the base of the gill-arch skeleton of a fish.

BASICONIC RECEPTOR. An olfactory or other sense-organ on an insect consisting of a small projecting cone whose base is connected to a nerve fibre.

BASIDORSALS. A pair of cartilaginous or bony elements from which the neural arch of a vertebra develops. In primitive vertebrates such as the *Cyclostomata* and in the embryos of higher vertebrates they are merely small pegs in front of the dorsal roots of each pair of spinal nerves. Later they fuse together to form the neural arch (*cf. Basiventrals*).

BASIHYAL. A bone or cartilage forming the middle part of the Hyoid Arch in a fish, amphibian etc. lying between the two rami of the lower jaw in the floor of the mouth.

BASILAR PLATE. See *Gnathochilarium*.

BASI-OCCIPITAL BONE. A bone forming the base of the hind-part of a vertebrate cranium. The whole hind-part or occipital region consists of the basi-occipital, two exoccipitals and the supra-occipital.

BASIOST. The bony or cartilaginous base to which the radialia or skeletal fin-rays are attached in a fish's fin.

BASIPODITE. The segment next to the coxopodite and bearing the two branches, the endopodite and the exopodite in the limb of a crustacean.

BASI-PTERYGIAL PROCESSES. Bony processes on the Basisphenoid for articulation with the Pterygoid bones in some vertebrates.

BASIPTERYGIUM. The basal cartilage to which the radial cartilages are attached in some fishes' fins.

BASISPHENOID. A bone in the middle part of the base of a vertebrate cranium between the basi-occipital and the presphenoid.

BASISPHENOID ROSTRUM. A splint of membrane-bone projecting forwards from the basisphenoid bone in some reptiles and replacing the presphenoid in the floor of the cranium.

BASITEMPORAL BONES. A pair of membrane bones on either side of the basisphenoid and uniting with it in some reptiles and birds.

BASIVENTRALS. A pair of cartilaginous or bony elements giving rise to the ventral part of the centrum of a vertebra and sometimes forming a haemal arch in the tail region (*cf. Basidorsal*).

BASOMMATOPHORA. Pulmonate molluscs with eyes at the bases of the posterior tentacles. They include most of the water snails such as *Limnaea* and *Planorbis*.

BASOPHIL. A cell in which the cytoplasmic granules can readily be stained with basic dyes such as haematoxylin. Examples are certain types of white blood corpuscles.

BASTARD WING (ALULA). A tuft of feathers attached to the 'thumb' of a bird, distinct from the other wing-feathers which are attached to the 'hand' and to the ulna.

BATESIAN MIMICRY. The resemblance of a harmless animal to a poisonous or dangerous one giving protection to the former, since predators will tend to avoid both.

23

BATHYERGIDAE. Burrowing rodents of Africa having short legs and tail, small eyes and ears and very little hair; the Cape Mole-rat is an example.

BATOIDEI (BATOIDEA). An alternative name for the *Raii* (sometimes spelt *Raji*): cartilaginous fish such as skates and rays which have the body flattened dorsoventrally and the pectoral fins greatly enlarged.

BATRACHIA. An alternative name for *Anura* or tailless amphibia (frogs and toads).

BATRACHIDAE. Toad-fish: carnivorous fish of warm seas having a large flattened head.

BDELLOID. Having the appearance of a leech.

BDELLOID ROTIFERS. Rotifers which move by a looping action like that of a leech.

BELEMNOIDEA (BELEMNITES). A group of Cephalopod molluscs, possibly ancestors of present-day octopuses and cuttle-fish, whose internal bullet-shaped shells are plentiful as fossils in Mesozoic rocks.

BENTHOS. Animals and plants living on the bottom of a sea or lake.

BERYCIDAE (BERYCOMORPHI). Perch-like deep-sea fish with a compressed body covered by circular or comb-like scales; having the pelvic fins in the thoracic position and having an air-bladder with a pneumatic duct.

BI-. Prefix from Latin *Bis*: twice.

BIBIONIDAE. Flies of the group *Nematocera* or 'long-horned' Diptera having a black or brown body nearly an inch long and large compound eyes sometimes meeting on the top of the head. The greyish white larvae live in the soil and are serious plant pests.

BICUSPID TOOTH. A tooth having two cusps or raised parts of the crown.

BICUSPID VALVE. The mitral valve or left auriculo-ventricular valve in the mammalian heart; so-called because of the two cusps or flaps by which it can close.

BIDDER'S ORGAN. A vestigial ovary, sometimes containing a mass of immature ova, in front of the testis or ovary in certain toads of both sexes.

BILE DUCT. The duct through which bile flows from the gall bladder to the duodenum in a vertebrate.

BILIRUBIN. A brown or orange bile pigment formed from the breaking down of red blood corpuscles.

BILIVERDIN. A green bile pigment formed from the breaking down of red blood corpuscles.

BILOPHODONT TEETH. Grinding teeth with two transverse ridges across the crown, *e.g.* in the Tapir.

BIMANA. Two-handed: a name sometimes used to distinguish Man from the apes or *quadrumana* by reason of the fact that Man has lost the power of opposing the large toe but retains the power of opposing the thumb.

BINARY FISSION (SIMPLE FISSION). Reproduction by division of a cell into two equal parts as in the case of *Amoeba*.

BINOCULAR VISION. Stereoscopic vision occurring when both eyes face the front and can be directed at the same object together. Such vision is a help in judging distances and occurs particularly in tree-dwellers.

BIO-. Prefix from Greek *Bios*: life.

BIOGENESIS. The generalization that living organisms must come from parents similar to themselves. In former times the idea of spontaneous generation from non-living matter was held.

BIOLOGICAL INDICATORS. Protozoa or other micro-organisms which can be used as indicators of chemical activity. Some kinds for instance will migrate to regions of low or high oxygen concentration and so may be used to locate the seat of respiratory activity in an aquatic animal.

BIOLUMINESCENCE. See *Luminescence*.

BIOTIC POTENTIAL. An estimate of the maximum rate of increase of any species of animal if left to itself and isolated from its natural enemies, disease or other inhibiting factors. Normally this potential reproductive rate is very great (many million per year in the case of insects for instance), but in the struggle for existence a balance is maintained and the population of any species remains roughly constant.

BIRAMOUS LIMBS (OF CRUSTACEA). The typical forked appendages of a crustacean, sometimes modified for various functions but essentially consisting of two branches arising from a basal portion. This basal part is called the *protopodite* and usually consists of two segments, *viz.* the coxopodite attached to the body and the *basipodite* which bears the two branches. The latter are called the *endopodite* and the *exopodite* and may be flattened (*phyllopodia*) or narrow (*stenopodia*) or may form legs, pincers, mouth-parts etc.

BLADDER WORM. The bladder-like larva of a tapeworm which, since it is found in a different host from the adult worm, was originally regarded as a distinct species (see *Cysticercus*).

BLASTO-. Prefix from Greek *Blastos*: a bud.

BLASTOCOEL (EMBRYOLOGY). A fluid-filled cavity within a blastula (*q.v.*).

BLASTOCYST (EMBRYOLOGY). A blastula or hollow ball of cells formed by repeated cleavage of the fertilized mammalian ovum. It differs from the blastula of a lower vertebrate in that it has an inner cell-mass enclosed in a hollow sphere. The outer sphere is the *trophoblast* by which the embryo becomes implanted on the uterus of the parent. The inner cell-mass soon becomes an embryonic disc from which the embryo develops.

BLASTODERM (BLASTODISC). A superficial layer of cells formed by partial cleavage of a large egg such as that of a bird. The presence of a large amount of yolk prevents complete cleavage and the disc is produced by the rapid and repeated division of cells on the side away from the yolk.

BLASTODISC. See *Blastoderm* above.

BLASTOIDEA. Echinoderms of the Carboniferous Period, sometimes classed as Crinoids, having an ovoid stalked body and a complex exoskeleton of plates.

BLASTOKINESIS. Movements of an embryo in an egg during development.

BLASTOMERES. The cells of a blastula (*q.v.*). (See *Cleavage*.)

BLASTOPORE (EMBRYOLOGY). An opening by which the cavity of a gastrula (the *archenteron* or primitive gut) communicates with the exterior. It may become the anus but more often becomes roofed over by the neural folds and forms the *Neurenteric Canal*.

BLASTOSPHERE. An alternative name for the Blastula (*q.v.*).

BLASTOSTYLE. A polyp modified for reproduction in certain colonial

C   25

coelenterates such as *Obelia*. It is usually a hollow elongated rod from which separate individual 'jelly-fish' or *medusae* bud off, detach themselves and swim away. These medusae are the sexual generation bearing gonads from which new asexual colonies are produced.

BLASTOZOITE. See *Blastozooid.*

BLASTOZOOID. A zooid which arises by budding from the parent, as in coelenterates, tunicates etc.

BLASTULA (EMBRYOLOGY). A hollow ball of cells formed by repeated cleavage of a fertilized egg-cell.

BLATTOIDEA. A superfamily of Orthoptera including the Blattidae (cockroaches) and the Mantidae (Praying Mantises). They are fast-running insects whose head is flexed at right angles to the body and whose fore-wings are thickened to form tegmina. The eggs are laid in a special horny capsule or *ootheca* (see also *Blattidae* below).

BLATTIDAE. Cockroaches: a family of insects having dark brown, oval, flattened bodies, long legs and long antennae. Mouth-parts are of the biting type and the insects, which are omnivorous, are regarded as vermin in most parts of the world.

BLENNIIDAE. Blennies: littoral or fresh-water fish found in most parts of the world; characterized by having the pelvic fins well forward in the jugular position and by having a dorsal fin formed entirely of spines.

BLENNIIFORMES. An alternative name for the *Jugulares*: a bony fish having the pelvic fins well forward in the jugular position.

BLEPHAROPLAST (PARABASAL BODY). A small structure of unknown function near the base of a flagellum in certain protozoa; connected with the basal granule of the flagellum by threads known as *rhizoplasts*. The name *Blepharoplast* has been used for the basal granule itself, but it is more accurately synonymous with the *parabasal body*.

BLIND SPOT. The part of the retina of a vertebrate eye where the optic nerve passes out on its way to the brain and where consequently there are no sensory cells.

BLOOD. A fluid which circulates throughout the animal body, carrying oxygen, food substances, hormones, waste products etc., usually contained in arteries, veins and capillaries or in larger cavities known as sinuses. It is kept moving by the pumping action of the heart. In all vertebrates and some invertebrates it contains the red pigment *haemoglobin* which readily combines with oxygen enabling the blood to carry much more of this gas than would otherwise be the case. In some invertebrates haemoglobin is replaced by the blue-green pigment *haemocyanin*.

BLOOD CORPUSCLES. Cells suspended in the fluid (plasma) of the blood. In vertebrates there are two main types, *viz.* red corpuscles or *erythrocytes* containing haemoglobin, and white corpuscles or *leucocytes* which help to prevent disease (see also *Erythrocyte* and *Leucocyte*).

BLOOD GILLS. Tubular outgrowths from the bodies of some aquatic gnat larvae (notably Chironomidae). They contain blood and were formerly thought to be respiratory organs, but it has been shown recently, using Protozoa as biological indicators, that respiration of these insects takes place over the whole surface of the body.

BLOOD GROUP. A group of people or animals whose blood may be mixed without any harmful interaction such as clumping together of the corpuscles. This clumping or agglutination is due to an incompatibility of the corpuscles of one group with the plasma of another. The plasma contains a substance known as *agglutinin* and the corpuscles a substance known as *agglutinogen*. In humans there are two agglutinogens 'A' and 'B' and two agglutinins 'α' and 'β'. People of group A contain 'A' corpuscles and 'β' plasma; people of group B contain 'B' corpuscles and 'α' plasma. Group AB contains both factors in the corpuscles and neither in the plasma. Group O contains none in the corpuscles and both in the plasma. When transfusing blood, 'A' must not come into contact with 'α', nor 'B' with 'β'. There are also other factors such as the Rhesus Factor (Rh) which may cause incompatibility.

BLOOD ISLANDS. Masses of cells which later coalesce to form a network of small blood-vessels: the beginning of the vascular system in a vertebrate embryo.

BLOOD PIGMENTS. See *Respiratory Pigments*.

BLOOD PLASMA. The fluid part of the blood in which corpuscles are suspended; containing dissolved salts, gases, food and waste substances, hormones, antitoxins etc.

BLOOD PLATELETS (THROMBOCYTES). Minute disc-shaped particles contained in mammalian blood; believed to be partly responsible for clotting by the production of *thrombokinase*.

BOIDAE. Boas, anacondas, pythons and similar large snakes of tropical forests which crush their prey in the coils of the body. They have vestiges of the hind limbs appearing as spurs on each side of the anus.

BOMBIDAE. Humble-bees: bees with large hairy bodies usually black or brown with yellow markings; living in less well organized colonies than those of the hive bee. Nests are often made in holes left by field mice or other creatures and consist of cells made from wax like those of the hive bee. The colony, however, usually dies off in the winter leaving only the young queens to hibernate and found new colonies the following year.

BOMBYCIDAE. A family of Lepidoptera which includes the silk-moth *Bombyx mori*. The larva or silk-worm secretes the thread for its cocoon from modified salivary glands through a median spinneret near the mouth.

BOMBYLIIDAE. Bee-flies: two-winged flies of the group Brachycera having a superficial resemblance to bees; feeding on nectar and helping in pollination. The larvae are parasitic on other insects.

BONE. The chief skeletal substance of most vertebrates consisting largely of a matrix of calcium phosphate having numerous small holes and passages for cells, blood vessels, nerves etc. Bones are classified according to their mode of origin as cartilage-bone (formed from cartilage); dermal or membrane bones (formed in the skin) and sesamoid bones (formed within tendons).

BOTRYOIDAL TISSUE. A peculiar type of mesenchyme or 'packing tissue' found in the bodies of leeches; consisting of a network of darkly pigmented tubular cells having intracellular ducts filled with red blood-like fluid.

BOVIDAE. Cattle, sheep, goats etc. Artiodactyla or even-toed ungulates of ruminating habit; having a complex four-chambered stomach and usually having horns which are not shed like the antlers of a deer.

BOWMAN'S CAPSULES. Small hollow cup-shaped capsules in the kidneys of vertebrates. Each capsule surrounds a coiled blood vessel or *glomerulus*, the two together forming a *Malpighian Body*. Waste liquid from the blood passes out of the glomeruli into the capsules and hence down uriniferous tubules to the ureters and so to the bladder or the cloaca.

BRACHI-, BRACHIO-. Prefix from Greek *Brachion*: the arm.

BRACHIAL. Appertaining to the arm.

BRACHIAL ARTERY AND VEIN. The vessels taking blood to and from the arm or front limb of a vertebrate.

BRACHIAL OSSICLES (OF ECHINODERMS). A row of calcareous ossicles in each arm of a Crinoid.

BRACHIAL PLEXUS (PECTORAL PLEXUS). A network of spinal nerves supplying the anterior limbs or fins of a vertebrate.

BRACHIOLES. Small arm-like processes, *e.g.* the reduced 'arms' of some Crinoids.

BRACHIOPODA. Marine invertebrates having a number of ciliated tentacles round the mouth and having a bivalve shell with the two valves situated dorsally and ventrally. The shell resembles that of a Lamellibranch mollusc but the anatomy of the animal is quite different. Brachiopoda were very common in Palaeozoic and Mesozoic times; millions of their fossil shells are to be found in the Jurassic limestone. There are, however, very few species surviving at the present day.

BRACHY-. Prefix from Greek *Brachus*: short.

BRACHYCEPHALIC. Short and broad; referring to the human head when the breadth is at least four fifths of the length, *e.g.* in Mongolians.

BRACHYCERA. Two-winged flies such as the robber-fly and the horse-fly: a suborder of Diptera having the antennae shorter than the thorax and often with a terminal bristle. The larvae are legless and have an incomplete head which may be partly retracted into the thorax. The pupae are exarate, *i.e.* with free appendages.

BRACHYDACTYLY. Shortness of fingers: a feature sometimes found in man and other animals; inherited as a Mendelian dominant.

BRACHYODONT TEETH. Teeth with short crowns which cease to grow when fully formed because the channel conducting blood through the roots to the pulp-cavity becomes closed or nearly so. Human teeth are of this type.

BRACHYURA. Crabs: decapodous crustacea having the carapace flattened and greatly expanded laterally while the abdomen is reduced and folded underneath.

BRACONIDAE. A family of small parasitic insects related to the Ichneumon flies. Eggs are laid on or in the body of some other insect such as a caterpillar and hatch out into small parasitic maggots which destroy the body of their victim before pupating.

BRADY-. Prefix from Greek *Bradus*: slow.

BRADYODONTI. Extinct sharks of the Devonian and Permian Periods, having teeth whose crowns consisted of numerous vertical parallel tubes of dentine. Included in this group are also the Holocephali represented at the present-day by *Chimaera*, known as the 'King of the Herrings'.

BRADYPODIDAE (TARDIGRADA). Sloths: primitive mammals of Central and South America belonging to the order Edentata; having poorly developed teeth with no enamel and having coarse hair usually coloured green by parasitic algae. They have long curved claws by which they walk slowly upside down on the branches of trees. On the ground they move very awkwardly. The skeletons of ancestral sloths of very great size have also been found in South America.

BRAIN. An anterior enlargement of the central nervous system made necessary by the fact that most of the sense organs are usually in the front.

BRANCHI-. Prefix from Greek *Branchion*: a gill.

BRANCHIAE. Respiratory organs of water-breathing animals: gills, ctenidia.

BRANCHIAL ARCHES. Arches of cartilage or bone extending round the pharynx and supporting the gills of a fish. There are usually five pairs of arches joined to a median ventral bar called the *basibranchial* cartilage. Each arch usually consists of four skeletal elements named from the base upwards the *hypobranchial, ceratobranchial, epibranchial* and *pharyngobranchial*.

BRANCHIAL BASKET. See *Branchial Arches*.

BRANCHIAL CHAMBER. (1) Of Crustacea: a cavity formed by part of the carapace enclosing the gills.

(2) Of Tunicates: the large pharynx whose lateral walls are perforated by numerous gill-slits.

BRANCHIAL CLEFTS. A series of apertures, usually from four to seven pairs, in the wall of the pharynx of fish and young amphibia, by which the throat communicates with the exterior and allows water which has entered by the mouth to pass out over the gills.

BRANCHIAL HEARTS. Accessory 'hearts' by which blood is pumped through the gills in cuttle-fish, octopuses etc.

BRANCHIOBDELLIDAE. Ectoparasitic annelids intermediate between earth-worms and leeches.

BRANCHIOPODA. The most primitive class of crustacea, mostly living in fresh water, having a varied number of body-segments and very little specialization of limbs. They include the well-known 'fairy-shrimp' *Chirocephalus* and the more specialized water-fleas such as *Daphnia*.

BRANCHIOSAURI. Small amphibians of the Carboniferous and Permian Periods resembling salamanders with rows of scales on the ventral part of the body.

BRANCHIOSTEGAL MEMBRANE. See *Branchiostegal rays*.

BRANCHIOSTEGAL RAYS. Horn-like processes backwardly directed from the ceratohyal bone, acting as supports for the branchiostegal membrane which forms part of the operculum of a fish.

BRANCHIOSTEGITE. Part of the carapace of a crab, lobster etc. projecting down on each side to enclose the gills within a branchial chamber.

BRANCHIURA. Carp-lice: crustaceans allied to the common *Cyclops*; having a large flattened head and suctorial mouth-parts by which they attach themselves to fish and live as ectoparasites.

BREGMA. The area in the mammalian head where the frontal and parietal bones meet.

BRONCHIOLE. One of the smaller breathing tubes branching off from the two large bronchi in the lungs of a higher vertebrate.

BRONCHUS. One of the two wide tubes leading from the trachea to the lungs in a higher vertebrate. As in the case of the trachea they are strengthened with rings of cartilage.

BROOD POUCH. A pouch on the body where eggs are retained and hatched, *e.g.*:

(1) Fish: in Sea-horses and Pipe-fish (*Syngnathidae*) the male carries the eggs in a ventral pouch formed of two folds of skin.

(2) Amphibia: in some frogs and toads the eggs are carried in pouches of skin on the back of the parent.

(3) Crustacea: the brood pouch may be a space under the carapace (as in *Daphnia*) or formed from overlapping plates called *oostegites* attached to the limbs (as in *Gammarus*).

BROTULINA. Fish living in the darkness of subterranean waters or at great depths in the sea, having an elongated body, united median fins and a tapering tail. Pelvic fins, if present, are attached to the pectoral girdle. The eyes may be reduced or absent.

BROWN BODY (OF POLYZOA). A polypide of a polyzoan colony which has degenerated and broken down to form a brown mass lying within the zooecium or body wall. A new polypide may grow from the old zooecium and the brown body then comes to lie inside the new stomach.

BRÜNNER'S GLANDS. Glands at the bases of the villi in the small intestine of a vertebrate. They secrete a number of enzymes and open into the *Crypts of Lieberkühn*.

BRUTA. An old name for the mammalian order *Edentata* which includes sloths and anteaters.

BRYOZOA. Another name for the *Polyzoa* (*q.v.*).

BUBULINAE. Large African antelopes with horns on both sexes; they include the Hartebeest, the Blessbok and the Gnu.

BUCCAL CAVITY. The cavity of the mouth usually containing the teeth and tongue.

BUCEROTIDAE. Hornbills: large birds of Africa and S.E. Asia with very large downwardly curved beaks and usually a horn-like head-ornamentation.

BUFONIDAE. Toads: tailless amphibia usually without teeth and having a drier skin than that of a frog, but able to secrete a thick white fluid from dermal glands. They include terrestrial, burrowing, arboreal and aquatic types.

BULBILS. See *Contractile Bulbils*.

BULBUS ARTERIOSUS. An enlargement of the ventral aorta of some vertebrates at the point where it leaves the heart. Its wall is of smooth muscle and there are no valves. In these respects it differs from the *Conus Arteriosus* which is composed of cardiac muscle, is contractile and contains valves. The bulbus tends to be prominent in bony fish but in other fish and in amphibia the conus predominates.

BULBUS CAROTICUS. See *Carotid Labyrinth*.

BULBUS CORDIS. See *Bulbus arteriosus*.

BULLA OSSEA. An alternative name for the *Tympanic Bulla* (*q.v.*).

BUNO-. Prefix from Greek *Bounos*: a small hill.

BUNODONT TEETH. Teeth like those of a pig in which there are distinct and separate conical elevations of enamel on the crown.

BUNO-LOPHODONT. A stage in the development of the teeth of many animals (*e.g.* Ruminants) intermediate between the *Bunodont* state (with small hillocks) and the *Lophodont* state (with transverse ridges).

BUPRESTIDAE. Metallic green or blue beetles whose legless larvae bore beneath the bark of trees and can be recognized by their greatly widened prothorax.

BURSA COPULATRIX. An invagination of the body-wall around the genital aperture of insects etc. adapted for receiving the intromittent organ of the male.

BURSA ENTIANA. The short swollen anterior part of the intestine of a fish, sometimes called the 'duodenum'.

BURSA FABRICII. A sac-like organ of uncertain function opening into the cloaca of a bird.

BUTEONIDAE. Buzzards and kites: carnivorous birds of the Old World belonging to the Falcon family; having a curved beak, powerful claws and a feathered head and neck.

BUTTERFLIES. Lepidoptera characterized by having clubbed antennae. The old system of classification divided the Lepidoptera into two groups: *Rhopalocera* or butterflies and *Heterocera* or moths. This classification has now been superseded by a division into *Homoneura* and *Heteroneura* based on the wing-venation. The former have an almost identical arrangement of veins in the two pairs of wings; they include a few of the larger moths and most of the tiny ones formerly known as *Microlepidoptera*. In the *Heteroneura*, on the other hand, the venation of the two pairs of wings differs markedly; all the butterflies and most of the larger moths are included in this group.

BYSSUS. A mass of threads formed by the hardening of a viscous secretion from the *Byssus pit* in marine mussels and similar molluscs. They enable the animal to become firmly attached to stones or to other mussels, a great advantage in tidal waters.

C

CACATUINAE. Cockatoos: parrots of Australasia and the Philippines, having a movable crest on the head.

CADOPHORE. A dorsal process of certain Tunicates (Doliolida) on which newly formed individuals, which have broken off from the parent, attach themselves to be carried about.

CAECUM (COECUM). A blind gut or diverticulum from the alimentary canal branching off, in the case of mammals, at the junction of the ileum and the colon; at its distal end is the *vermiform appendix*. Herbivorous animals have a long caecum in which fibres of wood or cellulose are broken down by bacterial action, but in carnivorous animals the caecum is usually absent or small.

CAENO-, CAINO-. Prefix from Greek *Kainos*: recent.

CAENOGENETIC. CAENOGENESIS. Features of an embryo larva which are adapted specially to the needs of the young and disappear in the adult.

CAINOZOIC. Belonging to the third geological epoch; that is from the Eocene to the present day.

CALAMUS. The hollow semi-transparent quill or stalk of a feather.

CALC-. Prefix from Latin *Calx, calcis*: Lime.

CALCANEUM. An alternative name for the *Fibulare* or heel-bone in many tetrapod vertebrates (see *Fibulare*).

CALCAR-. Prefix and roo from Latin *Calcaria*: a lime kiln.

CALCAR. Latin *Calcar*: a spur.

(1) The spur on the leg of a bird.

(2) A bony or cartilaginous process on the calcaneum of a bat; projecting backwards and inwards from the heel and helping to support the part of the wing-membrane which is between the legs.

CALCAREA. A class of sponges whose skeletons consist solely of calcareous and non-siliceous spicules.

CALCAREOUS NODES. See *Swammerdam's Glands.*

CALCIFEROUS GLANDS. Glands which secrete calcium carbonate or other calcium compounds, as for instance the oesophageal glands of an earthworm.

CALCIGEROUS GLANDS (OF AMPHIBIA). See *Swammerdam's Glands.*

CALLIPHORINAE. Blow-flies or bluebottles: a subfamily of Diptera superficially resembling the common house-fly but belonging to a different family, the *Tachinidae*. The larvae, which are without legs, pupate in a dark coloured barrel-shaped capsule or puparium. Most of these flies breed in carrion but some attack sheep and other animals; the maggots boring into the living flesh.

CALYMMA. The highly vacuolated outer gelatinous layer of Radiolarians.

CALYPTO-, CALYPTRO-. Prefix from Greek *Kaluptra*: a veil, *kaluptein*: to cover.

CALYPTER. The enlarged alula or proximal wing-lobe which in some of the Diptera forms a cover for the haltere (see *Alula*).

CALYPTOBLASTEA (LEPTOMEDUSAE). Coelenterates of the class Hydrozoa in which sedentary hydroid colonies alternate with free swimming medusae. A horny covering or *perisarc* extends over the hydroid colony forming cups or *hydrothecae* round the nutritive polyps and *gonothecae* round the reproductive individuals.

CALYPTOMERA. Water fleas (Cladocera) in which the carapace is well developed forming a dorsal brood pouch and extends down on each side to cover the body and limbs, *e.g. Daphnia.*

CAMELIDAE. Camels, dromedaries and llamas: ungulates whose feet generally have two toes although some fossil forms had four. The stomach has three chambers of which the first, the rumen, has a number of diverticula or pouches with narrow openings and sphincters so that a large amount of water can be stored.

CAMPAN-. Prefix from Latin *Campana*: a bell.

CAMPANIFORM RECEPTORS. Sense organs of insects, probably chemosensitive, in the form of bell-like projections from the skin with bipolar nerve cells beneath them.

CAMPANULA HALLERI. A vascular swelling where the lens is attached to the *retractor lentis* muscle in the eyes of some fish.

CAMPODEIDAE. Small wingless insects belonging to the Diplura or 'bristletails'. They are without eyes, have little body-pigment and have a tail-fork formed from a pair of long abdominal cerci.

CAMPODEIFORM LARVAE. Oligopod larvae: the common form of larvae among beetles and Neuroptera. They are usually predatory; have well-developed sense organs and legs but no abdominal appendages other than cerci. The name is given on account of their resemblance to *Campodea* and its allies.

CAMPTO-. Prefix from Greek *Kamptos*: flexible.

CAMPTOTRICHIA (KAMPTOTRICHIA). Fibrous jointed fin-rays such as are found in the lung-fish.

CANALICULI. Fine channels such as those permeating a bone.

CANCELLOUS BONE (CANCELLATED BONE). Spongy bone in which the blood vessels run in the interstices and not in definite Haversian Canals.

CANIDAE. A family of Carnivora which includes dogs, foxes and wolves.

CANINE TEETH. Long, usually pointed teeth between the incisors and the premolars in mammals; particularly prominent in the Carnivora.

CANNON BONE. A bone formed from the union of two metacarpals or meta-tarsals in many mammals, particularly in the Artiodactyla or even-toed ungulates.

CANOIDEA. A superfamily of Carnivora including dogs, bears, badgers etc., a group which, in the latest system of classification, comprises the two groups formerly known as Cynoidea and Arctoidea.

CANTHARIDAE. Carnivorous beetles which digest slugs and snails etc., by injecting gastric juice into them. The brightly coloured 'soldier' and 'sailor' beetles belong to this group and glow-worms are also sometimes included.

CANTHARINA. Sea-bream: carnivorous shore-fish without molar or vomerine teeth and having spinous fin-rays.

CAPILLARIES. The smallest blood-vessels which form a network linking arteries with veins.

CAPIT-. Prefix and root from Latin *Caput, capitis*: a head.

CAPITELLUM (CAPITULUM). A rounded bone-articulation, *e.g.* the head of a rib.

CAPITONINAE. Barbets: brilliantly coloured birds of tropical forests, having short legs with two toes pointing backwards and two forwards.

CAPITULUM (OF RIB). One of two processes (the other being the *tuberculum*) by which the rib of a mammal or bird articulates with the thoracic vertebrae. The capitulum articulates with the centrum of the vertebra and the tuberculum with the transverse process.

CAPR-. Prefix and root from Latin *Caper*: a goat.

CAPRIMULGIDAE (CAPRIMULGI). Night-jars or goat-suckers: small nocturnal woodland birds having a very wide mouth with bristles at the edge. The plumage is mottled and of various colours; the legs are short and weak. They feed chiefly on moths which they catch while flying. Eggs are laid on the ground.

CAPRINAE. A subfamily of the Bovidae comprising the sheep and goats.

CAPROIDAE. Boar-fish: spiny-rayed fish living in deep water of the Atlantic and Mediterranean; having the body compressed and with minute teeth on the jaws and vomer.

CAPSIDAE. Capsid Bugs: slender green or brown plant-bugs of the group Heteroptera; pests of apple, willow and other trees.

CAPTACULA. Extensible filaments or tentacles used both as sensory organs and for catching food, *e.g.* on many molluscs, annelids, etc.

CAPUT EPIDIDYMIS. An enlargement at the beginning of the epididymis or tube leading from the testis to the vas deferens in mammals.

CARABIDAE. Ground beetles: a large and varied family often having a bronze tinge or a violet iridescence. They have a narrow head, long thin antennae and long legs. In some species the elytra are fused together making flight impossible. The larvae, which are predatory, are of the usual campodeiform or long-legged type.

CARABOIDEA. A superfamily of beetles which are nearly all predators both in the adult and the larval stage. They include the *Carabidae* or ground-beetles, the *Cicindelidae* or Tiger-beetles and some aquatic forms such as the Great Brown Water Beetle (*Dytiscus*) and the Whirligig Beetle (*Gyrinus*).

CARAPACE. (1) Part of the exoskeleton of a crustacean made from a number of fused terga spreading like a roof over several segments of the head and thorax.
(2) The dorsal covering of a tortoise or turtle formed of bony plates fused together and usually covered by a horny epidermal layer.

CARCHARIIDAE. A family of sharks in which the mouth is crescent-shaped and ventral and the eyes have nictitating membranes. They include among others the Hammer-headed Shark whose eyes are situated at the extremities of large lateral head-lobes.

CARDIAC. Appertaining to the heart.

CARDIAC GLAND. A large digestive gland near the cardiac end of the stomach in certain mammals.

CARDIAC MUSCLE. Muscular tissue of a type found only in the wall of the heart. Its fibres somewhat resemble those of striated muscle but, unlike them, branch and rejoin to form a network. Its chief property is that of rhythmic contraction even when isolated from the body.

CARDIAC SPHINCTER. The sphincter at the entrance to the stomach of a vertebrate, *i.e.* at the end nearer to the heart.

CARDINAL VEINS. Large paired longitudinal veins or sinuses in fish and in the embryos of higher vertebrates. An anterior and a posterior pair collect blood from most parts of the body returning it by way of the *Cuvierian Ducts* to the heart.

CARDO. Latin *Cardo*: a hinge; any hinge-like part of an organ, *e.g.* the basal segment of the maxilla of an insect by which it articulates with the head.

CARETTOCHELYDAE. Turtles from New Guinea having no horny plates but only soft skin covering the bony carapace.

CARIAMIDAE. Long-legged marsh birds of South America resembling the secretary bird of Africa; easily domesticated.

CARID-. Root word from Greek *Karis, karidos*: a shrimp or prawn.

CARIDEA. Prawns and shrimps: a group of decapod crustaceans related to crayfish and lobsters but having flattened limbs adapted for swimming rather than walking.

CARIDOID FACIES. The structural arrangement of body and limbs which is found in the *Malacostraca, i.e.* shrimps, crayfish, lobsters etc. These crustaceans have twenty segments of which six belong to the head, eight to the thorax and six to the abdomen; a carapace encloses the thorax at the sides; there are stalked compound eyes, walking legs on the thorax, small biramous swimmerets on the abdomen and a broad flattened tail-fan.

CARIN-. Root word from Latin *Carina*: a keel.

CARINA. (1) The large median part of the sternum of a bird.
(2) The dorsal plate of a barnacle: one of the five calcified plates which form the carapace or mantle enclosing and protecting the body.

CARINATAE (CARINATES). Birds with a sternal keel or *carina, i.e.* most flying birds as opposed to the *Ratitae* or large running birds such as the ostrich which have no keel.

CARN-. Prefix and root from Latin *Caro, carnis*: flesh.

CARNASSIAL TEETH. The cutting teeth of carnivorous mammals: the last premolar in the upper jaw and the first molar in the lower. These have flattened cusps with sharp cutting edges.

CARNIVORA. A large order of mammals which, with few exceptions, are carnivorous in habit and are characterized by large canine teeth and sharp molars and premolars. There are two distinct groups sometimes regarded as separate orders: the *Fissipedia* or land species and the *Pinnipedia* or aquatic species. The former include cats, dogs, bears etc., whilst the latter include seals, sea-lions and walruses.

CARNIVORAE. Carnivorous pike-like fish with the head and body covered by scales; having teeth in both jaws and having the two halves of the mandible firmly united. The group includes fish from many parts of the world, some in lakes and rivers 14,000 ft. up in the Andes; others in the Dead Sea. The peculiar 'four-eyed' fish *Anableps* with two pupils in each eye belongs to this group.

CAROTID ARCH. The third aortic arch (*q.v.*) in the embryo of a vertebrate, giving rise to the common carotid artery in the adult.

CAROTID ARTERY. The main artery supplying blood to the head of a vertebrate. The common carotid artery bifurcates into an internal and an external branch.

CAROTID CANAL. A passage by which the internal carotid artery enters the skull.

CAROTID GLAND. See *Carotid Labyrinth*.

CAROTID LABYRINTH (BULBUS CAROTICUS, CAROTID GLAND). A network of capillaries forming a swelling at the junction of the external and internal carotid arteries in amphibia.

CAROTENE (CAROTIN). An orange coloured pigment allied to xanthophyll; synthesized by green plants and changed into Vitamin A in the animal body.

CARP-. (1) Prefix from Latin *Carpus*: a wrist.
(2) Prefix from Greek *Karpos*: a fruit.

CARPAL BONES. Small bones between the fore-arm and the hand of a tetrapod vertebrate. In the generalized fore-limb there are nine of these bones but in many animals there is reduction and fusion. The first or proximal row consists of the *Radiale* (adjacent to the Radius), the *Intermedium*, and the *Ulnare* (adjacent to the Ulna). Next comes a single bone, the *Centrale* and lastly a distal row of five bones corresponding with the five digits of the hand.

CARPOIDEA. Fossil echinoderms of the Cambrian and Silurian periods, resembling Crinoids but laterally compressed giving a bilateral symmetry.

CARPO-METACARPUS. The distal carpals fused with some of the metacarpals in the wing of a bird.

CARPOPODITE. The third of the five segments of the leg in such crustaceans as the crayfish, crab or lobster; in order from the base outwards the segments are known as the *Ischiopodite, Meropodite, Carpopodite, Propodite* and *Dactylopodite*.

CARPUS. The wrist (see *Carpal Bones*).

CARTILAGE. A skeletal material consisting largely of a matrix of white translucent *chondrin* sometimes containing white collagen fibres or yellow elastic fibres. In cartilaginous fish and in many lower chordates the whole skeleton is composed of cartilage, but in higher vertebrates the embryonic cartilaginous skeleton becomes largely replaced by bone in the adult.

CARTILAGE-BONE. Bone which develops from cartilage by a process in which the soft chondrin is gradually replaced by harder calcareous compounds. Two kind of cells take part in this action: namely *Osteoclasts* and *Osteoblasts*. The former break down the cartilaginous matrix and the latter deposit the harder bony substance in its place.

CARTILAGINOUS FISH (CHONDRICHTHYES OR ELASMOBRANCHII). Sharks, dogfish, skates, rays etc., whose skeletons are entirely composed of cartilage and whose skin is covered by dermal denticles. They also have open gill-slits which are not covered by an operculum. Reproduction is either viviparous or by large eggs in horny capsules produced after internal fertilization

CARUNCULA LACRIMALIS. A small conical fleshy projection at the inner. junction of the upper and lower eyelids in certain vertebrates.

CASTORIDAE. Beavers: large semi-aquatic rodents which burrow in river banks and, in the case of the North American species, construct dams.

CATA-, CAT-, CATH-. Prefixes from Greek *Kata*: down.

CATABOLISM. See *Katabolism*.

CATADROMOUS. Migrating down rivers to spawn in the sea (*cf. Anadromous*).

CATARRHINI. A group of Primates comprising the Old World apes and monkeys and also humans. They differ from the *Platyrrhini* or New World monkeys in having a narrow internasal septum and a tubular bony external ear-passage. Many are without tails but if a tail is present it is never prehensile.

CATHARTIDAE. King-vultures, condors and turkey-buzzards: carnivorous falcon-like birds of the Western Hemisphere, having hooked beaks and powerful claws. Many of them are very large birds, the female usually being larger than the male.

CATOSTEOMI (GASTROSTEIFORMES). Bony fish having an air-bladder without an open duct, or in some cases having no air-bladder; having no mesocoracoid bone and having the ventral fins, if present, in an abdominal position. They include such varied forms as sticklebacks, pipe-fish, sea-horses and flying-fish.

CAUDA EPIDIDYMIS. An enlargement at the distal end of the epididymis or tube leading from the testis to the vas deferens in mammals (see also *Caput epididymis*).

CAUDA EQUINA. The terminal spinal nerves of a mammal which together with the *Filum terminale* form a bunch. The name is derived from its resemblance to a horse's tail.

CAUDAL. Appertaining to the tail.

CAUDATA. An alternative name for *Urodela* or newt-like amphibians with tails and limbs.

CAUL. The amniotic membrane which encloses a foetus and is occasionally expelled without breaking at birth.

CAVICORN. Having hollow horns.

CAVIIDAE. Terrestrial or aquatic rodents with short incisors; tail short or absent; three digits on the hind feet and four on the front feet. The group includes cavies and Guinea-pigs and also the *Capybara*, the largest known rodent.

CEBIDAE. South American arboreal monkeys having both the thumb and the large toe opposable; having long prehensile tails, small heads and bare rounded ears.

CECIDOMYIDAE. Gall midges or gall gnats: minute two-winged insects, seldom exceeding one eighth of an inch in length, belonging to the suborder *Nematocera* and having long thin antennae, long narrow wings and very long legs. The larvae are yellow or brown, without legs and are the cause of galls on willow, oak, beech and many other trees.

CELL. A unit of living material consisting of a small mass of protoplasm bounded either by a thin membrane or in the case of plant cells by a thicker cell-wall of cellulose, lignin or other material. Embedded in the protoplasm is a nucleus consisting largely of chromatin surrounded by a nuclear membrane. The protoplasm around the nucleus may contain one or more vacuoles or fluid-filled spaces and frequently also a number of granular or rod-shaped bodies of varying and sometimes doubtful function (*mitochondria, chromoplasts, Golgi bodies* etc.). At the time of cell-division the chromatin of the nucleus takes the form of a definite number of rod-shaped *chromosomes* which are believed to carry the inherited qualities in units called *genes*. In normal cell-division or *mitosis* each chromosome splits lengthwise so that the two new cells each contain the same number of chromosomes as the parent cell. In reduction division or *meiosis* (*i.e.* during the formation of gametes) half the number of chromosomes goes into each new cell. The full number is known as the *diploid* number and the half number as the *haploid*.

CEMENT. A substance like spongy bone surrounding the roots of mammalian teeth and holding them in the sockets.

CENTETIDAE. Primitive insectivorous mammals represented by the *Tenrec* of Madagascar which has no tail and is about a foot long. It is regarded as primitive and put into a distinct family for the following reasons: the skull has no zygomatic arch; the testes are in the abdomen near the kidneys; the female urinogenital organs and anus open together.

CENTI-. Prefix from Latin *Centum*: a hundred.

CENTIMORGAN. A hypothetical unit, named after Morgan, for measuring the distance apart of genes in a chromosome. One centimorgan represents a cross-over value of 1%; one decimorgan a 10% cross-over value etc. (see *Cross-over value*).

CENTIPEDE. See *Chilopoda*.

CENTR-. Prefix and root from Latin *Centrum*: centre.

CENTRAL CANAL. The name commonly given to the cerebro-spinal canal running through the centre of the spinal cord of a vertebrate.

CENTRALE. The central tarsal or carpal bone in a pentadactyl limb situated between the proximal and distal rows of small bones which collectively form either the tarsus or the carpus.

CENTRAL NERVOUS SYSTEM. The brain and spinal cord of a vertebrate or the corresponding parts of an invertebrate: the part of the nervous system where most of the nerve cells are congregated and from which afferent and efferent nerves proceed.

CENTRARCHIDAE. Sun-fishes: North American fresh-water, spiny-rayed fish of perch-like form and able to build nests.

CENTRIOLE. A small granular body within the centrosome (*q.v.*), just outside the nuclear membrane in many cells. Before and during cell division it doubles and the two new centrioles move apart to form the poles of the spindle (*q.v.*). It is present in animals and lower plants but not in higher plants.

CENTROLECITHAL. A term used to denote eggs whose yolk is confined to the centre, as in some arthropods.

CENTROSOME. This name is sometimes used synonymously with *Centriole* (*q.v.*) and sometimes used to refer to the *Centrosphere*.

CENTROSPHERE. A clear area near the nucleus of a cell, containing the *centriole* (*q.v.*).

CENTRUM. A mass of bone or cartilage forming the centre or chief part of a vertebra; ventral to the spinal cord and surrounding the embryonic notochord.

CEPHAL-, CEPHALO-. Prefix and root from Greek *Kephale*: head.

CEPHALACANTHIDAE. Flying gurnards: spiny-rayed fish with large pectoral fins enabling them to glide through the air, but not as efficiently as the true Flying-fish.

CEPHALAD. Towards the head or front part of an animal.

CEPHALIC. Pertaining to the head.

CEPHALIC INDEX. Used in anthropology: a number obtained by dividing the maximum width of the head by the maximum length and expressing the result as a percentage.
Dolichocephalic (long-headed): up to 75%.
Brachycephalic (short-headed): over 80%.
Mesocephalic (medium): 75–80%.

CEPHALIZATION. Specialization of the anterior of an animal to form the head by the concentration in this region of feeding organs, sense organs, nerves etc.

CEPHALOCHORDA (CEPHALOCHORDATA). A primitive class of chordates containing only *Amphioxus* (the lancelet) and a few related forms which are all marine. Apart from certain highly specialized features they are considered to be an approximation to a hypothetical ancestor of all vertebrates. Their primitive features include the following: the presence of a notochord in the adult; a tubular dorsal nerve cord; many gill-slits in the pharynx; segmentally arranged myotomes or muscle-blocks all along the body. Unlike fish they have no definite head, no true brain or heart, no jaws and no paired fins. The skeleton consists only of simple cartilaginous bars supporting the gill-arches.

CEPHALO-PHARYNGEAL SKELETON (OF INSECTS). Chitinous skeletal structures forming secondary mouth-parts in the 'headless' larvae of certain flies. The head is in fact deeply sunk into the thorax; the original mouth-parts are atrophied and their place is taken by these secondary toothed or hooked structures.

CEPHALINAE. Small or medium sized African and Indian antelopes with horns on the males only, *e.g.* the *Duikerboks* of South Africa.

CEPHALOPODA. A class of mollusca including the cuttle-fish, the octopus, the chambered Nautilus and the fossil Ammonites; characterized by long prehensile tentacles surrounding the mouth. Present-day cephalopoda, except the nautilus and the argonaut, either have no shell or else it is small and internal. In the ammonites and nautilus, however, the shell is spiral with a number of chambers in the last of which the animal lives.

CEPHALOTHORAX. The fused head and thorax of many crustaceans and certain other arthropoda.

CEPHIDAE. Stem saw-flies: a small family of symphytous Hymenoptera whose larvae tunnel within the stems of grasses and other plants.

CERAMBYCIDAE. Longicorn or long-horned beetles: a large and world-wide family of beetles having very long antennae. Their larvae, which are yellowish grubs with hard brown heads and rudimentary legs, do immense damage to timber by excavating winding tunnels in which they pupate. They may live for four or five years in the larval state.

CERATA. Brightly coloured stinging processes on the body of the sea-slug *Eolis* and similar shell-less molluscs of the group *Nudibranchiata*. They contain pouches branching from the alimentary canal and the stings are in fact the undigested stinging cells of sea-anemones and jelly-fish which have been eaten. The sea-slug is thus protected by the stinging cells of its victims.

CERATO-. Prefix from Greek *Keras, keratos*: a horn.

CERATOBATRACHINAE. Frogs from the Soloman Islands with swollen tips on the fingers and toes and with well developed internal vocal sacs in the male.

CERATOBRANCHIAL CARTILAGES. Bars of cartilage forming the sides of the gill-arches of a fish; usually having a number of horn-like processes directed backwards.

CERATODUS (NEOCERATODUS OR EPICERATODUS). The Australian lung-fish: a large fish with many primitive features and many characters shared by amphibia. The skin is covered with overlapping cycloid scales; the tail is pointed and symmetrical; the paired fins are elongated and leaf-like. The swim bladder, which has a highly vascular wall, acts as a lung so that the fish can breathe out of water and move about in a clumsy fashion in the mud of a dried up lake or stream.

CERATOHYAL. Bones or cartilages forming the lateroventral part of the hyoid arch in the skeleton of a fish.

CERATOPOGONIDAE. A family of minute blood-sucking midges closely allied to the gnats.

CERATOPSIA. Gigantic herbivorous reptiles of the Cretaceous Period having three horns on the head and probably having dermal armour over most of the body. The best known genus *Triceratops* of Wyoming was 20 ft. long with a head 7 ft. long.

CERATOTRICHIA. Horny dermal fin-rays of fish.

CERC-, CERCO-. Prefix from Greek *Kerkos*: a tail.

CERCAL. Relating to the tail, as in *Homocercal, Heterocercal* etc.

CERCARIA. The fourth and last larval stage of the liver fluke (*Fasciola*) following the *redia* stage; like the redia it lives in the fresh-water snail (*Limnaea*). When mature, however, the cercaria passes out of the water snail, secretes a cyst round itself and only develops into an adult fluke if swallowed by the primary host (*e.g.* a sheep). A cercaria is like an immature liver fluke with a short tail.

CERCI (CERCI ANALES). Filamentous appendages at the posterior tip of the abdomen on many insects and other arthropods. Sometimes they are very long giving the appearance of a forked tail as on the mayfly.

CERCOPIDAE. Frog-hoppers: Homopterous insects whose name is suggested by the broad head, general frog-like appearance and jumping powers. The nymphs of many species produce and live in a frothy secretion of 'Cuckoo-spit' formed by blowing air into a fluid produced by glands near the anus.

CERCOPITHECIDAE. African and Asian monkeys, baboons, mandrills and Barbary Apes etc. usually having a non-prehensile tail and four nearly equal limbs; the thumb and the large toe are opposable and there are properly formed nails on all the digits.

CERE (CEROMA). A soft swelling at the base of the upper beak of a bird.

CEREBELLUM. An enlargement of the hind-brain of a vertebrate; lobed and convoluted in birds and mammals but a simple transverse commissure in some lower vertebrates. It is particularly concerned with unconscious co-ordination of muscles.

CEREBRAL CORTEX. The outer layer of the cerebral hemispheres consisting of 'grey matter' rich in nerve cells and synapses; in mammals it is particularly extensive and convoluted.

CEREBRAL GANGLIA. A term sometimes used for the large supra-oeso-phageal ganglia in certain invertebrates.

CEREBRAL HEMISPHERES. Paired enlargements of the fore-brain of a vertebrate behind the olfactory lobes and in front of the thalamencephalon. In mammals they are very large, extending back to cover most of the middle and hind-brain. The cortex or outer layer, known as 'grey matter', is concerned mainly with the co-ordination of functions and with intelligence.

CEREBRAL VESICLES (PRIMARY CEREBRAL VESICLES). Three hollow enlargements of the spinal cord in a vertebrate embryo, giving rise to the fore-, mid- and hind-brain.

CEREBRO-. Prefix from Latin *Cerebrum*: the brain.

CEREBROSPINAL CANAL. A fluid-filled passage running through the length of the spinal cord and enlarging to form the cavities of the brain.

CEREBROSPINAL FLUID. The lymph-like fluid filling the cavities of the brain and cerebrospinal canal.

CEREBRUM. See *Cerebral Hemispheres*.

CERTHIIDAE. Tree-creepers and wall-creepers: small insectivorous sparrow-like birds having a long slightly curved beak, a horny tongue and sharp claws, the longest on the hind-toe.

CERVIC-, CERVICO-. Prefix and root from Latin *Cervix, cervicis*: the neck.

CERVICAL. Appertaining to the neck.

CERVICAL RIBS. Ribs which are attached to the vertebrae of the neck and are usually much shorter than thoracic ribs. In mammals they are very much reduced and are fused to the vertebrae so that a small arch known as the *vertebrarterial canal* is formed on each side.

CERVICAL VERTEBRAE. The vertebrae of the neck: in nearly all mammals there are seven of these, the first being the *Atlas* and the second the *Axis*.

CERVICAPRINAE. African water-bucks: a subfamily of the Bovidae with horns on the males only.

CERVICUM. The neck-region of an insect or other arthropod.

CERVIDAE. Deer: ruminants in which the males have antlers which are shed after the mating season and rapidly grow again the following year. Antlers grow from the frontal bones and are covered by a soft velvety skin. They become more branched and complicated each season until complete maturity is reached.

CESTODA. Tapeworms etc.: Platyhelminthes with no mouth or gut and usually with a *scolex* or head from which numerous segments or *proglottides* are budded off. Each segment when ripe contains a complete set of hermaphrodite reproductive organs. As new segments form at the front, those at the back break off and are cast out with the faeces of the host. The eggs are eaten by a secondary host where they develop into larvae and encyst. A new primary host becomes infected by feeding on the secondary host.

CESTRACIONTIDAE. Bull-headed sharks: small sharks of the Pacific and S.E. Asia, having the upper lip divided into seven lobes and having small obtuse teeth in front and large pad-like teeth at the sides.

CETACEA. Whales: large aquatic fish-like mammals with no hind limbs; with the front limbs modified to form flippers; with no apparent neck and with a horizontal tail-fin or fluke. Hair is almost entirely absent but beneath the skin is a thick layer of fat or blubber. There are two main groups, *viz.* the toothless whales or *Mystacoceti* and the toothed whales or *Odontoceti.* The former include the Greenland Right Whale, the Blue Whale, the Rorqual and many others whilst the latter include the Sperm Whales, Bottle-nosed whales, dolphins and porpoises.

CHAET-, CHAETO-. Root and prefix from Greek *Chaite*: hair.

CHAETAE. Bristles, formerly known as *Setae,* of earthworms and other annelids; chitinous structures each secreted by a single large cell at the base of an ectodermal pocket called a *Chaetigerous Sac.*

CHAETIGEROUS SAC. See *Chaetae.*

CHAETODONTIDAE. Brilliantly coloured fish of coral reefs; having a compressed elevated body and spiny fins sometimes covered with small scales.

CHAETOGNATHA. Marine worms whose body is shaped like a torpedo with tail-fins and lateral fins. There are large chitinous jaws for seizing small crustacea or other prey.

CHAETOPODA. Annelids or segmented worms having chaetae or bristles. The group is divided into two orders: the *Polychaeta* which are mostly marine and may be free-swimming, burrowing or tube-dwelling, and the *Oligochaeta* or earthworms which have fewer chaetae, no sense organs, tentacles or parapodia and are usually hermaphrodite.

CHAETOPTERIDS. Complex tube-dwelling polychaete worms whose parapodia (outgrowths of the body wall) are modified for various functions: some bear hooks and bristles for anchoring the worm in its U-shaped tube; others are wing-like for pumping water through the tube; others again are ciliated and pass food particles to the mouth.

CHALAZA. A dense mass of albumen at each end of a bird's egg. It is spirally wound as the egg descends the oviduct and the two chalazae act as suspensors for the yolk-sac holding it in position within the membranes of the egg.

CHALCIDAE (CHALCIDOIDEA, CHALCIDS). Minute Hymenopterous insects nearly all parasitic or sometimes attacking the larvae of other parasites in which case they are said to be *hyperparasites*; a few species form galls and some develop within seeds. A special peculiarity is the phenomenon of polyembryony in which one egg may give rise to fifty or more separate embryos by a complex process of budding.

CHAMAELEONTIDAE. Chameleons: lizard-like reptiles with a laterally compressed body, projecting eyes which can be moved independently and a very long tongue able to shoot out and catch small insects. They are noted for their remarkable powers of changing colour to match their surroundings. This

D

is brought about by the rapid change in size and shape of certain specialized colour-cells or chromatophores in the skin.

CHAMBERED ORGAN (ECHINODERMS). A hollow space with five radial compartments situated within the large central ossicle of a Crinoid (sea-lily). It contains a plexus of nerves and is the nearest approach to a central nervous system in these creatures.

CHAMPSOSAURIDAE. Large aquatic lizard-like reptiles of the Cretaceous and early Eocene Periods.

CHARADRIIDAE (CHARADRIINAE). Plovers, lapwings, peewits etc. (see *Charadriiformes*).

CHARADRIIFORMES. Terrestrial, arboreal or marine birds with schizognathous skull (palatine bones not fused); with eleven primary wing quills but having the fifth feather on the ulna missing (aquintocubitalism). The tribe is divided into four groups and includes plovers, gulls, grouse and pigeons.

CHEILOSTOMATA. Polyzoa in which the aperture of the *zooecium* or capsular body-wall can be closed by an operculum.

CHEIROPTERA. Bats (see *Chiroptera*).

CHEIROPTERYGIUM. A pentadactyl or five-digited limb.

CHEL-. Prefix and root from Greek *Chele*: a claw.

CHELA. The pincer of a crab or lobster formed from the enlarged *propodite* on which the *dactylopodite* is hinged.

CHELICERAE. Sharp or clawed appendages in front of the mouth on spiders and other arachnida. They are used for attacking prey and may contain a poison gland at the base.

CHELIPED. A limb of a crustacean bearing a chela or pincer.

CHELONEMYDIDAE. Extinct turtles of the Jurassic Period.

CHELONIA. Tortoises and turtles: reptiles having the body encased in a large flattened bony capsule composed of a number of dermal bones forming the dorsal *carapace* and the ventral *plastron*. These may or may not be covered by dermal horny plates roughly corresponding to the underlying bones.

CHELONIDAE. Turtles: aquatic chelonians of warm seas and rivers in many parts of the world. They swim by means of paddle-like limbs and lay their eggs on sandy shores. Some of the edible turtles are as much as 4 ft. long.

CHELYDIDAE. Fresh-water tortoises of South America, Australia and New Guinea, having webbed feet and a long neck which bends laterally when the head is retracted. Most of them are carnivorous.

CHELYDRIDAE. Snapping turtles of Ecuador and alligator-turtles of North America: fierce reptiles having a powerful hooked beak and webbed feet with sharp claws.

CHEMO-RECEPTORS. Sense-organs or sense-cells which are stimulated by the chemical action of definite substances in contact with them, as for instance the organs of taste and smell.

CHERMESIDAE. See *Adelginae*.

CHEVRON BONES. Y-shaped haemal arches on the caudal vertebrae of reptiles.

CHIASMA. (1) A crossing over and interchange of parts of homologous chromosomes during meiosis (see *Crossing over*).

(2) The Optic Chiasma: a crossing of parts of the optic nerves beneath the mid-brain of a vertebrate.

CHILARIA. Flattened processes on the pregenital segment of the King Crab (*Limulus*). Their function is unknown.

CHILO-. Prefix from Greek *Cheilos*: a lip.

CHILODIPTERIDAE. Cardinal-fishes: small brilliant red perch-like fish often only about 1 in. long living in coral reefs.

CHILOPODA (OPISTHOGONEATA). Centipedes: carnivorous Myriopoda, distinguished from millipedes by having fewer segments and by having a single pair of legs on each. Although less specialized than insects, they resemble them in having antennae, mandibles and maxillae and in breathing by means of spiracles which lead into a complex tracheal system. There is a pair of poison claws on the first segment.

CHIMAERIDAE. Fantastically shaped fish with a cartilaginous skeleton but having an operculum and several other features usually only found in bony fish. They belong to the group *Holocephali* which contains many extinct genera but only three living ones. *Chimaera*, the so-called 'King of the Herrings', found in many parts of the world, has a shark-like body with a compressed head, a small mouth and very large pectoral fins.

CHIONIDIDAE. Sheath-bills: wading fish-eating birds of the Antarctic.

CHIRO-. Prefix from Greek *Cheir*: a hand.

CHIROMYIDAE. Squirrel-like animals of Madagascar, related to the lemurs, having long ears, bushy tail and eyes facing forwards, a characteristic which gives stereoscopic vision and is an advantage to arboreal animals when jumping from bough to bough. As in other lemurs and monkeys the thumb and the large toe are both opposable; this also being an advantage in tree-climbing.

CHIRONOMIDAE. Small two-winged flies whose aquatic larvae may be red, green, blue, yellow or colourless. The small red larvae of *Chironomus* itself are known as blood-worms from the fact that they are among the few insects which contain haemoglobin. At the hind end are two tubes or 'blood-gills' whose function is uncertain.

CHIROPTERA. Bats: flying mammals with wings formed from a membrane of skin, known as a *patagium*, stretched across from the front to the hind limbs and from these to the tail if one is present. The largest part of the wing extends over the very long fingers on each side, with the exception of the thumbs which are free and bear claws. Most bats are insectivorous but some are fruit-eaters and a few blood-suckers. They usually sleep hanging upside down during the day and fly out at night. The eyes are small but bats are able to emit and hear extremely high pitched notes, the echoes of which enable them to avoid objects when flying.

CHITIN. A horny protective substance forming the cuticle of insects and other arthropoda. It is an amino-polysaccharide and resists most solvent agents.

CHLAMYDO-. Prefix from Greek *Chlamus, chlamudos*: a cloak.

CHLAMYDOSELACHIDAE. Eel-like viviparous fish, five to six feet long having a cartilaginous skeleton and six pairs of gill-openings. They are found in Japanese seas and in the Atlantic and Arctic Oceans.

CHLAMYDOSPORES. A term used both in botany and zoology to denote spores which are individually enclosed in spore-cases, as in the Mycetozoa.

CHLORAGOGENS. Yellow pigmented cells which form around the gut of an earthworm or other annelid. They break loose and float about in the body cavity where they absorb nitrogenous waste substances from the coelomic fluid. They then break up and their remains are usually engulfed by amoeboid cells which either carry them to the nephridial tubules to be excreted or into the tissues of the body where they deposit their pigment.

CHLORO-. Prefix from Greek *Chloros*: green.

CHLOROCRUORIN. A green respiratory pigment containing iron, found in the blood of certain polychaete worms.

CHLOROMONADINA. Fresh-water protozoa of the class Mastigophora (Flagellata). They may be regarded as a link between plants and animals as they contain numerous green chloroplasts and are able to feed by photosynthesis.

CHOAN-, CHOANO-. Prefix from Greek *Choane*: a funnel.

CHOANAE. Internal nostrils: openings from the nasal cavity into the roof of the mouth in all air-breathing vertebrates.

CHOANATA (CHOANATAE). Vertebrates which have internal nostrils, *i.e.* all land vertebrates as well as lung-fish.

CHOANICHTHYES. Fish such as the lung-fish which have nostrils inside the mouth. Although there are only a few living genera the group is well represented in the Devonian and Carboniferous Periods and is believed to have given rise to Amphibia and hence to all land-living vertebrates.

CHOANOCYTES. Collared, flagellated cells covering large parts of the inner chambers of sponges and helping to keep the water circulating. They are similar in appearance to protozoa of the group Choanoflagellata and it is possible that sponges may have evolved from colonies of these.

CHOANOFLAGELLATA (CHOANOFLAGELLIDAE). Stalked protozoa, either single or in branching colonies, having a flagellum projecting from inside a protoplasmic collar or cup. Food particles are wafted by the flagellum towards the base of this collar where they are absorbed. The structure of these organisms is essentially similar to the flagellated cells known as *choanocytes* which line the inner chambers of sponges.

CHONDR-, CHONDRO-. Prefix and root from Greek *Chondros*: gristle or cartilage.

CHONDRICHTHYES (ELASMOBRANCHII). Cartilaginous fish: those whose skeletons are entirely of cartilage and which have open gill-slits not covered by an operculum. The group is at present small and included sharks, dogfish, skates and rays.

CHONDRIN. The white translucent substance forming the matrix of cartilage.

CHONDRIOSOMES. An alternative name for Mitochondria (*q.v.*).

CHONDROBLASTS. Cells which give rise to cartilage.

CHONDROCRANIUM (NEUROCRANIUM). The primary cranium: part of the skull composed of cartilage or cartilage-bone to which membrane-bones may later be added.

CHONDROSTEI. Fish with a cartilaginous skeleton but having bony plates on the head and parts of the body. They are represented at the present day only by the sturgeons (see *Acipenseroidei*).

CHONOTRICHA. Flask-shaped or cup-shaped protozoa usually attached by their bases to the bodies of crustacea etc. They are classed among the Ciliophora but the only cilia are those on the wide spiral funnel leading to the mouth. The body-wall is naked, this probably being a secondary feature associated with the sedentary habit.

CHORDA CELLS. Cells of the vertebrate embryo which give rise to the notochord. They are similar to mesodermal cells and move in from the dorsal lip of the blastopore until they come to lie beneath the neural plate.

CHORDAE TENDINEAE. Chords of tissue connecting the flaps or cusps of a valve to the muscular walls of the ventricle in the heart of a vertebrate.

CHORDAL VERTEBRAE. Vertebrae whose skeletal elements in the course of development invade and perforate the notochord instead of merely surrounding it. Such vertebrae are found in sharks, sturgeons and lung-fish.

CHORDATA. A phylum comprising all animals which at any stage in their development possess a notochord. They include all the vertebrates as well as the lower groups Hemichordata (Acorn-worms), Urochordata (Tunicates) and Cephalochordata (*Amphioxus*, the lancelet). (See *Notochord.*)

CHORDA TYMPANI. A branch of the facial or seventh cranial nerve of a vertebrate passing over the tympanic cavity and down the side of the face to the submaxillary salivary glands. It forms part of the parasympathetic system.

CHORDO-. Prefix from Greek *Chorde*: string.

CHORDOCENTROUS VERTEBRAE. Vertebrae whose centra are formed entirely or mainly from the sheath of the notochord as for instance in cartilaginous fish.

CHORDOTONAL RECEPTORS. Sense organs of insects probably detecting changes of tension in the muscles and possibly enabling the insect to balance itself while flying. Each consists of a thin fibre attached to the cuticle, passing through a rod called a *scolopale* to a sense cell beneath the surface.

CHORIOID. See *Choroid.*

CHORION. (1) The outermost of the foetal membranes in reptiles, birds and mammals, consisting of a layer of ectoderm with a layer of mesoderm beneath. In mammals it helps to form part of the placenta (see also *Amnion*).
(2) The 'shell' of an insect's egg.

CHOROID. The middle layer in the wall of the eye of a vertebrate, between the retina and the sclerotic coat. It contains blood-vessels and pigment.

CHOROID GLAND. A vascular plexus of unknown function round the optic nerve at the point where it leaves the eye in some fish.

CHOROID PLEXUS. A part of the roof of the vertebrate brain consisting of a highly vascular membrane with a large number of villi projecting into the cavity beneath; its principal function is to secrete cerebrospinal fluid. The anterior choroid plexus covers the third ventricle and the posterior covers the fourth ventricle.

CHROM-, CHROMO-, CHROMATO-. Prefixes from Greek *Chroma*, *chromatos*: colour.

CHROMAFFIN TISSUE. Tissue which can readily be stained, *i.e.* containing much chromatin.

CHROMATID. One of the two halves of a chromosome formed by longitudinal fission during cell division.

CHROMATIN. A constituent of chromosomes which readily stains with basic dyes such as haematoxylin.

CHROMATOPHORES. (1) Chromoplasts (*q.v.*).
(2) Pigment cells of the skin in frogs, chameleons etc. which by rapidly changing their size or shape can cause the animal to change colour and so match its surroundings. Such changes are under the control of the pituitary gland.

CHROMIDES. See *Cichlidae.*

CHROMIDIUM. A mass of chromatin granules of uncertain function in the cytoplasm of certain protozoa; they may appear occasionally or may be present throughout the greater part of the life-cycle.

CHROMOMERES. Granules running along the length of a chromosome, visible during meiosis and mitosis; believed to contain the genes.

CHROMOPLASTS (CHROMATOPHORES). Coloured protoplasmic structures found in plant cells and in the plant-like protozoa (Phytomastigina). They may contain chlorophyll, xanthophyll, carotin and various red or other coloured pigments.

CHROMOSOMES. Small elongated bodies in the nuclei of most cells, existing in a definite number for each species and believed to be the carriers of hereditary qualities. Normally they are in homologous pairs, one member of a pair being derived from each parent. The two in a pair are similar in size and shape. During meiosis homologous chromosomes separate so that gametes contain only one member of each pair, or half the total number of chromosomes present in the parent cell. Gametes are therefore said to be *haploid* and parent cells *diploid*. In addition to normal chromosomes there may be special ones which have been given the names 'X' and 'Y'. These carry the factor for sex determination. In humans all the egg-cells contain an X-chromosome; sperm cells an X or a Y. When these come together, therefore, the offspring may be 'X + X' (female) or 'X + Y' (male). Chromosomes normally only become visible during cell division. In the resting stage a nucleus appears only to contain a mass of chromatin.

CHRYS-, CHRYSO-. Prefix from Greek *Chrusos*: gold.

CHRYSALIS. The pupa of a butterfly (see *Pupa*).

CHRYSIDIDAE. Ruby-tail wasps: a family of wasps having bright shining green or red bodies. They lay their eggs in the burrows of other solitary wasps, bees or sawflies and their larvae feed on those of the other species.

CHRYSOCHLORIDAE. Mole-like insectivora of South Africa, *e.g.* the Cape Golden Mole.

CHRYSOMELIDAE. Leaf-beetles: small brightly coloured beetles such as the well known Colorado Potato Beetle, the Flea-beetles and the Asparagus beetle. The larvae are soft, fleshy, with short legs and usually brightly coloured. Many do considerable damage by eating the leaves of trees or farm and garden crops. They pupate in the soil.

CHRYSOMONADINA. A primitive order of flagellate protozoa having green, yellow or brown chromoplasts. There is no gullet and feeding is largely holophytic. Some of them pass through an amoeboid stage and some secrete calcareous or siliceous skeletal plates.

CHRYSOPIDAE. Green lacewing-flies: insects of the order Neuroptera having pale green bodies, iridescent wings and prominent reddish-golden eyes. The small white eggs are laid on long stalks affixed to leaf surfaces. Larvae are carnivorous and usually feed on plant-lice. Often these larvae camouflage themselves by carrying small plant fragments and the empty skins of their victims.

CHYLE. The milky emulsion of fatty substances, enzymes and water absorbed into the lacteals or lymphatic cavities in the villi of the small intestine.

CHYME. The thick creamy acid fluid formed of partly digested foods mixed with gastric juice in the stomach.

CICADELLIDAE. Leaf-hoppers: homopterous plant-sucking insects closely related to frog-hoppers and to aphides. They have small elongated bodies and can jump and fly with great speed. They probably spread many virus diseases among plants.

CICADIDAE. Cicadas: Homopterous insects of warm regions, having four membranous wings and long piercing mouth-parts. The males, which live only for a few days, can make a very loud rattling noise with drum-like membranes on the sides of the body; the females are silent. They lay their eggs in small holes which they bore into twigs. Larvae live, sometimes for many years, in the soil.

CICHLIDAE (CHROMIDAE, CHROMIDES). Spiny-finned fish which carry their young in the mouth. They are found particularly in Lake Tanganyika.

CICINDELLIDAE. Tiger-beetles: active predatory insects having large jaws with which they hunt their prey at great speed in sandy places. The larva is a sluggish animal living in a vertical burrow with its head out ready to seize any passing insect.

CICONIIDAE. Storks, Ibis and Spoonbills (see *Ciconiiformes*).

CICONIIFORMES. Long legged aquatic or marsh birds including herons, bitterns, storks, ibis, flamingoes etc.

CILI-, CILIO-. Prefix and root from Latin *Cilium*: an eyelash.

CILIA. Protoplasmic filaments growing out from a cell, like flagella but shorter and more numerous. By their rhythmic lashing they cause movements of water or other fluids or of a whole animal if small enough.

CILIARY FEEDING. A method of feeding by which the movement of cilia causes a stream of water and microscopic organisms to enter the mouth and pass down the alimentary canal. Such a method is used in *Amphioxus* and in many invertebrates.

CILIARY MUSCLES (CILIARY BODY). A ring of small muscles at the junction of the cornea and the choroid in the vertebrate eye, by means of which the shape of the lens can be slightly changed to bring about accommodation or alteration of focus.

CILIATA. Protozoa of the class Ciliophora excluding the small subclass known as Suctoria. They possess cilia throughout the whole of their adult life but do not have suctorial tentacles.

CILIATED EPITHELIUM. A layer of cells each usually bearing a number of cilia whose rhythmic beating can keep fluids moving, *e.g.* in the mammalian trachea and oviduct and the frog's oesophagus.

CILIOPHORA. Protozoa which at some stage in their life possess cilia and which usually have a micro- and a meganucleus. The class is a very large one and includes some of the most highly specialized of all protozoa.

CILIOSPORES. Ciliated spores such as are produced by certain protozoa.

CIMICIDAE. Bed-bugs: parasitic insects of the order Hemiptera living intermittently on their hosts and sucking blood from time to time. They have a flattened body with no wings, a tough skin and a suctorial proboscis.

CINCLIDAE. Dippers or water-ousels: small sparrow-like birds living by mountain streams. They feed on aquatic insects and are able to walk into and under water and to swim and dive.

CINGULUM. Any belt-like organ, *e.g.*
    (1) A ridge round the base of the crown of certain mammalian molar and premolar teeth.
    (2) The outer of two ciliated rings round the mouth in many rotifers.

CINOSTERNIDAE. American tortoises having the temporal region of the skull not roofed over and having eight plastral bones (ventral plates).

CIRCULUS CEPHALICUS. A circular blood vessel formed by the two latera dorsal aortae joining in front of and behind the gills in certain fish such as the cod.

CIRCULUS VENOSUS. A circular vein running round the rim of the visceral hump in snails and other pulmonate molluscs. It receives blood from the haemocoels of the body and communicates with the lung and heart.

CIRCUM-. Latin prefix: round, about, surrounding.

CIRCUM-OESOPHAGEAL BAND. See *Peripharyngeal band.*

CIRCUM-OESOPHAGEAL COMMISSURES. The two halves of the nerve collar surrounding the oesophagus and linking the ventral nerve cord with the dorsal 'cerebral ganglia' in earthworms, arthropoda and many other invertebrates.

CIRCUM-PHARYNGEAL BAND. See *Peripharyngeal band.*

CIRCUM-PHARYNGEAL COMMISSURES. See *Circum-oesophageal commissures.*

CIRCUMVALLATE PAPILLAE. Papillae of the mammalian tongue in the shape of small mounds each surrounded by a circular depression. Taste-buds are situated round the bases of these papillae.

CIRR-, CIRRH-. Prefix and root from Latin *Cirrus*: a lock of hair.

CIRRHAL OSSICLES (OF ECHINODERMS). Calcareous plates covering the cirri or tentacles of sea-lilies (crinoids).

CIRRHITIDAE. Large carnivorous perch-like fish such as the 'Trumpeter' of Australasia and New Zealand. They are among the most important food-fishes of the Southern Hemisphere.

CIRRHI. Slender filaments or tentacles such as those which surround the mouth of many marine worms and other invertebrates.

CIRRIPEDIA. Highly specialized and sometimes extremely degenerate crustaceans including the barnacles and the parasitic *Sacculina*. The former attach themselves to rocks by means of an anterior sucker and have a carapace or mantle strengthened with calcareous plates which completely enclose the body. The latter, which are usually parasitic on crabs, have root-like outgrowths permeating the body of the host and causing the phenomenon known as parasitic castration.

CITHERININA. Carp-like fish of Tropical Africa having a very long dorsal fin and minute labial teeth.

CLAD-, CLADO-. Prefix from Greek *Klados*: a twig.

CLADISTIA. See *Polypterini.*

CLADOCERA. Water-fleas: crustacea of the class Branchiopoda and the order Diplostraca having a laterally compressed carapace enclosing the trunk-limbs, two large forked antennae used for swimming, and a single sessile compound eye formed by the fusion of a pair (see also *Calyptomera*).

CLADOSELACHII (PLEUROPTERYGII.) Shark-like cartilaginous fish of the Devonian Period, with a fusiform body, heterocercal tail and a pair of lateral horizontal fins at right angles to the body.

CLASMATOCYTES (HISTIOCYTES). Large irregularly shaped phagocytes found in the lymph, connective tissue and blood of mammals. They are similar to monocytes but are distinguished from them by their ability to take up vital stains.

CLASPERS. (1) Of fish. The copulatory organs of a male elasmobranch fish, consisting of a pair of stiff rod-like projections, one on each side of the cloaca between the pelvic fins.

(2) Of insects. A pair of processes on the hind-end of the abdomen of a male insect serving to grasp the female when mating.

CLAVICLES. The collar bones: a pair of membrane bones on the ventral side of the pectoral girdle of many vertebrates. In some mammals they are reduced or absent, but in most they form the chief ventral part of the girdle and have superseded the coracoids which are predominant in birds and reptiles.

CLAVICULAR GIRDLE. The dermal part of the pectoral girdle in certain fish, consisting of the clavicle, cleithrum, supra-cleithrum and post-temporal bones. The latter are attached to the hind-part of the skull (*cf. Scapular girdle*).

CLEAVAGE. The repeated division of a fertilized ovum along definite planes at right angles to one another, usually forming first two cells, then four, eight, sixteen etc., eventually forming a *morula* or solid ball of cells and later a *blastula* or hollow ball of cells. If the egg has little or no yolk, cleavage is complete and equal. Eggs with more yolk may undergo complete but unequal cleavage forming larger cells on the yolky side, since the yolk inhibits cell division. Eggs with a very large amount of yolk such as those of birds undergo partial or *meroblastic* cleavage forming a disc of cells (blastodisc) on the side farthest from the yolk.

CLEIDOIC EGG. An egg enclosed within a protective shell.

CLEITHRUM. A dermal bone superimposed upon the original cartilaginous bones on each side of the pectoral girdle in many fish and some amphibia

CLERIDAE. Ant-beetles: brightly coloured predatory beetles commonly found feeding on the smaller bark-beetles which infest pine trees. The common British ant-beetle is black and red with white bands on the wing-cases. Its larva is scarlet with a dark head.

CLINOID RIDGE. That part of the vertebrate chondrocranium which forms the hind wall of the pituitary fossa. Embryologically it marks the junction of the parachordal and the trabecular regions.

CLITELLUM. The 'saddle' of an earthworm: a region in which the epidermis is thickened, highly vascular and glandular. Its function is to secrete the 'slime-tube' which binds two worms together during copulation and also the cocoon into which the eggs are later deposited.

CLITORIS. A small rod-like body projecting into the entrance of the vagina in female mammals. It is erectile and homologous with the penis of the male.

CLOACA. A cavity in the pelvic region into which the rectum and the urino-genital ducts open. An animal with a cloaca may therefore eject faeces, urine, eggs or seminal fluid through the single wide opening. A cloaca is present in most fish, amphibia, reptiles and birds and in the more primitive mammals, but not in the *Eutheria* or placental mammals.

CLOACAL PLATE. A plate of ectoderm and endoderm covering the end of the hind-gut in the embryos of most vertebrates. It afterwards becomes perforated to form the cloaca which in mammals is later divided to form the anus and the urino-genital aperture.

CLUB-SHAPED GLAND. The first gill-cleft to develop in the embryo of

Amphioxus. It forms a coiled glandular tube on the right of the pharynx near the position of the mouth before the latter is opened. It disappears at metamorphosis.

CLUPEIDAE (CLUPEOIDEA, CLUPEIFORMES). Herrings and similar fish with soft fin-rays, a scaly body and four pairs of gills; having the pectoral girdle attached to the skull and the pelvic girdle well back. They are extremely prolific and are found in coastal waters and rivers over most of the world.

CLYPEASTROIDA. 'Cake-urchins' or 'sand-dollars'. Flattened burrowing sea-urchins having many small spines and tube-feet.

CLYPEUS. A cuticular plate of an insect's head immediately anterior to the *Frons* and above the *Labrum.*

CNEMIAL CREST. A ridge on the tibia to which the main tendon of the knee is attached.

CNIDARIA. Coelenterates which possess *nematocysts* or stinging thread-cells, *i.e.* all hydroids, corals, jelly-fish and sea-anemones but not including the *Ctenophora.*

CNIDO-. Prefix from Greek *Knide*: a nettle.

CNIDOBLAST. A stinging thread-cell (see *Nematoblast* and *Nematocyst*).

CNIDOCIL. The 'trigger' of a nematoblast: a small protoplasmic projection which when touched stimulates the discharge of a stinging thread.

CNIDOSPORIDIA. Parasitic protozoa of the group Neosporidia having at one stage structures resembling nematocysts.

CO-, CON-, COM-. Latin prefix: with, together.

COARCTATE PUPA. A type of pupa found in many Diptera (*e.g.* the house-fly) in which the last larval skin forms a hard barrel-shaped case or *puparium.* When the adult fly emerges, the puparium splits and the top is pushed away by means of an eversible head-sac known as the *ptilinum.*

COBITIDINA. Loaches: small fresh-water fish with short dorsal and anal fins and with six barbels in the jaws.

COCC-, COCCO-. Prefix from Greek *Kokkos*: grain.

COCCIDAE. Mealy bugs and scale-insects: highly specialized and somewhat degenerate insects of the group Homoptera. Mealy bugs are covered with a fine waxy secretion; scale insects appear like small white scales on the bark of trees. The 'scale' is formed of several cast off skins which have shrunken and remain stuck together by a glutinous secretion. Many species are very injurious to trees; others have commercial value yielding such products as cochineal and shellac.

COCCIDIA. Protozoa, generally intestinal parasites, which undergo a complex life-history. Cells known as gamonts, precursors of gametes, adhere together by a sticky secretion forming a *syzygy.* From these gamonts there arise male and female gametes which unite to form zygotes. Each zygote then encysts and divides by multiple fission to give four sporoblasts each in a secondary cyst within the larger cyst. From the cysts emerge sporozoites which complete the life-cycle.

COCCIDIOMORPHA. Protozoa of the class Sporozoa and subclass Telosporidia having complex life-cycles as intracellular parasites. They include the Coccidia which are gut-parasites and the Haemosporidia which are blood-parasites.

COCCINELLIDAE. Ladybirds: small brightly coloured hemispherical beetles

50

with well defined spots and other markings on the wing-covers. The larvae are of the *campodeiform* type (*q.v.*) but differ from those of ground beetles in having the appendages and sensory organs much reduced. They feed voraciously on aphides and other small insects and have been used effectively for controlling these pests.

COCCOLITH. See *Coccolithophoridae*.

COCCOLITHOPHORIDAE. Marine planktonic protozoa having one or two flagella and two green or yellow chromatophores. The body is enclosed either in calcareous plates known as *coccoliths* or in rods known as *rhabdoliths*. Since these organisms can perform photosynthesis they may be regarded either as animal or plant and are included in the subclass Phytomastigina.

COCCOSTEOMORPHI. Extinct fish, sometimes up to thirty feet in length, belonging to the class *Arthrodira* (*q.v.*); having bony plates on the head, articulating with a strong bony cuirass on the front of the body. The hind-part was naked and ended in a whip-like tail.

COCHLEA. The spiral part of the labyrinth or inner ear containing the complex system of receptors by which various sounds are perceived. It is extremely well developed in mammals but in lower vertebrates it may be absent or may be in the form of a simple straight or slightly curved sac called a *lagena*.

COCOON. (1) A protective covering for eggs, *e.g.* of spiders or earthworms.
(2) A protective covering for the pupa of an insect; constructed of silk secreted by the larva and often mixed with particles of soil, wood etc.

COECILIIDAE. The chief family of Gymnophiona or limbless worm-like amphibians usually having cycloid scales embedded in the skin and arranged in the form of transverse rings.

COEFFICIENT OF DESTRUCTION. The percentage of the offspring of any animal which must normally be destroyed by disease, parasites or natural enemies in order to keep the population constant. If the number of offspring is very large, as for instance in insects, then the coefficient of destruction must also be large in order to maintain a balance (see also *Biotic Potential*).

COEL-. Prefix from Greek *Koilos*: hollow.

COELACANTHIDAE (COELACANTHINI). Coelacanths: primitive fish belonging to the group Crossopterygii whose paired fins are fleshy lobes covered with scales and fringed with delicate fin-rays arranged as in a fan. The tail is symmetrical and usually three-lobed. The swim-bladder is often ossified and therefore sometimes well preserved in fossils. The neural spines are only superficially ossified and appear hollow when fossilized (the name *Coelacanth* means 'hollow-spined'). The group was thought to have been long extinct, but recently several living specimens have been found off East Africa.

COELENTERATA. A large phylum of sedentary or free-swimming animals having a radially symmetrical sac-like body with a mouth at one end, usually surrounded by a ring of stinging tentacles. The digestive cavity, known as the *coelenteron*, has only this one opening, but repeated budding may give rise to a colony in which all the body-cavities are connected with each other. The body-wall is *diploblastic*, *i.e.* made of two layers of cells, ectoderm and endoderm with a gelatinous layer known as the *mesogloea* between them. The individuals of a colony are known as *polyps* or *hydroids*. They form the asexual sedentary generation which alternates with a free-swimming sexual phase known as the *medusa*. The latter is a disc-shaped jelly-fish which breaks away from the parent colony and swims away. It produces gametes and after fertilization has taken place a new polyp colony develops.

Included in the Coelenterata are corals and anemones in which the hydroid stage is dominant, jelly-fish in which the medusoid stage is dominant, and complex forms such as the 'Portuguese Man-of-war' formed of a large floating colony of diverse polyps.

COELENTERON. See *Coelenterata above*.

COELIAC ARTERY. The principal artery going to the stomach, liver and associated organs of a vertebrate.

COELIACO-MESENTERIC ARTERY. An artery with branches to the stomach, spleen, liver and intestines in some vertebrates.

COELIAC PLEXUS. See *Solar Plexus*.

COELOCONIC RECEPTORS. Olfactory or other sense-organs of an insect, more advanced than basiconic receptors and differing from them in having the terminal process of the sense cell sunk in a small pit (see *Basiconic*).

COELOM. The body-cavity of most triploblastic animals, formed by splitting of the mesoderm into an outer and an inner layer.

COELOMATE, COELOMATA. Those animals which have a coelom, *i.e.* all triploblastic animals other than Platyhelminthes, Nematoda, Rotifera and a few other minor groups.

COELOMODUCT. A duct leading from the coelom to the exterior, *e.g.* the oviduct and the excretory ducts of many animals.

COELOMOSTOME. The inner opening of a coelomoduct.

COEN-. Prefix from Greek *Koinos*: common.

COENENCHYMA. The tissue of the coenosarc: that which connects together all the polyps of a colony, as for instance in coral.

COENOSARC. A system of branching tubes connecting all the polyps of a colony and enabling water and food particles to be shared by all.

COFFIN BONE. The third or last phalanx of a horse's foot.

COFFIN JOINT. The joint between the second and third phalanges of a horse's foot.

COLD BLOODED ANIMALS. See *Poikilothermous*.

COLEOPTERA. Beetles: an order of insects comprising about a quarter of a million species, the largest order in the whole animal kingdom. The fore-wings are thickened to form *elytra* or covers which protect the membranous hind-wings. The mouth-parts are of the biting type.

Beetles are holometabolous, *i.e.* they undergo complete metamorphosis. The larvae are very variable: some are predatory insects of the *campodeiform* type with long legs and antennae; others such as the wireworm are entirely herbivorous and have the legs and antennae reduced. The larvae of chafer beetles are of the *scarabaeoid* type with a large soft inflated abdomen.

COLII. Mouse-birds: small long-tailed mouse-coloured birds of creeping habit, living in the forests of Africa and feeding on fruit.

COLL-. Prefix from Greek *Kolla*: glue.

COLLAGEN. A protein which forms the white fibres of animal connective tissue and which gives gelatine when boiled.

COLLATERAL GANGLIA. Ganglia of the vertebrate autonomic system, situated medially in the mesenteries and connected with the spinal nerves by the splanchnic nerves.

COLLEMBOLA. Spring-tails: very small primitively wingless insects having biting mouth-parts, short antennae and legs without tarsi. The abdomen has six segments which are sometimes fused together. On the fourth segment is a *furcula* or springing organ which, when not in use, is held beneath the abdomen by a 'catch' or *retinaculum*. When this is released the furcula springs back propelling the insect through the air (see also *Ametabola* and *Apterygota*).

COLLETERIAL GLANDS. A pair of glands opening into the vagina or oviduct of many insects. Their function is to secrete a sticky fluid which may be used to fasten eggs to surfaces or in some cases may harden into an *ootheca* or egg-case.

COLLOBLASTS. Lasso-cells: specialized cells with a long coiled thread ending in a sticky head. They are borne on the tentacles of Ctenophora (sea-goose-berries) and are used for catching prey.

COLLUM. (1) The neck (Latin): any neck-like structure.

(2) The tergite of the post-labial segment of a millepede, consisting of a large dorsal plate between the head and the first pair of legs.

COLON. The large intestine of a vertebrate between the small intestine and the rectum; having the function of absorbing water from the faeces. The name is also used to denote the corresponding part of the intestine in some invertebrates.

COLONICI. Insects which live in colonies.

COLOUR CHANGES. Colour changes which help an animal to match its surroundings are of two kinds, *viz.*

(1) Temporary changes brought about by chromatophores (*q.v.*) as in the chameleon and to a less degree in the frog.

(2) Seasonal changes due to moulting of fur or feathers and to the growing of a new coat of the appropriate colour. Examples of this are the Arctic fox, Arctic hare, stoat, ptarmigan etc. all of which turn white in winter.

COLUBRIDAE. A large family of snakes all of which have movable facial bones and have both jaws and palate toothed; many are poisonous.

COLUBRINAE. Harmless non-poisonous snakes having overlapping scales on the skin; having teeth on the entire length of the maxillary and dentary bones but having no grooves or ducts in any of the teeth and no poison sacs. There are about a thousand species of which the common British grass snake is one.

COLUMBAE. Pigeons and doves: medium sized birds having a short neck and legs and a fairly long beak with naked scaly skin at its base. In addition to all true doves and pigeons, the extinct flightless Dodo of Mauritius is sometimes included in this group.

COLUMBIDAE. Doves and Pigeons (see *Columbae* above).

COLUMELLA. Diminutive of Latin *Columna*: a column.

(1) See *Columella Auris*.

(2) The pillar about which the whorls of a spiral shell are coiled.

(3) The axis of the cochlea.

COLUMELLA AURIS. A single auditory ossicle connecting the tympanic membrane with the inner ear in amphibia, birds and some reptiles. It is homologous with the *Hyomandibula* of fish and with the *Stapes* (stirrup) of the mammalian ear.

COLUMELLA CRANII. See *Epipterygoid*.

COLUMELLA MUSCLE. The muscle by means of which the foot of a snail or similar mollusc can be retracted into the shell.

COLUMNAE CARNEAE. Ridges of muscular tissue projecting from the thick walls of the ventricles into the ventricular cavities of the vertebrate heart.

COLYMBIDAE. Divers: marine birds breeding on the shores of inland waters in periarctic regions (see *Colymbiformes*).

COLYMBIFORMES. Divers and grebes: water birds with webbed or lobed toes, flattened metatarsus and short tail feathers. The body is carried upright and the feet are far back. The young birds are hatched with a complete covering of down.

COMATULA. A 'feather-star' or free-swimming Crinoid.

COMMENSAL, COMMENSALISM. Two organisms of different species living in close association and sharing the same food, but not dependent on each other sufficiently to be called symbionts or parasites.

COMMISSURE. A line or band of tissue linking two parts or organs, *e.g.* nerve strands connecting paired parts of the brain; the two halves of the nerve collar in annelids and arthropods; commissural blood-vessels linking dorsal and ventral vessels in an earthworm etc.

COMMUNIS SYSTEM. The visceral afferent nervous system consisting of sensory nerves running from the mucous membrane of the pharynx, mouth and viscera and converging on the visceral lobe ('taste-brain') in the medulla oblongata. Such fibres form parts of the fifth, seventh, ninth and tenth cranial nerves.

COMPACT BONE. Bone in which the blood-vessels run in Haversian canals surrounded by concentric bony rings.

COMPASSES. Slender radial arch-like ossicles in the base of the structure known as *Aristotle's Lantern* which forms the skeleton of the mouth-parts of sea-urchins (Echinoidea). (See *Aristotle's Lantern*.)

COMPENSATING HYPERTROPHY. The abnormal growth of a part of the body when another part has been removed.

COMPENSATION SAC. A membranous sac with an opening next to the mouth in some Polyzoa. It fills with water when the tentacles shoot out and empties when they are withdrawn, thus keeping the volume of the organism unchanged. This is necessitated by the fact that these polyzoa are almost entirely covered by a calcified capsule or *zooecium* and therefore cannot change their shape. In other species which are less rigid, compression of the body brings about eversion of the tentacles.

COMPLEMENTAL MALES. Small or degenerate males which live on or in the body of the female. In some barnacles the normal individuals are hermaphrodite but the complemental males have the ovaries suppressed; they may also lose their alimentary canal and lead a semiparasitic existence in the mantle cavity of the larger partner.

COMPOUND EYES. Eyes of insects and crustaceans consisting of numerous visual units or *ommatidia*. The number of these ranges from many thousands in dragon-flies and beetles to less than a dozen in some ants. Each consists of a cuticular lens beneath which is a crystalline cone and a retinula or group of light-sensitive cells. The whole eye is convex or hemispherical with the apices of the cones converging towards the optic nerve in the centre. A black pigment normally surrounds each cone and it is thought that an insect perceives a 'mosaic image', *i.e.* one made up of a large number of dots. In some cases however, the layer of pigment surrounding each cone is capable of retraction and the image, although brighter, will be blurred owing to its being formed by overlapping points of light. This is the case with many nocturnal insects.

CONARIUM. (1) An alternative name for the pineal organ.
(2) The transparent larva of *Velella*.

CONCAVE AND CONVEX VEINS. In primitive insects such as the mayfly and in some fossil insects the veins of the wings run either on ridges or in furrows. The former are known as convex veins; the latter concave.

CONCH-. Prefix and root word from Latin *Concha*, Greek *Konche*: a shell-fish.

CONCHA. A part of the nasal cavity in certain reptiles where a shelf projects inwards from each side wall, increasing the surface of the nasal epithelium.

CONCHOSTRACA. A suborder of the Diplostraca: crustaceans having a large and variable number of limbs covered by a laterally compressed carapace formed of two halves hinged together and movable by means of adductor muscles. This hinged carapace gives them a superficial resemblance to bivalve molluscs.

CONDITIONED REFLEX. A reflex action modified by association with a particular experience so that instead of responding to the original stimulus, the action takes place in response to the new stimulus. The best known example is that of the dog in Pavlov's experiment whose salivary glands were made to react in response to the ringing of a bell.

CONDYL-. Prefix from Greek *Kondulos*: a knuckle.

CONDYLARTHRA. Extinct plantigrade mammals of the Eocene period having pentadactyl limbs without hoofs; having an elongated skull with a very small brain and having all four types of mammalian teeth.

CONDYLE. A knob of bone fitting into a corresponding socket or groove in another bone and allowing movement in definite directions, *e.g.* the occipital condyles of the skull fitting into the sockets on the Atlas vertebra.

CONES (OF THE EYE). Retinal cells, roughly cone-shaped, which perceive colour in the vertebrate eye. They are not present in all vertebrates and, where present, are usually concentrated in the region of the *Fovea Centralis*.

CONGLOBATE GLAND. A median gland of unknown function opening into the ejaculatory duct of the male reproductive system in certain insects.

CONGROGADINA. Marine fish allied to the blennies having no pelvic fins and having a tapering tail without a caudal fin.

CONIOPTERYGIDAE. Mealy-wings: small insects of the order Neuroptera, having very simple wing-venation and having the body covered by a powdery exudation from epidermal glands. The larvae infest trees and prey upon minute insects, mites etc.

CONJUGANT. One of a pair of gametes undergoing conjugation.

CONJUGATION (SYNGAMY). The union of two gametes to form a zygote. Some authorities prefer to restrict the use of the term to the fusion of iso-gametes and to use the word fertilization for that of heterogametes.

CONJUNCTIVA. The thin transparent membrane covering the cornea of a vertebrate eye: a continuation of the epidermis of the eyelids.

CONNECTIVE TISSUE. Tissue which fills spaces and holds the organs to-gether in the animal body. In vertebrates it consists of a jelly-like substance containing white collagen fibres or yellow elastic fibres or both, together with the cells which make these fibres and which are known as *fibroblasts*. There may also be fatty cells (adipose tissue) and blood vessel (see *Areolar tissue, Adipose tissue, Collagen*).

CONODONTS. Minute cone-shaped teeth, possibly of annelids or molluscs, found as fossils in the Silurian and Carboniferous limestone.

CONOPIDAE. Two-winged insects having a remarkable protective resemblance to wasps and bees and sometimes starting life as parasites on them.

CONTINUITY OF THE GERM PLASM. See *Weismannism*.

CONTINUOUS VARIATIONS. Variations within a species, merging insensibly into one another and fluctuating about a mean. Such variations if treated statistically exhibit graphically a 'normal' curve. Variations of stature and of colour are common examples. It is now believed that the phenomenon is due to the interaction of a large number of genes.

CONTOUR FEATHERS. Small feathers arranged in regular lines or 'contours' on the surface of a bird's body; they often have an aftershaft or duplicate vane near the base.

CONTRACTILE BULBILS. Pulsating swellings, sometimes containing valves, in some of the principal arteries of animals which have no true heart (*e.g.* in *Amphioxus*).

CONTRACTILE VACUOLES. Cavities or vacuoles acting as osmo-regulators and excretory organs in many protozoa. They appear to be clear spaces in the protoplasm which gradually expand as they fill with water and periodically collapse expelling their contents to the exterior. In *Amoeba* they are solitary spherical cavities but in some of the more complex protozoa such as *Euglena* and *Paramecium* they are accompanied by a number of accessory vacuoles leading into them.

CONUS ARTERIOSUS. A muscular cone where the ventral aorta leaves the ventricle in some fish and amphibians. Since it is made of cardiac muscle and contains valves it is regarded as a part of the heart and should be distinguished from the *Bulbus arteriosus* or *Bulbus cordis* which may be in front of the conus or may replace it in some fish.

CONVERGENCE. CONVERGENT EVOLUTION. A phenomenon whereby animals in different groups come to resemble each other through similarity of habit or environment. The marsupial moles and marsupial ant-eaters, for example, resemble the corresponding types among the higher or Eutherian mammals, although in no way related to them.

COPEPODA. Free or parasitic crustaceans including the well known freshwater *Cyclops*; they have no carapace; have six pairs of thoracic legs and none on the abdomen, and are characterized by a single median eye.

COPRO-. Prefix from Greek *Kopros*: dung.

COPRODAEUM. That part of the cloaca of a bird into which the rectum opens.

COPROZOIC. Inhabitants of dung, mostly protozoa, merging on the one hand into intestinal parasites and on the other into inhabitants of damp earth.

COPULATION. Sexual intercourse between two animals.

COPULATORY BURSA. See *Bursa copulatrix*.

COPULATORY ORGAN. An organ such as the penis which enables the male animal to discharge seminal fluid into the vagina or oviduct of the female.

COPULATORY SPICULES. A pair of chitinous rods used by male nematodes to keep open the genital apertures of themselves and their partners during copulation.

CORACIAE. A group of birds comprising six families and including king-fishers, bee-eaters, hoopoes and hornbills (see *Coraciiformes*).

CORACIIDAE. Rollers: brightly coloured birds with long wings and a curved beak with sharp edges and a wide gape. They are found in Europe, Africa and India; a few species are peculiar to Madagascar.

CORACIIFORMES. A large tribe of short-legged arboreal birds having blind and helpless young and often nesting in holes. The group includes many common birds such as kingfishers, owls, nightjars, swifts and humming-birds.

CORACOIDS. A pair of cartilage-bones forming the ventral part of the pectoral girdle and connecting the scapulae with the sternum in many vertebrates. In mammals the clavicles, which are membrane bones, have superseded the coracoids, the latter being reduced to small processes on the scapulae. In the lower vertebrates, however, both coracoids and clavicles are usually present.

CORALS. Colonial coelenterates which secrete a calcareous matrix often of complex structure and great beauty. This may be formed in one of two ways. In the 'precious coral' (*Corallium*) and the 'Organ pipe coral' (*Tubipora*) the skeleton is formed by the fusion of spicules in the mesogloea. These species belong to the order Alcyonaria. In the larger reef corals on the other hand which belong to the Zoantharia, the calcareous mass is ectodermal in origin and is formed first from the basal discs of the polyps, afterwards extending up into numerous folds and grooves to form cups or *thecae* in which the polyps live. These cups join to form a solid mass and the actual living organism forms merely a thin film over the surface of a calcareous structure which may be very large.

CORBICULA. A pollen basket formed by long curved hairs on the hind leg of a bee.

CORIUM. The dermis or underlayer of the skin of a vertebrate.

CORIXIDAE. Water-bugs also known as 'Lesser water boatmen' belonging to the order Hemiptera and the suborder Heteroptera. Unlike the large water boatman (*Notonecta*) which swims on its back, the Corixidae swim in the normal way with their backs uppermost. They are plentiful among pondweeds and carry an air supply down in a hollow of the back covered by the wings. Most are herbivorous.

CORM. (1) General: although primarily a botanical term, this word is occasionally used to denote any enlarged axis or stem-like structure.

(2) Of Crustacea: the main axis of the limb; in biramous limbs which have one branch longer than the other the corm consists of the base or *protopodite* and the longer of the two branches (usually the *endopodite*).

CORMIDIUM. A complex colony of zooids or polyps of various types found in such coelenterates as the 'Portuguese Man-of-war' and others of the group Siphonophora. On one cormidium there may be leaf-like *hydrophyllia* for protection and respiration; *gasterozooids* for feeding; *dactylozooids* for stinging and *gonozooids* for reproduction. The whole colony is often kept afloat by a large balloon-like *pneumatophore*.

CORN-. Prefix and root from Latin *Cornu*: a horn.

CORNEA. The transparent horny part forming the front of the eye of a vertebrate.

CORNEAGEN LAYER. Transparent epidermal cells beneath the cuticular lenses in the compound eyes of insects and crustaceans. The lenses are shed with the rest of the cuticle at each moult and are renewed by means of these corneagen cells.

CORNEOSCUTES. Horny epidermal scales of reptiles.

CORNICLES. Two short dorsal tubes near the hind-end of the abdomen on Aphides and other plant-bugs. They secrete the sticky *honey-dew* which is much sought after by ants, or the white waxy 'wool' of the woolly aphis.

CORNUA. (1) Horns or any horn-like structures.

(2) Hyoid cornua: elongated processes directed obliquely backwards and in some cases also forwards from the Hyoid cartilage in the floor of the mouth of mammals, birds and some reptiles and amphibia.

CORONA. (1) Any crown-like structure. Latin *Corona*: a crown.

(2) Of Echinoderms: the globular shell of a sea-urchin or echinoid consisting of dermal plates firmly sutured together.

CORONA RADIATA (EMBRYOLOGY). A layer of follicle cells which sometimes encloses the mammalian ovum for a short time when first released from the ovary.

CORONARY SYSTEM. The arteries and veins supplying blood to the muscular walls of the heart in a vertebrate. The coronary arteries lead from the aorta just above the semi-lunar valve; the veins drain back into the right auricle.

CORONARY BONE (CORONOID BONE). One of the membrane bones of the lower jaw in reptiles, projecting upwards to form a crown-like peak immediately behind the teeth. It is absent in mammals.

CORPORA. Plural of Latin *Corpus*: a body.

CORPORA ALLATA. A pair of small ductless glands behind the brain in insects and their larvae. It is believed that they secrete hormones connected with metamorphosis and ecdysis.

CORPORA BIGEMINA. The two optic lobes on the dorsal surface of the mid-brain in all lower vertebrates, *i.e.* all except mammals which have four lobes, the *corpora quadrigemina*.

CORPORA CAVERNOSA. A pair of laterally placed masses of erectile tissue in the mammalian penis. Together with the median *Corpus spongiosum* they contain numerous spaces which can be filled with blood, bringing about an increase of turgor and hence erection of the penis prior to copulation.

CORPORA MAMMILLARIA. Two prominences projecting downwards from the floor of the mid-brain behind the pituitary body in mammals.

CORPORA QUADRIGEMINA. The four optic lobes of the mammalian brain corresponding to the two *corpora bigemina* of lower vertebrates.

CORPORA STRIATA. Thickened lower portions of the side walls of the Telencephalon in the neighbourhood of the Optic Chiasma of the vertebrate brain.

CORPUS. Latin, a body (Pl. *Corpora*).

CORPUS ALBICANS. A mass of fibrous tissue in the mammalian ovary formed from the degeneration of the *Corpus luteum* (*q.v.*).

CORPUS CALLOSUM. A transverse band of nervous tissue connecting the two cerebral hemispheres in the brain of a mammal.

CORPUSCLES. See *Blood Corpuscles*.

CORPUS LUTEUM. The Yellow Body: an endocrine organ formed by enlargement of a Graafian Follicle after the release of an ovum in mammals. In the normal course of events these bodies develop for a short time and then die down. If pregnancy occurs, however, they continue to grow and to secrete the hormone *progesterone* which brings about enlargement of the uterus and other changes.

CORPUS SPONGIOSUM. Spongy erectile tissue running along the middle of the mammalian penis. The urethra runs through the centre of it and on each side there are sheets of muscle forming the boundaries of blood sinuses. When the latter are filled with blood they become turgid and cause erection of the penis prior to copulation (see *Corpora cavernosa*).

CORTEX. The outer layer of any organ, *e.g.* of the brain, kidneys or adrenal glands.

CORVIDAE. Ravens, crows, rooks, jackdaws, magpies and jays: fairly large birds of the Passerine group with strong thick beaks slightly curved and notched.

CORYPHAENIDAE. Large mackerel-like fish with brilliant colours.

COSMOID SCALES. COSMIN. One of the three main types of fish-scales; the others being the *Ganoid* and the *Placoid* type. Cosmoid scales are found at the present time only in lungfish and coelacanths. Each scale has an outer layer of *cosmin* similar to dentine, a middle layer of spongy bone and a lower layer of closely packed bony lamellae.

COSSIDAE. Goat Moths: some of the largest species of British moths having a wing expanse of about three inches. The caterpillars do considerable damage to oak, willow, poplar and other trees. Eggs are laid at the base of the tree and the caterpillar bores an irregular tunnel between the bark and the wood. Later it enters the wood and makes winding passages often several inches wide. It lives for several years in this way and eventually pupates in the soil.

COSTAL. Appertaining to the ribs, *e.g.* Costal arteries, muscles, cartilages etc.

COSTAL MARGIN. The anterior margin of an insect's wing.

COSTAL VEIN. COSTA. The first vein of an insect's wing, running along the anterior margin (see *Venation*).

COTTIDAE. Bull-heads and Miller's Thumbs: small spiny-rayed fish of rock-pools and shores in North Temperate regions. They are found in both salt and fresh water. The body is oblong or cylindrical and there are only four pairs of gill-slits.

COTYL-. Prefix from Greek *Kotule*: a cup or cup-shaped hollow.

COTYLEDONS. Parts of a placenta in which the villi are aggregated to form small tufts such as are found in many ruminants.

COTYLOID. A small bone at the junction of the ilium, ischium and pubis, forming part of the acetabulum in some mammals.

COTYLOPHORA. Ruminants having a cotyledonary placenta: deer, giraffes, cattle, sheep and goats.

COTYLOSAURS. The most primitive of all reptiles, the best known being *Seymouria*, a fossil of the Permian period. They are characterized by a skull in which the temporal cavity is completely roofed over (anapsid type).

COWPER'S GLANDS. A pair of glands of uncertain function opening into the urethra in the male mammal.

COXA. COXOPODITE. The first segment of the leg of an insect or of any appendage of a crustacean.

COXITE. The base of a rudimentary appendage such as those on the abdomens of certain insects. They may form lateral plates fused to the sterna.

CRACIDAE. Curassows: arboreal birds of tropical America having large air-spaces in the skeleton. Some species are easily domesticated but rarely breed in captivity. Hybrids with domestic fowls have been recorded.

CRAMPTON'S MUSCLE. A striated muscle in the eye of a bird, able to decrease the diameter of the eyeball in the neighbourhood of the junction between the cornea and the sclerotic. This causes the surface of the cornea to become more convex and helps the eye to accommodate itself for nearer vision.

CRANI-. Root from Latin *Cranium*, Greek *Kranion*: a skull.

CRANIAL FLEXURE. A bend in the neural tube of a vertebrate embryo near the part which later becomes the mid-brain.

CRANIAL NERVES. Nerves which lead directly to or from the brain. Fish and amphibia have ten pairs; reptiles, birds and mammals twelve. They are always in the same order and are as follows:
I, *Olfactory*; II, *Optic*; III, *Oculomotor*; IV, *Pathetic* or *Trochlear*; V, *Trigeminal*; VI, *Abducens*; VII, *Facial*; VIII, *Auditory*; IX, *Glossopharyngeal*; X, *Vagus*; XI, *Spinal Accessory*; XII, *Hypoglossal*.

CRANIAL SEGMENTS. The somites or segments which go to form the head in a vertebrate embryo: the *Premandibular*, *Mandibular* and *Hyoid* segments in front of the auditory capsule and two or more *Metotic* segments behind it.

CRANIATA. All chordates having a definite head, *i.e.* all vertebrates and cyclostomes but not the protochordata.

CRANIO-FACIAL ANGLE. An angle used in the measurement of skulls, chiefly in anthropology, to indicate the relative positions of the face and the cranium. A line known as the *basicranial axis* is drawn from the centre of the foramen magnum to the front of the presphenoid bone. Another line, the *facial axis*, is drawn from the anterior of the first line to the front tip of the upper jaw. When the face projects straight out in front of the cranium (*prognathism*) this angle is nearly 180°. When the cranium is large and the face is below it, as in humans, the angle may approach 90°. This condition is known as *orthognathism*.

CRANIUM. The principal part of a skull: the part which encloses the brain. A complete skull is built up of the cranium, sense-capsules, jaws and gill-arches, the latter being reduced and modified in land animals.

CREMASTER. A hooked appendage on the posterior tip of a chrysalis. It may be used to hang the chrysalis up or to attach it to the inside of a cocoon.

CREODONTIA. Primitive carnivorous mammals of the Eocene and Miocene periods, showing affinities with both the Insectivora and the carnivorous marsupials. They had large canine teeth but no carnassials. Some were wholly terrestrial; others semi-aquatic or seal-like.

CRIBELLUM. A sieve-like plate forming part of the spinneret in spiders and a few insects.

CRIBRIFORM PLATE. The hind-part of the mesethmoid bone of a mammal separating the cranial cavity from the nasal cavity; perforated for the passage of the olfactory nerves.

CRICOID CARTILAGE. One of several cartilages strengthening the wall of the larynx in mammals (see also *Arytenoid*).

CRINOIDEA. Sea-lilies and feather-stars: Echinoderms with branching arms and having a general structure like an inverted starfish on a long stalk with the mouth in the centre of the upper side; the whole body is covered with calcareous plates. They may remain anchored or may break away from their stalk and swim freely. Crinoids are an extremely ancient group dating from the Silurian and Cambrian periods.

CRISTA ACUSTICA. A ridge marking the end of a bundle of sensory nerve fibres in each ampullia of the vertebrate ear.

CROCIDURINAE. Primitive mouse-like insectivora of Africa and Asia; sometimes aquatic, sometimes burrowing.

CROCODILIA. Large lizard-like reptiles with powerful jaws, immovable quadrate bones, an elongated hard palate and teeth set in sockets. The nostrils are situated at the end of the long snout and can be closed. The body is covered with very heavy bony plates as well as by horny epidermal scales which are pitted and sculptured. The heart is four-chambered, a feature unique among reptiles.

CROCODILIDAE. A family comprising most of the present-day crocodiles: they include the caimans of Central and South America, alligators of North America and China and the true crocodiles of Africa, S.E. Asia, North Australia and Tropical America (see also *Crocodilia*).

CROP. A dilatation of the oesophagus where food may be kept for a time before passing on to the gizzard or stomach. The term not only applies to birds but also to insects, earthworms and other invertebrates.

CROSSING OVER (CHIASMA). An interchange of parts between pairs of homologous chromosomes which have come together during meiosis. This leads to separation of genes which had hitherto been linked and it follows that groups of hereditary factors, which are normally inherited together, may on occasion become separated and linked to other qualities. In the fruit-fly *Drosophila*, for example, long wings are usually linked with grey body and short wings are linked with black body. A small proportion, however, show a cross-over so that long wings go with black body and short wings with grey body. The percentage of such interchanges is called the *Cross-over Value* or *C.O.V.* and is proportional to the distance apart of the genes on the chromosomes. The further apart two genes are, the more likely they are to become separated (see also *Centimorgan*).

CROSS-OVER VALUE. See *Crossing over*.

CROSSOPTERYGII. A group of bony fish represented chiefly by fossils but including at the present time the newly discovered *Coelacanths* which were thought to have been long extinct. In some systems of classification the lung-fish (Dipnoi) are also included in the group but there are many points of difference between these and the coelacanths. Both groups, however, have fleshy lobed fins covered with scales and with fin-rays arranged round them as in a fan. The body is covered with cosmoid scales (see also *Choanichthyes* and *Coelacanthidae*).

CROSS-VEINS. Transverse veins linking the principal longitudinal veins in an insect's wing and giving it greater strength. In Mayflies and Dragonflies they are numerous and variable but in the more specialized orders they tend to become few and located in fixed positions so that they may be used as a basis for species identification (see *Venation*).

CROTCHETS. Curved double-pointed chaetae of certain Annelids.

CRUCIFORM CENTRA. Centra (of vertebrae) in which calcified fibro-cartilage extends from the perichondrium outwards on each side between the neural and haemal arches to give a cross-shaped section, as for instance in some cartilaginous fish.

CRUMEN. A cutaneous gland similar to a sebaceous gland in a depression of the lachrymal bone of antelopes and deer.

CRURAL. Appertaining to legs.

CRURA CEREBRI. Thickened ventral portions of the mid-brain of a vertebrate, consisting of nervous tracts linking the thalamencephalon with the hind-brain.

CRUSTACEA. A large and varied group of Arthropods, mostly aquatic, ranging from such well known creatures as crabs, shrimps and lobsters to such highly specialized types as the barnacle. Besides having the exoskeleton and jointed limbs typical of arthropods they have two pairs of antennae, a pair of mandibles and usually many other appendages variously modified for walking, grasping, swimming etc. They may have a carapace or large plate of exoskeleton covering and protecting the head and thorax.

CRYPT-, CRYPTO-. Root and prefix from Greek *Kruptos*: hidden.

CRYPTS OF LIEBERKÜHN. Small pits at the bases of the villi of the small intestine, into which flow digestive juices.

CRYPTOCERATA. Insects of the order Hemiptera and the suborder Heteroptera having very small inconspicuous antennae. They include many aquatic species such as the well known 'water-boatmen' *Notonecta* and *Corixa*.

CRYPTODIRA. Tortoises and turtles covered with horny plates and able to bend the neck in a vertical plane; they include the majority of terrestrial and aquatic forms.

CRYPTOMITOSIS. A process of nuclear division which takes place in certain parasitic and coprozoic protozoa. There are no distinct chromosomes but the mass of chromatin becomes concentrated round the equator and then behaves in a similar manner to normal chromosomes, dividing and moving to opposite poles of the cell.

CRYPTOMONADINA. Flagellated protozoa of fresh or salt water, belonging to the subclass Phytomastigina, that is plant-like organisms which can do photosynthesis. They have a gullet but it is doubtful whether it can be used for feeding. The colour ranges from green to brown.

CRYPTURI. Another name for the Tinamiformes: a group of Central American birds which are particularly interesting from an evolutionary standpoint because, although good flyers, they appear to be closely related to the ostrich and other flightless birds.

CRYSTALLINE CONE. A transparent cone acting as an additional lens between the cuticular lens and the retinal cells in each *ommatidium* or visual unit of a compound eye.

CTEN-. Root and prefix from Greek *Kteis, ktenos*: a comb.

CTENIDIA. Comb-like gills usually situated within the mantle-cavity of a mollusc (*e.g.* in the octopus).

CTENODIPTERINI. Fossil lungfish with numerous small dermal plates on the skull and having the dorsal, ventral and caudal fins much better developed than those of the present-day lung-fish.

CTENOID SCALES. Fish-scales with one edge toothed like a comb; they are found chiefly in the spiny-rayed fish (Acanthopterygii).

CTENOPHORA. 'Sea-gooseberries': free-swimming marine organisms, sometimes classed with the Coelenterata; usually globular and transparent, without nematocysts but having a pair of long tentacles bearing 'lasso-cells'. The mouth is at one pole of the sphere and a sense organ is at the other. Movement is by means of cilia borne on meridional comb-like plates.

CTENOPODA. The most primitive group of water-fleas or Cladocera having a well developed carapace covering the six pairs of trunk-limbs which are all alike and are used for straining food particles from the water.

CTENOSTOMATA. Polyzoa in which the aperture of the cup-shaped zooecium can be closed by a folded membrane when the tentacles are withdrawn.

CUBICAL EPITHELIUM. Cells which are roughly cubical with a large spherical nucleus in the centre. They are found lining the ducts of glands and, like most forms of epithelium, are in a single layer attached to a basement membrane.

CUBITAL BONE. CUBITUS. The Ulna.

CUBITAL FEATHERS. Secondary quill feathers: those feathers of a bird's wing which are attached to the ulna; the primaries being attached to the *manus* or hand.

CUBITAL VEIN. CUBITUS. The fifth longitudinal vein of an insect's wing. The principal veins named from the anterior are: *costal, subcostal, radial, median, cubital* and *anal.*

CUBOID. A bone formed from fusion of the fourth and fifth tarsals in many mammals.

CUCULIDAE. Cuckoos: medium sized birds with a curved beak, long pointed wings and a blunt tail. The fourth toe, normally pointing backwards, can be directed forwards. The outstanding habit of some species is that of laying eggs in the nests of other birds. There are many, however, which build their own nests and at least one, known as the Black Witch, in which several females lay their eggs in a common nest.

CUCULIFORMES. Arboreal birds with a united palate (desmognathous) and with the first and fourth toes directed backwards. They include cuckoos and parrots.

CULICIDAE. A family of two-winged insects comprising most of the common gnats and mosquitoes. They have long piercing mouth-parts and long antennae with numerous whorls of bristles. Scales similar to those of butterflies are situated along the wing-margins and veins. The abdomen is also scaly except in the Anopheline or malaria-carrying group. The larvae and pupae are aquatic, the former breathing through a pair of tubes opening at the hind-end; the latter at the head-end.

CULMEN. A median longitudinal ridge on the upper part of a bird's beak.

CUMACEA. Small marine crustaceans of the group Peracarida which carry their young in a brood-pouch formed from flattened appendages known as *oostegites.* The carapace is greatly enlarged, extending in front of the head and inflated on each side to form a branchial chamber.

CUNEIFORM. Wedge-shaped: a name particularly applied to certain carpal and tarsal bones.

CURCULIONIDAE. Weevils: a family of beetles belonging to the suborder *Rhynchophora* of which there are about 60,000 species. The head is prolonged into a snout which is used for boring into grain or other parts of plants where the eggs are laid.

CURSORIA. Cockroaches, mantids, stick-insects and leaf-insects: a group of Orthoptera having all the legs of nearly equal length. They thus contrast with the *Saltatoria* (grasshoppers and locusts) which have the hind-legs much longer for jumping.

CUTANEOUS. Appertaining to the skin.

CUTICLE. A layer of horny non-cellular material covering the epidermis of many animals.

CUTIN. A horny lipoid substance forming the cuticles of plants and some animals.

CUTIS. The dermis or lower layer of the skin.

CUTIS LAYER. See *Dermatome*.

CUVIERIAN DUCTS. A pair of blood vessels leading from the sinuses and veins of the body into the *sinus venosus* of the heart in fish; they are homologous with the anterior venae cavae of higher vertebrates.

CUVIERIAN ORGANS (OF ECHINODERMS). Organs of defence in Holothurians or 'sea-cucumbers'; they consist of numerous branching tubes covered with a sticky substance and opening into the hind-gut. When the animal is attacked a violent contraction of the body-wall causes them to shoot out entangling an enemy in the sticky threads. The action is sometimes so violent that the cloaca may be ruptured and the whole alimentary canal may be forced out of the body.

CYATHOZOOID. The primary zooid of certain colonial Tunicates.

CYCL-, CYCLO-. Prefix from Greek *Kuklos*: a circle.

CYCLICAL VARIATION. A periodic fluctuation in the number of individuals of any plant or animal in a particular locality; sometimes correlated with periodic climatic changes.

CYCLOID SCALES. Fish-scales of a rounded or elliptical shape such as those of salmon and many common fish.

CYCLOMYARIA. See *Doliolida*.

CYCLOPHYLLIDEA. A suborder of Cestoda comprising the majority of tapeworms: those having a scolex with four cup-shaped suckers and a rostellum with a crown of hooks.

CYCLOPTERIDAE (DISCOBOLI). Lump-suckers: spiny-rayed fish whose pelvic fins are placed well forward and are modified to form the bony centre of a sucking disc for attachment to stones etc.

CYCLORRHAPHA. Diptera such as the house-fly whose larvae are without head or legs and whose pupae are co-arctate, *i.e.* contained in a barrel-shaped *puparium* (see *Puparium* and *Co-arctate pupa*).

CYCLOSIS. Circulatory movement of protoplasmic structures within a cell, as in the case of the food vacuoles in *Paramecium* and the chloroplasts in certain plant cells.

CYCLOSPONDYLOUS. A term used to describe the vertebrae of certain fish in which the centrum is calcified to form a double cone surrounding the notochord.

CYCLOSTOMATA. (1) Fish-like creatures such as the lamprey which have no true jaws but have a terminal suctorial mouth surrounded by a ring of teeth.
   (2) Polyzoa whose zooecium or capsular body-wall is tubular and without an operculum.

CYNIPOIDEA. Gall-wasps: narrow-waisted or apocritous Hymenoptera having straight antennae and a laterally compressed body. The majority produce galls on oak or other trees and the formation of these is thought to be due to the production of some type of auxin either by the insect or by the injured tissues of the plant.

CYNO-. Prefix and root from Greek *Kuon, kunos*: a dog.

CYNODONTIA. Extinct reptiles of the Triassic period having dog-like teeth and many mammalian characters (see *Theromorpha*).

CYNOIDEA (CANOIDEA). The group of Carnivora to which dogs, wolves and foxes belong (see also *Canidae*).

CYPHONAUTES LARVA. The typical larva of a polyzoon, like a trochosphere with a bivalve shell.

CYPRINOIDEA (CYPRINI, CYPRINIDAE). Carp-like fresh-water fish having a smooth head, a scaly body and pelvic fins placed well back. The mouth is toothless but there are teeth further back in the pharynx; barbels are often present. The air bladder is large and is connected with the ear by Weberian Ossicles.

CYPRINODONTIDAE (POECILIIDAE). Carnivorous pike-like fish inhabiting fresh water of tropical and subtropical regions. The head and body are covered with scales and the mouth is without barbels. Most species are viviparous and the females are larger than the males; the anal fins of the latter may be modified to form copulatory organs.

CYPRIS LARVA. A type of larva occurring in barnacles and others of the class Cirripedia. It resembles the adult of the genus *Cypris* in having a bivalve carapace with adductor muscles.

CYPSELI. Swifts and humming-birds (see *Cypselidae* and *Trochilidae*).

CYPSELIDAE. Swifts: migratory birds resembling swallows, having long narrow curved wings, strong claws and feathered metatarsi. Most of their time is spent on the wing catching insects. They breed in large numbers in caves and rocks, the nests being glued together by a sticky secretion from the mouth.

CYST. A thick protective membrane secreted by many organisms when in a resting stage.

CYSTI-, CYSTO-. Prefix from Greek *Kustis*: a bladder.

CYSTIC DUCT. That part of the bile duct of a vertebrate which leads from the gall bladder, receives branches from various parts of the liver and eventually forms the common bile duct leading into the duodenum.

CYSTICERCUS. The bladder-worm or larva of a tapeworm, found in the secondary host and only developing into an adult worm when eaten by the primary host. Bladder-worms in pork, for example, develop into tapeworms in the human body; those in rabbits develop to maturity in the dog. The name *Cysticercus* was given before the bladder-worm was known to be identical with the tapeworm.

CYSTOFLAGELLATA. See *Noctiluca*.

CYSTOGENOUS CELLS. Cells which secrete the material from which a cyst is formed.

CYSTOIDEA. Fossil crinoids having stalked sac-like bodies with tentacles round the mouth; the epidermis is covered with hexagonal plates perforated by pores either in pairs or in diamond patterns.

CYSTIPHORINAE. The Bladder-nosed seal and the Elephant-seal, both of which have a dilatable sac on the face or on the proboscis.

CYTO-. Prefix from Greek *Kutos*: a hollow (= appertaining to a cell).

CYTOCHROME. A respiratory pigment containing iron, present in most cells.

CYTOPLASM. The protoplasm surrounding the nucleus of a cell as distinct from the nucleoplasm within the nucleus.

CYTOSTOME. The 'cell-mouth' of certain protozoa such as the Ciliophora. It may open into a vestibule or groove of ectoplasm or into a deeper 'gullet' from the base of which food-vacuoles can form and pass into the organism.

CYTO-TROPHOBLAST. The part of the mammalian placenta nearest to the foetus, where the cells of the trophoblast are not broken down to form a syncytium (*cf. Plasmodi-trophoblast*).

D

DACTYLO-. Prefix from Greek *Daktulos*: a finger.

DACTYLOPODITE. The fifth or distal segment of the leg of a crab, lobster or similar crustacean. It may take the form of a claw or may be hinged to the enlarged propodite to form part of a pincer (see *Chela*).

DACTYLOZOOIDS. Stinging polyps in the form of long tentacles which often surround the gasterozooids or feeding polyps in colonial coelenterates. Their specialized structure enables them to catch prey and pass it on to the other polyps for digestion.

DALLIIDAE. Fresh-water pike-like fish from Alaska and Siberia.

DANININA. Small carp-like fish from the fresh waters of India and S.E. Asia.

DARTS (OF NEMATODA). Sharp projections of the lining of the buccal cavity in some parasitic nematode worms, especially those living in plants. They are used for penetrating the cell-walls of the host.

DART SAC. A sac branching off the vagina of a snail, containing a sharply pointed calcareous 'dart' which is discharged with great force prior to copulation. When two snails come together each shoots out the dart so forcibly that it penetrates the skin and tissues of the other. Some time after this the two snails which have thus been stimulated come together again and mutual fertilization takes place.

DARWINISM. See *Natural Selection*.

DARWIN'S POINT. A small swelling sometimes occurring near the top of the pinna of the human ear; thought to be a vestige of the former point of the ear.

DASY-. Prefix from Greek *Dasus*: dense.

DASYPAEDES. Birds whose young are covered with down at birth.

DASYPODIDAE. Armadillos: primitive South American mammals of the order *Edentata*, having the back and sides covered with an armour of bony plates over which lie horny epidermal scales.

DASYPROCTIDAE. Agoutis of South America and the West Indies: large rodents with long incisors, a short tail and hoof-like claws.

DASYURIDAE. Carnivorous marsupials of Australasia including the large 'Tasmanian wolf', the smaller 'Tasmanian Devil', the 'Marsupial cat' and a number of small insectivorous species.

DE-. Latin prefix: from, away, out, down.

DEALATION. Loss of wings by ants or other insects. In the case of ants, mating takes place during flight after which the fertilized queen descends and rubs or pulls her wings off. The loss of these appears to stimulate in her an instinct to go below ground and to start laying her eggs.

DE-AMINATION. The breaking down of excess of proteins or amino-acids in the liver to form urea which is afterwards excreted by the kidneys.

DECA-. Prefix from Greek *Deka*: ten.

DECAPODA. (1) Shrimps, lobsters, crabs etc. Ten-legged crustacea of the class *Malacostraca* having a carapace fused with all the thoracic segments.
(2) Cuttle-fish, squids and other Cephalopod molluscs having ten tentacles round the mouth.

DECIDUATA. Mammals in which the villi of the placenta are so intimately united with the uterine mucous membrane that the latter comes away with the foetus at birth (see also *Adeciduata*).

DECIMORGAN. See *Centimorgan*.

DECUSSATION. Crossing over, as for instance of some of the optic nerve fibres beneath the brain (see also *Optic Chiasma*).

DEEP PETROSAL NERVE. See *Petrosal Nerves*.

DEFOLIATORS. Insects which completely strip trees by feeding on the leaves, *e.g.* the larvae of certain moths and sawflies.

DEHYDROGENASE. A respiratory enzyme which catalyses oxidation of a substance by removal of hydrogen from it; the latter usually combining with some other substance which is called a hydrogen acceptor.

DELOBRANCHIATA. Primitively aquatic Arachnida breathing by gill-books: they include the extinct *Eurypterids* and the present day king-crabs (*Xiphosura*).

DELPHINOIDEA (DELPHINIDAE). Toothed whales or Odontoceti including sperm whales, porpoises and dolphins. Sperm whales have teeth in the lower jaw only; the others in both jaws.

DEMI-. Prefix from French *demi*, Latin *dimidius*: half.

DEMIBRANCH. See *Hemibranch*.

DEMIFACET. One of two depressions in adjacent thoracic vertebrae; the two together making a complete facet for the articulation of the capitulum of a rib.

DEMOSPONGIAE. Sponges whose skeleton may be composed of simple needle-like spicules or of four-rayed spicules. There is no calcium in these, but they may be made of silica or of the horny substance *spongin*. The common bath sponge *Euspongia* belongs to this class.

DENDR-, DENDRO-. Root and prefix from Greek *Dendron*: a tree.

DENDRITE. An arborization or branching structure at the end of a nerve fibre (see *Synapse*).

DENDRITIC TENTACLES. Much-branched tentacles, as, for instance, those of a Holothuroid (sea-cucumber).

DENDROBATINAE. Toothless frogs of Tropical America and of Madagascar, some of which secrete a strongly poisonous fluid from the skin.

DENDROCHIROTAE. Holothuroids, such as *Cucumaria*, the 'sea-cucumber' having dendritic or tree-like tentacles (see *Holothuroidea*).

DENDRON. A branched or tree-like nerve fibre conveying impulses towards the nucleus or cell-body of a neurone (*cf. Axon*).

DENDROPHRYNISCINAE. Aquatic frogs from Brazil and Peru without teeth, tympanum or Eustachian tubes.

DENT-. Prefix and root from Latin *Dens, dentis*: a tooth.

DENTAL FORMULA. A simple method of describing the teeth of a mammal. Using the initial letters for incisor, canine, premolar and molar and the number of each kind of tooth in the upper and lower jaws on one side only, the dental formula of a rabbit is I $\frac{2}{1}$, C $\frac{0}{0}$, P $\frac{3}{2}$, M $\frac{3}{3}$, or more simply $\frac{2.0.3.3}{1.0.2.3}$. That of a dog is $\frac{3.1.4.2}{3.1.4.3}$ and that of a human $\frac{2.1.2.3}{2.1.2.3}$.

DENTAL LAMINA. A long band of ectoderm which has sunk below the surface of the mouth and runs parallel to the edge of the jaw in an embryo vertebrate. It gives rise to the enamel of the teeth.

DENTARIES. A pair of membrane-bones bearing teeth and forming the front part of the lower jaw in most vertebrates. Bony fish, reptiles and birds have a number of bones behind the dentaries but in mammals the dentaries constitute the whole lower jaw.

DENTICLE. Any small tooth-like structure, *e.g.* the dermal denticles in the skin of a cartilaginous fish.

DENTINE. Ivory: a fairly hard bone-like substance forming the inner part of a tooth covered by a harder layer of enamel. The dentine originates in the dermis; the enamel in the epidermis.

DEPENDENT DIFFERENTIATION. A term used in embryology to describe the formation of an organ which will only develop in the presence of an organizer, that is, a chemical substance obtained from a neighbouring tissue. The lens of a frog's eye, for instance, develops under the influence of organizing tissue from the optic cup. The cup itself, however, is *self-differentiating* or independent of other tissues.

DERCETIDAE. Extinct eel-shaped fish of the Cretaceous Period having a long head, long jaws and several rows of triangular scutes on the body.

DERCETIFORMES. See *Heteromi* and *Dercetidae*.

DERICHTHYIDAE. Eel-like fish living at great depths in the Atlantic.

DERM-. Prefix and root from Greek *Derma, dermatos*: skin.

DERMAL. Formed in the dermis or lower layer of the skin, *e.g.* Dermal bones, dermal scales etc.

DERMAL BONES (MEMBRANE BONES). Bones which do not replace pre-existing cartilage, but are deposited independently in the lower layer of the skin. Those forming the roof of the mammalian cranium are of this type, as also are the clavicles.

DERMAL FIN-RAYS (DERMOTRICHIA). Horny or bony rays supporting the fins of fishes; distinct from the cartilaginous rays (radialia) which are usually at the base of a fin (see *Ceratotrichia* and *Lepidotrichia*).

DERMAPTERA. Earwigs: hemimetabolous insects in which the fore-wings are reduced to small thickened elytra and the membranous hind-wings are roughly semicircular in shape. There is a large forceps at the hind-end of the abdomen. They have biting mouth-parts and are omnivorous; a few wingless species are ectoparasites.

DERMATOME (CUTIS LAYER). The outer layer of each mesodermal somite of a vertebrate embryo, giving rise to the dermis.

DERMIS (CORIUM). The lower layer of the skin formed from the mesoderm and usually containing blood vessels, connective tissue, nerve endings etc.

DERMOPTERA (GALEOPITHECIDAE). Flying lemurs: a group of little-known mammals inhabiting the forests of S.E. Asia. They have a membrane or *patagium* stretched between the legs but it is much thicker and more hairy than that of a bat. They are usually classed as a distinct order of mammals but apparently have affinities both with the insectivora and the primates.

DERMOTRICHIA. See *Dermal fin-rays*.

DESM-, DESMO-. Prefix from Greek *Desmos*: a bond, chain.

DESMERGATE. An ant intermediate between a soldier and a worker.

DESMOGNATHOUS. A term used to describe the jaws of such birds as swans, geese, ducks, parrots and many birds of prey in which the maxillo-palatine bones are large and spongy, usually uniting with each other across the middle line ventral to the vomer.

DETERMINATE CLEAVAGE. Cleavage of a fertilized egg of the 'mosaic' type in which at a very early stage definite areas are set apart to form particular types of tissue or particular organs.

DETORSION. A phenomenon shown by certain Gasteropod molluscs in which the visceral hump, after undergoing torsion or twisting in the embryo, later reverses the process. In primitive or embryonic gasteropods the gills and the anus are directed backwards and the auricles of the heart are behind the ventricles. During their development most gasteropods go through a stage at which the whole visceral hump suddenly and rapidly rotates through an angle of 180° making these organs point forwards. This is known as *torsion*. In the group known as Opisthobranchiata, however, the visceral hump partly or completely turns back again usually causing the loss of certain organs and of the shell. The group includes such types as sea-slugs, sea-butterflies and sea-hares (see also *Torsion*).

DEUTO-, DEUTERO-. Prefix from Greek *Deuteros*: second, secondary.

DEUTOCEREBRUM (DEUTEROCEREBRUM, MESOCEREBRUM). The middle portion including the antennary lobes in the brain of an Arthropod.

DEUTOMERITE. The hind segment containing the nucleus in certain protozoa, such as *Gregarina*, a parasite of insects. Protozoa of this kind are divided by constrictions of the ectoplasm into three regions, *viz.* the *Epimerite* or fixing organ, the *Protomerite* or middle region and the *Deutomerite* or hind region.

DEUTOPLASM. The yolk or food material in an egg, frequently consisting largely of the protein *lecithin*.

DI-. Greek prefix: double, twin.

DIA-. Greek prefix: through, across.

DIAKINESIS. The stage of meiosis at which the chromosomes arrange themselves round the 'equator' of the cell and become attached to the spindle. It marks the transition from the prophase to the metaphase.

DIAPAUSE. A slowing down of metabolism and delay of development of insects etc. often connected with seasonal changes of atmospheric conditions but sometimes apparently due to an inborn rhythm. It may take the form of normal hibernation, but in some cases a diapause lasts for several years. This is probably due to the fact that the insect fails to produce the normal hormones necessary for continued growth or development. A prolonged diapause may, however, be ended by some sudden shock such as invasion by a parasite.

DIAPEDESIS. The passage of white corpuscles out of or into a blood vessel by changing shape and squeezing between other cells.

DIAPHRAGM. A transverse muscular partition separating the cavities of the thorax and the abdomen in mammals. By raising and lowering the diaphragm the pressure in the chest cavity is increased and decreased causing the lungs to expand and contract when breathing.

DIAPHYSIS. The central shaft of a bone which is formed separately from the epiphyses or end-pieces and becomes fused with them as the bone becomes completely calcified.

DIAPSIDA. Vertebrates having a diapsid skull, *i.e.* one with two temporal fossae on each side. These fossae or cavities are one above the other and are behind the orbit. They are separated from it by the post-orbital bar and from each other by the superior temporal bar. The Diapsida include birds and a number of reptiles of which crocodiles, dinosaurs and pterodactyls are examples.

DIASTATAXY. An alternative name for *Aquintocubitalism* in birds.

DIASTEMA. A gap in the teeth, as, for instance, between the incisors and the premolars of a rabbit.

DIASTER. The two asters at the extremities of the spindle in cell-division.

DIBAMIDAE. Worm-like lizards of Australia and New Guinea having no limbs, no sternum, concealed eyes and no ear-opening.

DIBRANCHIA. See *Dibranchiata* below.

DIBRANCHIATA. Cuttle-fish, squids and octopuses: Cephalopod molluscs with a single pair of gills, eight or ten tentacles and with the shell reduced or absent (*cf. Tetrabranchiata*).

DICHOCEPHALOUS RIB. The type of rib found in mammals, birds and some reptiles, having two heads known as the *tuberculum* and the *capitulum* articulating with the thoracic vertebrae.

DICTYO-. Prefix from Greek *Diktuon*: a net.

DICTYOSOME. A Golgi Body (*q.v.*).

DICYCLICAL. Having two sexual cycles a year, usually one in the spring and the other in the autumn.

DICYNODONTIA. Extinct mammal-like reptiles of the Triassic period, having as the only teeth a pair of tusks growing from persistent pulps in the maxillae.

DIDACTYLA (POLYPROTODONTIA). Marsupials not showing syndactylism: opossums, marsupial cats, marsupial moles and Tasmanian wolves.

DIDELPHIA. An alternative name for the Marsupialia.

DIDELPHYIDAE. Opossums: arboreal marsupials of North and South America having long prehensile tails. Some are carnivorous; some insectivorous; a few are aquatic.

DIDIDAE. Dodos: large, recently extinct, flightless birds of Mauritius, resembling but rather larger than a turkey. The last known specimen was living in 1681.

DIDUNCULIDAE. Dove-like birds of the Samoan Islands having a compressed beak with hooked extremity and with horny tooth-like structures on the lower half.

DIENCEPHALON. An alternative name for the *Thalamencephalon*: the posterior part of the fore-brain bearing the pituitary organ ventrally and the pineal organ dorsally.

DIFFERENTIATION. The process by which cells, tissues or organs gradually change their appearance and function during embryonic development.

DIFFERENTIATION CENTRE. An area containing organizing tissue from which differentiation of an embryo spreads.

DIGENEA. Liver flukes etc., endoparasites which spend their lives in two alternate hosts, one usually a vertebrate and the other an invertebrate. The common liver fluke, for example, spends its larval stages in the water snail *Limnaea* and its adult stage in the liver and bile duct of a sheep.

DIGENESIS. Alternation of sexual and asexual reproduction.

DIGESTION. The breaking down of complex food substances by enzymes to form simpler soluble substances which can be absorbed. In most animals this takes place in the alimentary canal or gut. Enzymes are secreted either by glands in the gut-wall or by larger glands, such as the pancreas and salivary glands, whose ducts open into the gut. In some simple organisms, however, digestion may be intracellular, the food particles in such cases being engulfed by amoeboid cells and broken down in food-vacuoles within these cells. Such a process takes place in protozoa, coelenterates and a few other invertebrates.

DIGIT-. Prefix from Latin *Digitus*: a finger.

DIGITAL FORMULA. A series of figures indicating the number of phalanges in each finger or toe. For example, that of the human hand is 2.3.3.3.3.

DIGITIGRADE. Walking on the toes or fingers with the rest of the foot off the ground: the normal walking method of dogs, cats and most fast running animals.

DIMORPHIC, DIMORPHISM. Existing in two distinct forms, as, for instance, male and female forms; hydroids and medusoids, etc.

DINERGATE. A soldier-ant having an enlarged head.

DINO-. Prefix from Greek *Deinos*: terrible.

DINOFLAGELLATA (DINOFLAGELLIDA). Green, yellow or brown protozoa protected by cellulose plates and having two flagella at right angles to one another. They belong to the group Phytomastigina, which forms a link between plants and animals. One of the best known examples is the phosphorescent marine genus *Noctiluca*.

DINOMYIDAE. Mouse-like rodents of Peru with long bushy tails and with a cleft upper lip.

DINOPHYSINAE. Pelagic Dinoflagellate protozoa having a shell in the form of two lateral plates.

DINORNITHIDAE. Moas: large, extinct, flightless birds of New Zealand, sometimes up to 10 ft. high. It is thought that they were hunted by primitive man and became extinct within recent times.

DINOSAURIA. Terrestrial or amphibious reptiles, often of great size, of the Triassic and Cretaceous Periods. Most had long limbs and long tails; sometimes the hind-limbs were bird-like. Although most were herbivorous, some of the later species in the Cretaceous Period were carnivorous and probably extremely fierce. These had relatively large brains which probably enabled them to hunt and prey on the others.

DIODONTIDAE. Porcupine-fish: tropical fish with inflatable body; with bifid divergent spines on the anterior vertebrae and with the teeth fused to form a beak.

DIOTOCARDIA (ASPIDOBRANCHIATA). Gasteropod molluscs which have two auricles in the heart and two fern-like gills. They include the Ormer (*Haliotis*) and the Limpet (*Patella*).

71

DIPHY-. Prefix from Greek *Diphues*: twofold, of double nature.

DIPHYCERCAL (PROTOCERCAL). A type of tail which is primitively symmetrical and pointed, as in many fossil fish and a few living ones.

DIPHYLLIDEA. Tapeworms in which the scolex has two suckers only and a long neck armed with spines. The only genus known is found in the spiral intestines of selachians, the larva being found in prawns.

DIPHYODONT. Having two sets of teeth: a condition characteristic of mammals. Lower vertebrates are often *polyphyodont*, *i.e.* they have a continuous succession of teeth.

DIPLEURA. A bilaterally symmetrical animal.

DIPLEURULA (ECHINOPAEDIUM, PLUTEUS). A bilaterally symmetrical larva of echinoderms.

DIPLO-. Prefix from Greek *Diplous*: double.

DIPLOBLASTICA. Organisms whose body-walls consist of two layers, the *ectoderm* and the *endoderm* with a gelatinous, non-cellular *mesogloea* between them. They are contained in a single phylum, the Coelenterata comprising such creatures as coral polyps, sea-anemones and jelly-fish.

DIPLOID. Having the full number of chromosomes, *i.e.* twice the haploid number. Normally the body-cells have the diploid number and the gametes the haploid number. In this way when two gametes combine to form a zygote the full number is restored.

DIPLOMONADINA. Symmetrical protozoa with two nuclei and with a number of flagella on each side. Most of them are parasites of vertebrates including man. They belong to the group Zoomastigina.

DIPLOPODA (PROGONEATA). Millepedes: herbivorous Myriopoda having a cylindrical body of about seventy segments and differing from centipedes in having two pairs of legs on each abdominal segment. The antennae are short and club-shaped.

DIPLOPORES. Pores in pairs perforating the dermal plates of some Crinoids (see *Cystoidea*).

DIPLOPORIDA. Crinoids having diplopores.

DIPLOSPONDYLY. Having two vertebrae in each body-segment, as, for instance, in the tails of many fish.

DIPLOSTRACA. Branchiopod crustaceans with a laterally compressed carapace; with large biramous antennae and with compound eyes which are fused together and sessile. The order includes the water-fleas (*Cladocera*).

DIPLOTENE. A stage in meiosis at which each chromosome begins to divide longitudinally into two chromatids. The separation is incomplete and along the length of the chromosomes may be seen *chiasmata* where the individual chromatids cross over and parts of each interchange with corresponding parts of the partner (see *Crossing over*).

DIPLURA. Bristle-tails: small, white, primitively wingless insects living in the soil. They have a forked tail formed from two long thin cerci and have biting jaws lying in pockets from which they are protruded when feeding.

Diplura come nearest to the probable ancestral type of insect and form a link between insects and Myriopoda.

DIPNOI (DIPNEUSTI, DIPNEUMONES). Lung-fish: scaly fish with external and internal nostrils and with an air-bladder modified to form a lung. The latter is very vascular and is connected with the back of the throat so that the fish can survive for long periods if the river or lake in which it lives dries up. The paired fins are fleshy and can be used as primitive limbs, enabling the fish to move clumsily when out of water. Lung-fish were common in Devonian and Carboniferous times but at present only three species survive. They are considered to be intermediate between fish and amphibians.

DIPODIDAE. Jerboas: long-legged, jumping, mouse-like rodents of North Africa etc.

DIPROTODONTIA (SYNDACTYLA). Marsupials having fewer than three incisor teeth on each side of each jaw and having the second and third toes bound together. The group includes kangaroos, wallabies, flying phalangers and koala bears.

DIPROTODONTIDAE. Extinct diprotodont marsupials, some as large as a rhinoceros, from the Pleistocene of Australia.

DIPSADOMORPHINAE. Long-tailed snakes with lateral nostrils and with well developed teeth and poison fangs. They are not usually harmful to man as the poison is often weak and the fangs too far back. There are about seventy genera, terrestrial, arboreal or aquatic, in most parts of the world except the northern regions.

DIPTERA. Two-winged flies: a large and varied order of holometabolous insects with a single pair of membranous wings and a reduced hind pair modified to form *halteres* or balancers. The mouth-parts are often lengthened into a proboscis for piercing and sucking. The larvae are legless and the pupae exarate or co-arctate. There are about 85,000 species of Diptera, many of which transmit diseases to animals and humans either by blood-sucking or by carrying the germs on their bodies.

DIRECTIVE EVOLUTION (ORTHOGENESIS). An apparently inherent trend in a particular group of animals to evolve in a particular direction, *e.g.* a tendency to increase or decrease in size; to develop lightness of bones etc.

DIRECTIVES (OF ACTINOZOA). Mesenteries or membranes in the body cavity of a sea-anemone which, by differing slightly from the normal, give an indication of a plane of bilateral symmetry in an otherwise circular organism.

DISCAL CELL. A large 'cell' or area of wing in some of the more specialized butterflies and moths, formed by the reduction or disappearance of intervening veins and cells.

DISCO-. Prefix from Latin *Discus*: a disc.

DISCOBOLI. Lumpsuckers (see *Cyclopteridae*).

DISCOCEPHALI (ECHENEIFORMES). Remoras: spiny-fish whose dorsal fin is modified to form a sucker for attachment to other fish or to floating objects.

DISCOGLOSSIDAE. Frogs and toads whose tongues are in the form of a round non-protrusible disc. They have teeth in the upper jaw; separate Eustachian tubes, and the males are without vocal sacs. There are short ribs on the anterior vertebrae.

DISCOIDAL PLACENTA. See under *Metadiscoidal Placenta*.

DISCOMEDUSAE. A subdivision of the Scyphozoa comprising most of the common disc-shaped jelly-fish having the following characteristics: a large mouth with ciliated tentacles underneath; a large stomach with four interradial pouches in which are the gonads; numerous short tentacles round the circumference. In this group of Coelenterates the polyp stage is reduced to a small larva called a *Scyphistoma*. This becomes segmented or strobilated and the young medusae or *Ephyrae* detach themselves from the top by transverse fission.

DISCONTINUOUS DISTRIBUTION. A phenomenon in which similar types of organism are found only in widely separated parts of the world. This is usually an indication that the group is a very ancient one and has become extinct in all but a few places. The three existing species of lung-fish found in Australia, Africa and South America are a good example.

DISCONTINUOUS VARIATION (MUTATION). The occurrence of an individual which differs markedly from the average type. Such a phenomenon may be due to a chemical or physical change in the chromosomes or to an alteration in the number of chromosomes (polyploidy).

DISCUS PROLIGERUS. A ring of cells lining the cavity of a *Graafian follicle* in the mammalian ovary.

DISSOGENY (DISSOGONY). Having two periods of sexual maturity, one during the larval; the other during the adult stage.

DISTICHODONTIA. Fresh-water fish of Tropical Africa having a long dorsal fin, a rounded belly and a chain of Weberian Ossicles connecting the air-bladder to the ear.

DITOKOUS. Producing only two eggs or two young at a birth.

DITREMATA. Marsupial and Eutherian mammals in which the anal aperture is separated from the renal and reproductive apertures by the perineum. They are thus distinguished from the *Monotremata* such as the Platypus which have a common opening.

DIVERSICORNIA. An ill-defined group of beetles which includes about forty families belonging to the suborder Polyphaga. Beetles of diverse types, which cannot be put into any more definite classification, are placed in this group for convenience. They include Coccinellidae (Ladybirds), Dermestidae (Larder beetles) and Elateridae (Click beetles).

DIVERTICULUM. A blind tube branching off from a chamber or canal, as, for instance, the coecum.

DOBIE'S DOTS. Darkly staining discs ('A-discs') distributed along the fibrils of striped muscle and separated from one another by lighter regions ('J-discs'). The close proximity of numerous light and dark discs in adjacent fibrils gives the appearance of transverse stripes across the muscle.

DOCOGLOSSA. Gasteropod molluscs such as the common limpet which have a radula or tongue with rows each consisting of a few strong teeth suitable for scraping algae off rocks.

DOLICHO-. Prefix from Greek *Dolichos*: long.

DOLICHOCEPHALIC (SCAPHOCEPHALIC). Long headed (referring to human races): having the breadth of the head 75% or less of the length (*cf. Brachycephalic*).

DOLICHOSAURIA. Small, extinct, aquatic reptiles with long necks and well developed limbs but otherwise rather snake-like.

DOLIOLIDA. Pelagic tunicates having a tail in the larval stage but losing it in the adult. They reproduce by buds or blastozooids which break away from the parent but re-attach themselves to a dorsal process or *cadophore* where they develop.

DOMINANT. A term used in genetics to denote the stronger one of a pair of contrasted characters, *i.e.* the one which shows itself in a hybrid (*cf. Recessive*).

DORSAL. On or near the back of an animal.

DORSALIA. See *Arcualia*.

DOUBLE CIRCULATION. The type of blood circulation found in mammals and birds in which the heart has four chambers and the pulmonary and systemic circulations are completely separated. Blood flows from the left ventricle to the tissues of the body and back to the right auricle, after which it goes from the right ventricle to the lungs and back to the left auricle.

DOWN FEATHERS (PLUMULAE). Small soft feathers covering the whole body of a bird and keeping it warm.

DRAIN-PIPE CELLS. Perforated cells usually placed end to end making a pipe-like structure as, for instance, in the excretory organs of earthworms, turbellarians etc.

DRIFT NET (COELENTERATES). A mass of long tentacles or dactylozooids hanging beneath a colonial coelenterate such as the Portuguese Man-of-war (*Physalia*). They trail through the water stinging any fish which may touch them, after which the feeding polyps or *gasterozooids* attach themselves to the fish and begin to digest it.

DROMAEO-, DROM-, DROMA-. Prefix from Greek *Dromein*: to run.

DROMAEOGNATHOUS. A term used to describe the jaw-structure of birds in which the maxillo-palatine bones are united with the broad vomer. Such an arrangement is found in the ostrich, emu, kiwi and other running birds.

DROMATHERIIDAE. Primitive marsupials, sometimes regarded as reptiles, from the Upper Trias.

DRONES. Male bees which develop by parthenogenesis, *i.e.* from unfertilized eggs. They are somewhat larger than workers and have a broader head and large eyes. Their life is short, as they either die soon after mating or are killed by the others at the beginning of the winter.

DUCTLESS GLANDS. See *Endocrine organs*.

DUCTUS ARTERIOSUS (DUCTUS BOTALLI, DUCT OF BOTALLO). A blood-vessel connecting the pulmonary artery with the aorta before birth and so short-circuiting the lungs. It closes up and atrophies about the time of birth so that more blood then goes to the lungs as these come into use.

DUCTUS CAROTICUS. A blood-vessel connecting the carotid artery with the aorta in some vertebrates. It is the vestige of part of the lateral dorsal aorta present in the embryonic state, but like the Ductus arteriosus it may atrophy as the animal develops.

DUCTUS CUVIERI. See *Cuvierian Duct*.

DUCTUS EJACULATORIUS (EJACULATORY DUCT). A duct leading from the male genital organs to the exterior, through which seminal fluid is expelled.

DUCTUS ENDOLYMPHATICUS (ENDOLYMPHATIC DUCT). One of a pair of ducts which, in fish, connect the sacculus of the ear to the exterior and keep the pressure of the endolymph the same as that of the surrounding water. In land vertebrates the aperture is closed and the duct becomes the blind sac known as the *Saccus endolymphaticus*.

DUCTUS THORACICUS. The main duct of the mammalian lymphatic system leading into the left anterior vena cava.

DUCTUS VENOSUS (EMBRYOLOGY). A vein formed from the fusion of the two posterior vitelline veins which run in the wall of the yolk-sac outside the developing embryo of a bird. It later gives rise to the hepatic vein and the base of the posterior vena cava.

DUCTUS VENOSUS ARANTII (EMBRYOLOGY). A blood-vessel connecting the allantoic vein in the placenta of a mammal to the posterior vena cava.

DULOSIS. Slavery among ants: the brown ant, for instance, living in the same nest and working for the red.

DUMB-BELL BONE (PARADOXICAL BONE). A bone in the roof of the mouth of Monotremes, considered to be homologous with the prevomer of a reptile, but which has disappeared in higher mammals.

DUODENUM. That part of the alimentary canal of a vertebrate which immediately follows the stomach and which receives the pancreatic and bile ducts.

DUPLICIDENTATA. Rabbits and hares: rodents having in the upper jaw a second pair of incisor teeth behind the first. They are now classed as a separate order, the *Lagomorpha*.

DURA MATER. A tough fibrous membrane surrounding the brain and spinal cord of a vertebrate. It is the outer of the two *meninges*, the inner being the more delicate and vascular *Pia mater*.

DUVERNOY, GLANDS OF. Another name for the *Glands of Bartholin (q.v.)*.

DYSCOPHINAE. Climbing frogs of Madagascar and Burma, having discs on the ends of the digits.

DYSGENESIS. Infertility, especially between hybrids which are fertile with members of either parent-type.

DYTISCIDAE. Large water-beetles: a family of aquatic carnivorous beetles, most of them with a large oval body and powerful back legs which are flattened and fringed with bristles for swimming. When submerged they carry a store of air beneath the wing-covers. The male is distinguished by the broadened front tarsi which form cup-like suckers for gripping the female when mating.

E

E-, EX-. Latin prefix: out of, without.

EAR. An organ combining the functions of hearing and of giving a sense of balance in vertebrates. In fish it is only an organ of balance but in amphibians, reptiles, birds and mammals, the organ of hearing is present in a varying degree of development. The mammalian ear consists of the following parts:

(1) The *Pinna* or outer funnel-shaped flap of skin.

(2) The *External Auditory Meatus* or passage.

(3) The *Tympanic Membrane*.

(4) The *Middle Ear*: an air-filled chamber connected with the throat by the *Eustachian Tube*, and containing three ossicles known as the *malleus, incus* and *stapes* (hammer, anvil and stirrup) which convey vibrations inwards from the tympanic membrane.

(5) The *Fenestra ovalis* and *Fenestra rotunda* (oval window and round window), two membranes which separate the middle from the inner ear.

(6) The *Inner Ear* or *Labyrinth* consisting of three Semicircular Canals (the organ of balance) and the *Spiral Cochlea* (organ of hearing).

Amphibians, reptiles and birds differ from mammals in having no pinna, in having only one ossicle (the *columella auris*) and in having a very poorly developed cochlea.

EAR-BRAIN (TUBERCULUM ACUSTICUM). A swelling on the dorsal side of the *Medulla oblongata* into which run the seventh, eighth, ninth and tenth cranial nerves, the afferent sensory components of which are known as the *Acustico-lateralis* system.

EC-. Prefix from Greek *Ek*: out of, off.

ECARDINES. Brachiopoda having shells without a hinge.

ECDYSIS. Moulting: the periodic shedding of the cuticle of an arthropod or the outer epidermal layer of a reptile.

ECHENEIFORMES (ECHENEIDAE). See *Discocephali*.

ECHIDNIDAE. Spiny ant-eaters: one of the two families of Monotremes, found only in Australasia. They are primitive reptile-like mammals covered with spines and with hairs. The snout is elongated and toothless; the tongue is very long and protractile.

ECHINO-. Prefix and root from Greek *Echinos*: a spine, prickle.

ECHINODERMATA (ECHINODERMS). The phylum of invertebrates which includes star-fish (Asteroidea), brittle stars (Ophiuroidea), sea-urchins (Echinoidea), sea-lilies (Crinoidea) and sea-cucumbers (Holothuroidea).

They are usually radially symmetrical although the larvae show bilateral symmetry. They are characterized by an exoskeleton of calcareous ossicles either in the form of plates or spines or tiny pincer-like *pedicellariae*. Locomotion is usually by means of numerous tube-feet – small sac-like organs which can be extended when filled with water from an elaborate water-vascular system.

ECHINOIDEA. Sea-urchins: globular or disc-shaped echinoderms like star-fish without arms. They have ambulacral grooves, tube-feet and a ventral mouth. The whole body is covered with calcareous plates, long spines and small pincer-like *pedicellariae*, the latter sometimes bearing poison glands. Locomotion is by means both of tube-feet and of spines.

ECHIUROIDEA. Marine worms having an extremely long ciliated proboscis or prostomium used for feeding. Although showing few signs of segmentation they are classed with the Annelids owing to their early development from a trochosphere larva similar to those of other annelids, and also owing to the presence of chaetae or bristles, although these are few in number.

ECTADENIA. Accessory glands arising from the ectoderm of the male ejaculatory duct (*e.g.* in insects).

ECTO-, ECT-. Prefix from Greek *Ektos*: outside.

ECTOBLAST. An alternative name for *Ectoderm* (*q.v.*).

ECTODERM. The outermost of the three primary layers of cells of an embryo, the others being the endoderm and the mesoderm. These three layers are formed at a very early stage of development and each always gives rise to the same types of tissue. Ectoderm produces epidermis and nervous tissue; mesoderm produces muscles, blood-vessels etc. and endoderm produces the lining of the gut.

ECTOGNATHA. Insects of the order Thysanura (Bristle-tails) having the mouth-parts projecting outwards and well developed (*cf. Entognatha*).

ECTONEURAL SYSTEM. The chief part of the nervous system of star-fish and other echinoderms, lying either in grooves in the skin or just beneath the surface. Included in this system are the nerve-ring round the mouth and the radial nerves along each of the five arms of the star-fish.

ECTOPARASITES. Parasites which live on the outside of their hosts or near the outside, as, for instance, in the mouth or rectum.

ECTOPLASM. A term used to denote the outer layer of protoplasm of a unicellular organism in cases where this is in any way different from the internal protoplasm or *endoplasm*. There are many forms which it may take; it may be clear as in *Amoeba*; highly vacuolated as in many planktonic protozoa, or may form a stiff pellicle as in Ciliophora, Sporozoa and many flagellates.

ECTOPROCTA. Polyzoa in which the oral tentacles can be retracted into a sheath, and in which the anus opens just outside these tentacles. The alimentary canal is thus U-shaped (see also *Endoprocta*).

ECTOPTERYGOID. The outermost of the three pterygoid bones in the skull of a bony fish, the others being the *endopterygoid* and the *metapterygoid*. It corresponds with the *transpalatine* bone in reptiles and forms part of the hindpart of the palate.

EDENTATA (EDENTATES). Primitive mammals whose teeth are either absent altogether or if present are brown, peg-like and without enamel. The group includes sloths, ant-eaters and armadilloes of South America as well as the two extinct forms, the *Megatherium* or Giant Ground Sloth and the *Glyptodon* or Giant Armadillo. All these belong to the group Xenarthra of the New World. There are also a few African and Asian species, such as the pangolin or scaly ant-eater, which have been put in a separate order, the Pholidota. These were formerly regarded as Edentates and were put in the subgroup Nomarthra, sometimes also called Effodientia.

EDENTULOUS. Without teeth: anodontoid.

EFFECTOR. Any organ such as a muscle or gland which is brought into action by stimulation from a nerve.

EFFERENT. A term used for nerves or blood-vessels leading away from any particular organ or part of the body.

EFFERENT BRANCHIAL VESSELS (EPIBRANCHIAL VESSELS). Vessels carrying blood away from the gills of a fish, towards the dorsal aorta.

EFFERENT NERVES. Nerves which conduct impulses away from the brain or spinal cord, as, for instance, the motor nerves which stimulate muscles.

EFFODIENTIA. Another name for the *Nomarthra*, the Edentates of Asia and Africa, *viz*. the Pangolin or scaly ant-eater and the Aardvark or Cape anteater. These are now usually placed in separate orders known as the Pholidota and the Tubulidentata respectively (see also *Edentata*).

EGESTION. Getting rid of faeces or unused food substances from the body.

EGG. A term usually used to denote an ovum or egg-cell with much yolk surrounded by several membranes, possibly albumen or jelly and a shell (see *Egg-membrane*).

EGG-CELL. See *Ovum*.

EGG-MEMBRANE. A general name denoting any membrane, shell or layer of jelly or albumen surrounding and protecting an egg. The chief kinds are:
(1) Primary or *Vitelline* membranes secreted by the ovum or the oocyte.
(2) Secondary membranes secreted by cells of the ovary.
(3) Tertiary membranes secreted by glands in the oviduct. The albumen and shell of a bird's egg are of this type.

EGG-SHELL. See *Shell.*

EGG-TOOTH. A small horny projection on the upper beak of a young bird, used for piercing the shell of the egg as the bird hatches.

EJACULATORY DUCT. The distal part of the sperm-duct or *Vas deferens* which by its muscular contractions can bring about the ejection of seminal fluid.

ELAIODOCHON (ELAEODOCHON). The oil-gland or preening gland of a bird.

ELAPINAE. A subfamily of snakes, all extremely poisonous, characterized by having a series of small solid teeth behind the poison fangs. The tail is cylindrical, a fact which distinguishes them from sea-snakes whose tail is laterally compressed. They include the majority of poisonous snakes in all continents, but are absent from Madagascar and New Zealand.

ELASMOBRANCHII. An alternative name for Cartilaginous fish or *Chondrichthyes.* The name refers to the fact that the gill-slits are exposed and not covered by an operculum.

ELASTICA (OF NOTOCHORD). More correctly named the *Membrana elastica interna* and the *Membrana elastica externa*: two sheaths of fibrous elastic tissue surrounding the notochord of a vertebrate. During embryonic development they may have cartilage or bone deposited in them to form the centra of the vertebrae.

ELASTIC CARTILAGE. Cartilage containing fibres of yellow elastic tissue, *e.g.* in the pinna of the ear.

ELASTIC TISSUE. The chief component of ligaments: connective tissue composed largely of yellow branching fibres of *elastin.*

ELASTIN. See *Elastic Tissue* above.

ELATERIDAE. Click-beetles: a family named on account of their habit of springing into the air with a clicking sound when placed on their backs. The larvae are well known as wireworms, agricultural pests living in the soil and attacking roots.

ELECTRIC ORGANS. Organs capable of giving an electric shock to any animal which touches them. They are found on the bodies and tails of such fishes as *Torpedo* and *Raia*, and consist of spots or areas where the potential difference accompanying muscular contraction is greatly increased. Each electric organ consists of numerous small compartments filled with a gelatinous substance through the middle of which runs a plate of finely granular nucleated substance having numerous nerve endings.

ELEUTHEROZOA. A general name for the non-sessile types of Echinoderm: star-fish, brittle stars, sea-urchins and sea-cucumbers, as distinct from the *Palmatozoa* or sessile types which include the stalked Crinoids (sea-lilies) and many fossil types which were sessile although not always stalked.

ELYTRA. Thickened horny fore-wings of beetles, etc., covering and protecting the membranous hind-wings. In a few species of beetle, whose flying wings are reduced or absent, the elytra are cemented together down the middle line.

EMBALLONURIDAE. Bats having a long freely moving tail extending beyond the inter-femoral membrane. They are found mostly in the tropics.

EMBIOPTERA. A small order of hemimetabolous insects with soft, elongated, flattened bodies; the males have two pairs of wings but the females are wingless. They are social insects living in silken tunnels beneath stones or under bark. The silk is produced from glands in the front legs.

EMBOLOBRANCHIATA. Terrestrial arachnids breathing by lung-books or tracheae or both; they include spiders, scorpions, ticks, mites and phalangids.

EMBOLOMERI. An extinct family of primitive amphibia included in the Labyrinthodontia, having vertebrae each with two centra, one behind the other. They thus give a clue to the evolution of the vertebral column.

EMBRYO. A plant or animal in the early stages of development before the principal organs are complete.

EMBRYONIC DISC. (1) The Blastoderm or blastodisc: a disc of cells on one side of the yolk-sac, from which the future embryo develops in the egg of a reptile, bird etc.

(2) The homologous disc of cells of the blastocyst in mammalian embryonic development.

EMBRYONIC KNOB. A knob-like mass of cells to one side of the mammalian blastocyst, beneath the *Cells of Rauber*; later giving rise to the embryonic disc and the embryo itself.

EMBRYONIC MEMBRANES (FOETAL MEMBRANES). Membranes extending outside the developing embryo of a reptile, bird or mammal, protecting it and enabling it to obtain oxygen and food material. They are the *Amnion*, *Chorion* and *Allantois* or a composite membrane formed from several of these.

EMBRYOPHORE. Any structure on which an embryo develops.

EMPIDAE. Empid flies: small carnivorous flies with interesting mating habits. The male kills a small insect, wraps it up in silk thread and presents it to the female. While she is unravelling it and feeding on it she is too contented and busy to devour the male, as she would normally do, and he is enabled to fertilize her safely.

EMPODIUM. A median bristle between the two claws on the foot of an insect.

EN-, EM-. Greek prefix: in, into.

ENAMEL. The hard outer layer of a tooth covering the softer dentine within (see also *Dental lamina*).

ENCHYLEMA. A name used to denote the more liquid constituent of cytoplasm which, according to one theory, is enclosed in the meshes of a network of more viscous material known as *spongioplasm*. According to another theory the structure is that of a foam and the meshwork is an optical section of the walls of bubbles or alveoli containing the liquid.

ENCYSTMENT. The formation of a cyst or thick membrane round an organism, usually to protect it from drought or other unfavourable conditions during a resting stage in its development.

END BULB. See *Tactile corpuscle*.

ENDITES. Lateral lobes on the inner or median side of some crustacean limbs, especially on the flattened type of limb known as a *phyllopodium*. The number of endites may be up to six but is usually less than this; one of them may be enlarged to form a *gnathobase* for manipulating food.

ENDO-. Prefix from Greek *Endon*: within.

ENDOBLAST. An alternative name for the endoderm, the innermost of the three primary germ-layers of an embryo.

ENDOCARDIUM. The inner lining of the heart.

ENDOCHONDRAL OSSIFICATION. The formation of bone from cartilage, commencing near the centre round the marrow cavity and gradually spreading outwards towards the periosteum.

ENDOCRINE GLANDS. Ductless glands such as the thyroid, pituitary, adrenals etc. producing hormones or internal secretions which go directly into the blood-stream and are thence carried round the body to have various stimulating effects on growth and metabolism.

ENDOCUTICLE. The flexible inner layer of the cuticle of an insect or other arthropod, composed of chitin and proteins. During moulting this layer is digested away by enzymes from special epidermal glands. It thus leaves the outer layers of cuticle loose and ready to be cast off.

ENDOCYCLICA. Sea-urchins or Echinoidea of the regular symmetrical type in which the mouth is central and underneath, while the anus is directly opposite to it on the upper side.

ENDODERM. The innermost of the three primary germ-layers of an embryo, giving rise to the lining of the alimentary canal (*cf. Ectoderm* and *Mesoderm*).

ENDOLYMPH. Fluid contained in the innermost parts of the membranous labyrinth of the vertebrate ear.

ENDOLYMPHATIC DUCT (DUCTUS ENDOLYMPHATICUS). A duct connecting the inside of the sacculus of the ear with the exterior in some fish; or ending as a blind sac in most land vertebrates. In amphibia the right and left endolymphatic ducts meet in the middle line and connect with the calcareous nodes lying alongside the vertebrae.

ENDOLYMPHATIC SAC. This name is sometimes given to the closed endolymphatic duct of a land vertebrate, but is also used to denote a widening or dilatation of the endolymphatic duct in fish, or in some cases a backward diverticulum from the point where the two ducts meet in the median line (see *Endolymphatic duct*).

ENDOMIXIS. A process which brings about a state of rejuvenation in certain protozoa, such as *Paramecium*, which have two nuclei. The meganucleus degenerates and disappears but is later replaced by a new one formed by division of the micronucleus.

ENDOPARASITE. A parasite living inside the host as distinct from an ectoparasite which lives outside the host.

ENDOPHRAGMAL SKELETON. An internal skeleton in many crustaceans, formed by ingrowths of the cuticle known as *apodemes* which serve for the insertion of muscles. They may be united to form a complex framework, as is notably the case in crabs, lobsters, shrimps etc.

ENDOPLASM. A name used to denote the internal cytoplasm of a cell where this differs markedly from the outer layer or *ectoplasm* (see *Ectoplasm*).

ENDOPLEURITES. Y-shaped ingrowths of cuticle from the sides of the thorax, forming part of the endophragmal skeleton of crustaceans.

ENDOPODITE. The inner branch of the biramous limb of a crustacean (*i.e.* that nearest to the median line).

ENDOPROCTA. Polyzoa in which the anus opens within the circle of oral tentacles, the alimentary canal being U-shaped (see also *Ectoprocta*).

ENDOPTERYGOID. The uppermost of the three pterygoid bones of a fish (see also *Ectopterygoid*).

ENDOPTERYGOTA. Insects which pass through a complete metamorphosis in which the larva is very different from the adult and in which there is always a pupal stage.

ENDOSKELETON. A skeletal structure wholly within an organism; the term

81

usually refers to the bony or cartilaginous skeleton of a vertebrate but may be used also to denote such structures as the spicules of a sponge or other invertebrate.

ENDOSOME. A nucleolus or karyosome, the latter name being more especially used to denote those which are basic staining.

ENDOSTERNITES. Apodemes or ingrowths from the ventral cuticular plates of a crustacean or an insect. They form part of the endophragmal skeleton to which important muscles are attached.

ENDOSTEUM. A thin membrane lining the cavity inside a bone.

ENDOSTYLE. A glandular ciliated groove running along the floor of the pharynx in *Amphioxus*, in some Tunicates and in the larvae of lampreys (Cyclostomata). It produces threads of mucus to which food particles adhere and which are passed dorsalwards round the pharynx and backwards into the gullet by ciliary action. From observations of the development of the lamprey it can be shown that the thyroid gland has evolved from the endostyle.

ENDOTHELIO-CHORIALIS. A type of placenta, found, for example, in the Carnivora, in which the uterine epithelium and the uterine connective tissue break down, but the maternal capillaries remain intact (*cf. Haemochorialis*).

ENDOTHELIUM. A layer of very thin cells lining the blood-vessels and heart of a vertebrate.

ENDYSIS. Development of new skin-structures (*cf. Ecdysis*).

ENGYSTOMATIDAE. Tropical frogs with a small snout-like mouth for feeding on ants.

ENT-, ENTO-. Prefix from Greek *Entos*: within.

ENTEPICONDYLAR FORAMEN. A foramen above the inner condyle of the humerus in some mammals and reptiles.

ENTERO-. Prefix from Greek *Enteron*: intestine.

ENTEROCEPTORS. Nervous receptors inside the alimentary canal.

ENTEROCOEL (EMBRYOLOGY). A portion of the coelom which has been in communication with the archenteron, *e.g.* in the mesodermal pouches of *Amphioxus* etc.

ENTEROKINASE. An enzyme formed in the walls of the duodenum and acting as a co-enzyme with *trypsinogen* from the pancreas to form *trypsin* which breaks down proteins.

ENTERON. The alimentary canal.

ENTEROPNEUSTA. An alternative name for the *Hemichordata*, a small group of primitive chordates including the burrowing 'acorn-worm' *Balanoglossus* and some sedentary ciliary-feeding worms. They possess gill-slits and a notochord in the anterior part of the body, and are therefore related to other chordates and to vertebrates, but their embryology suggests relationship also with the Echinoderms.

ENTOCOEL. The space between two mesenteries of a pair in the body-cavity of a sea-anemone or of certain coral polyps. In polyps with a large number of mesenteries the secondary and tertiary pairs are added, but always in the *exocoels, i.e.* the spaces between one pair and the next and never in the entocoels.

ENTODINIOMORPHA. Protozoa, sometimes having strange shapes, found in large numbers in the alimentary canals of sheep and cattle. They probably help to break down the vegetable matter in the food by feeding on it and being in turn digested further down. They belong to the *Oligotricha*, a group of ciliated protozoa having a 'gullet', an undulating membrane and an adoral wreath, but with the body-cilia absent or reduced.

ENTOGNATHA. See *Entognathous* below.

ENTOGNATHOUS. A term used to denote insects having the mouth-parts in pockets from which they can be protruded when feeding, as in the *Diplura* or Bristle-tails formerly called *Entognatha* (see also *Diplura*).

ENTOMOSTRACA. A name used formerly to include all crustaceans other than the Malacostraca. Although no longer used in formal classification it is convenient to keep the name for the lower crustaceans as a whole.

ENVELOPE CELLS. Cells occurring in protozoa of the order Cnidosporidia, containing sporoblasts, which later give rise to complicated multicellular spores. The envelope cells are usually found in pairs contained in a body known as a pan-sporoblast within the syncytial body of the parent protozoon.

ENZYME. A catalytic substance, secreted by living cells, which helps to promote or speed up some specific metabolic activity such as digestion oxidation etc. Enzymes usually act only in dilute solution with a particular pH value and within narrow limits of temperature.

EOHIPPUS. A small ancestral type of horse of the Eocene Period, having three toes on the hind foot and four on the front foot.

EP-, EPI-, EPH-. Various forms of the Greek prefix: upon.

EPANORTHIDAE. Marsupials of the Diprotodont type having a small vestigial pouch. The group includes a large number of fossils from Central and South America but is at present only represented by two species living in the mountains of Ecuador and Columbia.

EPAXONIC MUSCLES. In fish and some higher vertebrates each myotome or segmental muscle is divided into two by a longitudinal partition in which the ribs are formed. The muscles dorsal to this are called *epaxonic*; those ventral to it *hypaxonic*.

EPHEMEROPTERA. May-flies: a primitive order of insects having four membranous wings of which the front pair is considerably larger than the hind pair. The veins of the wings are in the form of a simple network like that of the dragon-fly. The eyes are large and the antennae short; at the hind-end are three long tail filaments. The larvae or nymphs are aquatic and herbivorous; they may live for several years in a pond or stream before metamorphosis. There is no pupa, but the first winged stage, known as the *sub-imago*, resembles the adult in size and shape. It is, however, of a dull brown colour. This casts its skin off almost immediately and reveals the brightly coloured shining adult which, being unable to feed, lives only for a few hours or a day.

EPHIPPIUM. (1) An egg-case formed from the thickened cast off carapace of certain crustacea (*e.g. Daphnia*).

(2) The pituitary fossa in the base of the vertebrate skull.

EPHYRA. The free-swimming larva of a jelly-fish, formed by strobilation or segmentation of the polyp-like *scyphistoma larva*. The latter is a sedentary individual resembling a hydra, anchored at the base and having a mouth with oral tentacles at the upper end. It becomes segmented by a series of constrictions and eventually forms a number of young ephyrae like a pile of saucers. These break away one by one and swim off to become new jelly-fish.

EPIBLAST. An alternative name for the ectoderm, the outermost of the three primary germinal layers of an embryo, giving rise to epidermis and nervous tissue.

EPIBOLY. An embryonic process by which a blastula (hollow ball of cells) is converted into a gastrula (two-layered sac) by overgrowth of the ectoderm which spreads over the endoderm. The net result is a gastrula similar to that obtained in some organisms by invagination of the endoderm.

EPIBRANCHIAL. Above the gills:

(1) Epibranchial cartilages or bones forming the dorso-lateral part of each branchial arch of a fish.

(2) Epibranchial arteries: the efferent branchial arteries of a fish, leading from the gills to the dorsal aorta.

(3) Epibranchial space of Mollusca: the space above the gill membranes in a Lamellibranch or bivalve mollusc. Water which has entered the mantle cavity by way of the ventral or inhalent siphon, continually passes through microscopic pores in the gill-lamellae into the epibranchial cavity and out through the dorsal or exhalent siphon.

(4) Epibranchial space of Crustacea: the upper part of the gill-chamber under the carapace of a crab or similar crustacean.

EPICARDIUM. The delicate covering layer of cells outside the wall of the heart.

EPICORACOIDS. A pair of cartilages or cartilage-bones superimposed on the ventral surface of the coracoids in frogs and other amphibia and in many reptiles.

EPICRANIAL PLATES. Two cuticular plates covering the dorsal surface of an insect's head.

EPICUTICLE. The outer layer of the cuticle of an insect or other arthropod; it is usually very thin and is made of *sclerotin* which is impervious to water.

EPIDERMIS. The outer layer of the skin derived from the embryonic *ectoderm*. In vertebrates it usually consists of stratified epithelium outside which is often a horny layer formed from dead cells. At the base of the epidermis is the *Malpighian Layer* consisting of actively dividing cells which constantly replace the dead cells above.

EPIDIDYMIS. The long narrow coiled tube connecting the testis with the *vas deferens* in higher vertebrates.

EPIGAMIC CHARACTERS. Secondary sexual characters brought about by the action of hormones. Examples are the pitch of the voice, the presence or absence of horns or antlers, the colour of feathers etc.

EPIGASTRIC VEIN. A vein in birds corresponding to the front part of the anterior abdominal vein of a reptile or amphibian. It runs ventrally forwards to the liver and joins the hepatic vein.

EPIGLOTTIS. A small flap of flesh and cartilage protecting the opening of the larynx in mammals. During swallowing it comes down over the glottis preventing food or drink from entering it.

EPIHYAL BONES. A pair of cartilage-bones which, together with the ceratohyals, hypohyals and basihyal, make up the hyoid arch of a vertebrate. These bones are well developed in many fish but tend to be reduced or absent in higher vertebrates.

EPIMERE. A somite: one of a series of segmentally arranged blocks of mesoderm in a vertebrate embryo, giving rise to myotomes (blocks of muscle), sclerotomes (which form the vertebrae) and dermatomes (which form the lower layer of the skin).

EPIMERITE. The fixing organ or anterior segment on certain parasitic protozoa such as *Gregarina* which frequent the alimentary canals of many insects.

EPIMERON. The posterior part of the subcoxa of an insect (see also *Episternum*).

EPIMYSIUM. The fibrous sheath of a muscle.

EPINEURAL CANALS. Canals or passages formed by the roofing in of the ambulacral grooves enclosing each radial nerve in a sea-urchin (echinoid). In star-fish and other members of the phylum the grooves are open, exposing the radial nerves to the water.

EPINEURIUM. The fibrous sheath of a nerve.

EPIOTIC BONE. One of the five bones forming the auditory capsule of a fish, the others being the *pro-otic, opisthotic, sphenotic* and *pterotic*. In higher vertebrates these are fused together to form the *periotic* bone.

EPIPHARYNGEAL GROOVE. A ciliated and glandular groove in the roof of the pharynx of *Amphioxus*, by which food particles are conveyed backwards to the oesophagus.

EPIPHARYNX. The membranous roof of the mouth of an insect, provided with taste receptors and sometimes elongated to form part of the proboscis.

EPIPHRAGMA. A membrane formed of calcium phosphate and mucilage secreted by snails before hibernating. The snail withdraws into its shell and then secretes this membrane forming a disc over the opening of the shell, except for a tiny hole left for breathing.

EPIPHYSIS. The end part of a bone which ossifies separately from the central part or *diaphysis* but later fuses with it.

EPIPHYSIS OF BRAIN. An alternative name for the *Pineal Body* (*q.v.*).

EPIPLEUR (EPIPLEURAL FOLDS). A pair of folds of skin forming the floor of the atrium in *Amphioxus*; they grow inwards from the lateral walls (*metapleural folds*) and meet in the mid-ventral line.

EPIPODITE. An external process on the base of the limb of a crustacean. It may be flattened, branched or folded and usually has a respiratory function.

EPIPUBIC BONES (MARSUPIAL BONES). Two ventral bones in front of the pubic bones in Monostremes and Marsupials but absent in higher mammals. In marsupials they help to support the pouch.

EPIPROCT. A small tergal plate above the anus of an insect.

EPIPTERYGOID (COLUMELLA CRANII). A bone forming an ascending process from the pterygoid or pterygo-quadrate in the skull of a reptile; homologous with the *alisphenoid* bone of a mammal.

EPISTERNUM. (1) The cartilaginous anterior part of the sternum in some vertebrates, *e.g.* a frog.
(2) A name sometimes used for the *interclavicle*.
(3) The anterior part of the subcoxa of an insect, the posterior part being called the *epimeron*. The two are separated from each other by the *pleural suture*.

EPISTOME. A plate in front of the mouth of a crab or similar crustacean, formed from the sterna of the mandibular and antennal segments.

EPITHELIO-CHORIALIS. A type of placenta found for example in cattle and pigs, where all the layers of the placenta are retained and no 'erosion' of tissues takes place (*cf. Haemochorialis*).

EPITHELIUM. Cells which form a thin layer covering and protecting the body, lining the alimentary canal or forming the secretory parts of a gland. They are classified according to their shape as columnar, cubical or squamous (pavement) and are usually packed tightly together and resting on a basement membrane. The term epithelium is sometimes wrongly applied to similar cells of mesodermal origin which are known as *endothelium* (lining blood vessels) or *mesothelium* (lining the coelom).

EPOÖPHORON (ORGAN OF ROSENMÜLLER). A vestige of the embryonic mesonephros (middle kidney) which may persist in an adult female mammal; it is homologous with the tubules of the epididymis in the male.

EQUIDAE. Horses, asses and zebras: a group of Perissodactyla or odd-toed ungulates having a single large hoof on each foot. Many ancestral types had three or more toes and there has been a gradual reduction of these so that only the middle one now remains.

EQUITANT. Overlapping saddlewise.

EQUITANT WHORLS. In the calcareous shells of Foraminifera numerous chambers are arranged in a spiral, each whorl overlapping the previous one at the sides and hiding it. The inner whorls of the spiral can therefore only be seen in a decalcified specimen. Such overlapping is said to be equitant.

EREPSIN. A proteolytic enzyme produced by glands in the wall of the small intestine.

ERETHIZONTIDAE. Tree porcupines of Central and South America with long prehensile tails.

ERGATANER. A male ant resembling a worker.

ERGATE. A worker ant.

ERGATOGYNE. A female ant having the characteristics of a worker.

ERGATANDROMORPH. A worker ant with male characteristics.

ERINACEIDAE. Hedgehogs: Insectivora with spiny skin, plantigrade feet and a flexible vertebral column enabling them to roll up when disturbed. There are about twenty species found in most countries of the Old World.

ERUCIFORM LARVA. A name formerly used for a polypod or caterpillar type of larva having a fleshy body, a thin skin and pro-legs or cushion-feet on the abdomen.

ERYTHR-, ERYTHRO-. Prefix from Greek *Eruthros*: red.

ERYTHRININA. Fresh-water carp-like fish of tropical America, having a scaly body without an adipose fin and a naked head with no barbels.

ERYTHROCYTES. Red blood corpuscles.

ESOCIDAE. Pikes: carnivorous and very voracious fresh-water fish with scaly body, soft fin-rays and no barbels on the mouth. The paired pelvic fins and the unpaired fins are far back on the body.

ESOCIFORMES (HAPLOMI). Pike-like fishes (see *Esocidae* above).

ETHMOID BONE (MESETHMOID BONE). A cartilage-bone forming the anterior boundary of the cranium in most vertebrates. It usually consists of three parts, *viz.*:
(1) The cribriform plate at right angles to the axis of the cranium and perforated for the passage of nerves.
(2) The median plate continuous with the nasal septum.
(3) The spongy lateral parts against the inner walls of the maxillae.

ETHMO-PALATINE LIGAMENTS. A pair of ligaments connecting the palatoquadrate cartilage (upper jaw) with the ethmoid region of the chondrocranium in most cartilaginous fish (see *Hyostylic*).

EU-. Greek prefix: well, good, genuine, true.

EUANOSTRACA. The present-day genera of Anostraca or 'fairy-shrimps' as opposed to the more primitive fossil forms (see *Anostraca*).

EUBLEPHARIDAE. Lizards of Africa, Asia and America, similar to geckos but having functional eyelids.

EUCARIDA. Crustaceans of the order Malacostraca, having the carapace fused to all the thoracic segments; having no oostegites and having compound eyes on stalks. The group includes the large order of Decapoda (crabs, lobsters etc.). (See also *Caridoid Facies*.)

EUCEPHALOUS. A term applied to insect larvae which have a definite head as distinct from the acephalous type in which it is absent or much reduced.

EUCILIATA. All the ciliated protozoa other than the small primitive group known as Prociliata.

EUGLENOIDIDA (EUGLENOIDINA). Flagellate protozoa of the group Phytomastigina containing numerous chloroplasts and forming a link between plants and animals. *Euglena*, the typical member of the order has one flagellum arising from the bottom of the 'gullet' and has a contractile vacuole fed by accessory vacuoles. Some species have pyrenoids and food reserves of *paramylum*, a variety of starch.

EUGLENOID MOVEMENT. The characteristic method of locomotion of *Euglena* and certain other protozoa. It consists of a contraction of one part of the body and a dilatation of another causing a kind of wriggling movement.

EUGREGARINARIA. Parasitic protozoa of the order Gregarinidea, distinguished from the Schizogregarinaria by the fact that they do not undergo schizogony or multiple asexual fission. They include the well known *Monocystis*, a parasite of the earthworm, and *Gregarina*, a parasite of insects.

EULAMELLIBRANCHIATA (EULAMELLIBRANCHIA). Lamellibranch (bivalve) molluscs in which the gill-filaments are united by connective tissue to form perforated ciliated membranes hanging longitudinally under the shell-valves on each side. The group includes the common fresh-water mussel (see *Lamellibranchiata*).

EUMITOSIS. Normal cell division or mitosis as distinct from *paramitosis* and *cryptomitosis* (*q.v.*).

EUPHAUSIACEA. A small group of pelagic shrimp-like crustacea forming the chief food of whales.

EURYPTERIDA. Aquatic arachnids of the Silurian period resembling scorpions but sometimes attaining a length of 6 ft.

EURYPYGIDAE. Sun-bitterns: crane-like birds of America with long neck and slender bill.

EUSELACHII. Sharks and rays: cartilaginous fish distinguished from all others by having numerous teeth developing in succession, and by having pectoral fins with three basal elements, *viz.* the *pro-*, *meso-* and *metapterygium*. Apart from these features they have the general characters of cartilaginous fish: the whole skeleton is made of cartilage and the gill-slits are not covered by an operculum.

EUSTACHIAN TUBE. A tube connecting the back of the throat with the cavity of the middle-ear behind the tympanic membrane in all land vertebrates. It is considered to be homologous with the first gill-slit in some fish and with the spiracle in dogfish. By its means the air pressure is equalized on both sides of the tympanic membrane.

EUSTACHIAN VALVE. A rudimentary sinu-auricular valve where the venae cavae enter the right auricle of the mammalian heart.

EUSTHENIIDAE. Stone-flies (Plecoptera) of Australia and New Zealand, having nymphs or larvae of a very primitive type with lateral gills on the abdominal segments. They may live for several years and pass through as many as thirty instars before the adult fly is formed.

EUSUCHIA. All crocodiles other than the two small extinct orders Parasuchia and Pseudosuchia (see *Crocodilia*).

EUTHERIA (MONODELPHIA, PLACENTALIA). The higher mammals comprising all except the Monotremes and Marsupials. The embryo of the Eutherian always has an allantoic placenta and reaches an advanced state of development before birth.

EUTHYNEURY. A secondary symmetrical arrangement of the nervous system in certain gasteropod molluscs. In the *Opisthobranchiata*, for instance, the embryonic nervous system is twisted and unsymmetrical but symmetry is attained by detorsion or untwisting of the visceral hump. In pulmonate snails, on the other hand, the twisted visceral loop of the nervous system becomes shortened and fused with the circumpharyngeal nerve collar. These two classes of molluscs, although not closely related, are sometimes put into a single group, the *Euthyneura* (see *Torsion*).

EVOLUTION. The gradual change of animals and plants from one generation to another, leading to new varieties, species, genera etc. (see *Natural Selection*).

EX-. Latin preposition and prefix: out of.

EXARATE. A type of insect pupa in which the legs and other appendages are free and clearly visible and the abdomen is movable. In some cases such a pupa is capable of limited locomotion. Exarate pupae are found in nearly all holometabolous insects other than Lepidoptera.

EXCITOR NEURONE. A neurone which directly stimulates a muscle, gland or other organ.

EXCRETION. The elimination from the body of unwanted materials: water, carbon dioxide and nitrogenous substances produced as a result of metabolism.

EXHALENT CANALS (OF SPONGES). Passages by which water flows from the flagellated chambers to the central chamber or paragaster in the most complex type of sponge, the *Leucon* type.

EXHALENT SIPHON. An opening from which water is forced outwards, *e.g.* in Lamellibranch molluscs, Tunicates etc.

EXITES. Lateral lobes on the outer side of the limb of a crustacean, more particularly on the flattened type of limb known as a *phyllopodium*. The distal exite is often enlarged and known as a *flabellum*; the more proximal ones, the *epipodites*, are sometimes modified to form gills.

EXOCCIPITALS. Two bones forming part of the hind-part of the vertebrate cranium. In mammals each bears a condyle or projection fitting into a socket in the atlas vertebra. This form of articulation makes possible the nodding movement of the head.

EXOCOEL (OF COELENTERATES). The space in a sea-anemone or a coral polyp where secondary mesenteries may develop between two adjacent pairs of primary mesenteries (see *Entocoel*).

EXOCUTICLE. The middle cuticular layer of an insect or other arthropod, outside the endocuticle and beneath the epicuticle. It is usually very thick and is composed of chitin and proteins impregnated with sclerotin and sometimes melanin.

EXOCYCLICA. Irregular flattened sea-urchins (echinoidea), either round, heart-shaped or roughly pentagonal, with the ambulacral grooves resembling five petals (see *Clypeastroidea*).

EXOGAMY. Syngamy or sexual union of two gametes derived from different parents, as distinct from *autogamy* in which they are both derived from one parent. The terms are generally used with reference to protozoa.

EXOPODITE. The outer branch of the biramous limb of a crustacean.

EXOPTERYGOTA (HEMIMETABOLA). Insects in which there is no pupal stage and in which metamorphosis involves very little change at each instar. The young, which are known as nymphs, are generally similar to the adult, but are smaller and have wing-buds at an early stage. Eleven orders are comprised in the Exopterygota. They include cockroaches, earwigs and all kinds of bugs and lice, as well as such insects as the dragonfly and the mayfly. The two latter have aquatic nymphs differing more markedly from the adults than do the nymphs of other orders.

EXOSKELETON. The thick cuticle of an insect, crustacean or other arthropod, usually made of chitin impregnated with calcium compounds making it very hard. As its presence hinders growth, the animal must of necessity cast it off from time to time and can only grow between the loss of one cuticle and the formation of the new one. This moulting is known as *ecdysis*.

EXPIRATION. The breathing out of air from the lungs.

EXTENSOR MUSCLES. Muscles which bring about the straightening of a limb.

EXTERNAL AUDITORY MEATUS. The outer ear-passage leading to the tympanic membrane in birds and mammals; absent in amphibia and reptiles.

EXTERNAL CAROTID ARTERIES. A pair of arteries branching off from the common carotids in the neck of a vertebrate and taking blood to the external parts of the face and to the tongue.

EXTERNAL EAR. The parts of the vertebrate ear which are external to the tympanum, *i.e.* the pinna and the outer ear passage.

EXTERNAL GILLS. Tuft-like outgrowths of the body-wall having a large surface through which dissolved oxygen can pass into the blood. They are found in the tadpoles of frogs, newts etc., as well as in many invertebrates.

EXTERNAL ILIAC VEINS. Veins bringing blood from the external parts of the legs to the posterior vena cava of a vertebrate.

EXTERNAL JUGULAR VEINS. Veins bringing blood from the external parts of the head to the anterior venae cavae of a vertebrate.

EXTERNAL NARES. Nostrils or openings connecting the olfactory cavities with the exterior.

EXTERNAL RECTUS MUSCLE (POSTERIOR RECTUS MUSCLE). The hindmost of the four rectus muscles attaching the eyeball to the orbit in all vertebrates. There are four rectus muscles at right angles to one another: the anterior, posterior, superior and inferior. In front of these are the superior and inferior oblique muscles.

EXTERNAL RESPIRATION. Breathing: the mechanism by which air enters and leaves the body, as distinct from internal or true respiration involving a process of oxidation with production of energy.

EXTERO-. Prefix from Latin *exter*: outer.

EXTEROCEPTOR. A sensory cell or organ which receives stimuli from outside (*cf. Interoceptor* and *Proprioceptor*).

EXTRA-. Latin prefix: on the outside.

EXTRA-EMBRYONIC COELOM. A term used in vertebrate embryology to denote an extension of the coelom outside the limits of the embryo, that is between the amnion and the chorion in reptiles, birds and mammals.

EXTRA-THECAL ZONE. The part of a coral-polyp colony overlapping the calcareous cups or *thecae* in which the polyps live. It is a continuous layer forming new buds all the time and depositing a mass of calcareous material below it.

EXTRINSIC EYE MUSCLE. See *Eye muscles*.

EXUMBRELLAR SURFACE. The upper or 'dorsal' surface of a jelly-fish, that is the side opposite to the mouth. The word dorsal is not generally used in the case of such radially symmetrical animals since most of them are derived from a flattened bilaterally symmetrical larva which has a true dorsal and ventral surface.

EXUVIAE. The cast-off skin of an insect etc. (see *Ecdysis*).

EYE. A single or composite receptor organ by which an animal can perceive light. The essential feature of all types of eye is that light, whether focused by a lens or not, impinges on the sensitive cells of the retina bringing about a chemical change which stimulates the optic nerve. In the human eye light enters through the cornea, passes through the aqueous humour and the aperture of the iris to the lens. The latter focuses it to form an image on the retina at the back of the eyeball. There are two types of light-sensitive cells in the retina: *cones* which perceive colour and *rods* which perceive black and white only. Further refinements of the human eye are the ability to regulate the amount of light entering it by varying the aperture of the iris, and the ability to alter the focal length of the lens by changing its shape (see also *Compound eye* and *Ocellus*).

EYE-BRAIN. The optic lobes formed from the enlarged roof of the mid-brain in vertebrates.

EYE MUSCLES. Six small muscles which move the eyeball of a vertebrate. They are the anterior, posterior, superior and inferior rectus muscles and the superior and inferior oblique muscles.

EYE-SPOT. A light-sensitive receptor consisting of a small spot of pigment able to absorb light, undergo some temporary chemical change and give a stimulus to the organism. Many of the simpler invertebrates, including a number of protozoa, possess such eye-spots.

F

FACIAL ANGLE. See *Cranio-facial angle*.

FACIAL NERVE. The seventh cranial nerve of a vertebrate having several branches innervating the face, mouth, palate etc.

FACILITATOR. A substance such as *adrenalin* or *acetyl choline* which is produced by a nervous impulse at a synapse. As the substance diffuses across the gap it activates the next neurone, so that an impulse seems to pass across from one nerve to the next although in fact a new impulse is produced in each.

FACULTATIVE PARASITE. An organism which is sometimes free-living and sometimes parasitic.

FAECES. Waste matter egested from the anus consisting of partly broken down food substances, dead bacteria, dead cells, mucus and bile pigments.

FALC-, FALCI-. Prefix from Latin *Falx, falcis*: a sickle.

FALCATE. FALCIFORM. Sickle-shaped, as for instance certain protozoa.

FALCIFORM BODIES. Sickle-shaped sporozoites of *Monocystis* and similar protozoa.

FALCIFORM LIGAMENT. A short double sheet of mesentery connecting the liver with the ventral wall of the abdomen in a vertebrate.

FALCONIDAE (FALCONINAE). Falcons, hawks, eagles and buzzards (see *Falconiformes* below).

FALCONIFORMES. Carnivorous birds with curved beak, hooked at the tip and with strong talons for seizing small mammals, birds or other prey. The group includes vultures, condors, hawks, eagles, buzzards, kites and falcons.

FALCULA. The curved claw of a bird.

FALLOPIAN TUBES. An alternative name for the oviducts of mammals.

FALSE PALATE. The roof of the mouth in mammals, formed by flat extensions of the maxillae meeting in the middle and separating the buccal from the nasal cavity.

FALSE RIBS. Ribs which, although extending round the chest, are not attached directly to the sternum. In many mammals for instance, the first seven pairs are true ribs articulating with the sternum but the next two or three pairs are false ribs attached to the cartilaginous parts of the seventh pair.

FAT BODY. (1) Of Amphibia: a yellow lobed body attached to the anterior of each kidney and testis. By means of this reserve of fat the gonads are nourished during the winter in readiness for the breeding season in the spring.

(2) Of Insects: a mass of fatty tissue filling most of the body cavity. The food stored in this tissue includes fats, proteins and glycogen which form a reserve for hibernation or pupation.

FAUCES. Two pillars forming the sides of the throat between which are the tonsils in a mammal.

FEATHERS. Epidermal structures forming the body-covering of birds. The chief kinds are: *remiges* or wing-quills, *rectrices* or tail-quills, *coverts* surrounding the bases of the quills, *contour feathers* arranged in rows, *down feathers* forming a soft covering for the whole body and *filoplumes* which are minute and hair-like.

FECHNER'S LAW. The intensity of sensation increases approximately as the logarithm of the strength of the stimulus: a generalization which is open to doubt.

FELIDAE. The cat family: carnivorous, mostly nocturnal mammals with very long canine teeth and sharp carnassials; having wide cheek-bones and large auditory capsules. They walk on the toes and these have sharp retractile claws used for tree-climbing or for holding their prey. The group includes cats, lions, tigers, leopards etc., and the extinct sabre-toothed tiger with its enormously developed upper canines.

FEMORAL ARTERIES. Arteries taking blood into the thighs of a tetrapod vertebrate.

FEMUR. (1) The thigh-bone or upper bone of the hind leg in a tetrapod vertebrate.
(2) The first long segment of an insect's leg, between the trochanter and the tibia.

FENESTRA METOTICA. A passage between the auditory capsule and the occipital part of the skull in higher vertebrates: an opening for the glosso-pharyngeal and vagus nerves and the internal jugular vein.

FENESTRA OVALIS. The Oval Window: a membrane separating the middle ear from the inner ear in land vertebrates. Sound vibrations are transmitted from the outer ear-drum or tympanum across the middle ear by means of three small bones, the *malleus, incus* and *stapes* (hammer, anvil and stirrup). The latter is in contact with the oval window and sets it in motion, so conveying vibrations to the liquid in the inner ear.

FENESTRA PRO-OTICA. A passage in front of each auditory capsule in the skulls of higher vertebrates: an opening for the trigeminal, facial and abducens nerves.

FENESTRA ROTUNDA. The round window: a membrane separating the middle ear from the inner ear in higher vertebrates. It is below the oval window and relieves the pressure on the liquid in the inner ear when the oval window vibrates (see *Fenestra ovalis*).

FERTILIZATION. The union of male and female gametes to form a zygote.

FERTILIZATION MEMBRANE. The vitelline membrane or primary egg-membrane which has become thickened and more clearly defined after the entry of the sperm cell.

FIBRIN. A protein in the form of insoluble threads which help to make a clot and so prevent loss of blood when any blood vessels are injured. Fibrin is formed by the interaction of the soluble protein *fibrinogen* with the enzyme *thrombin*, the latter being produced only in the presence of calcium ions.

FIBRINOGEN. See *Fibrin* above.

FIBROBLASTS. Cells which give rise to fibres of connective tissue.

FIBRO-CARTILAGE. Cartilage in which white fibres of collagen are embedded, as for example in the intervertebral discs.

FIBULA. A bone which, together with the tibia, forms the lower part of the hind-leg of a vertebrate. It is usually more slender than the tibia and is often partly or wholly fused with it.

FIBULARE (CALCANEUM). The proximal tarsal bone adjacent to the fibula in a vertebrate. Frequently it is elongated and projects backwards to form a heel bone.

FIERASFERIDAE. Small eel-like fish of tropical seas, often living harmlessly in cavities of star-fish or molluscs.

FIGITINAE. A subfamily of gall-wasps which parasitize aphides, scale-insects and flies.

FILI-, FILO-. Prefix from Latin *Filum*: a thread.

FILIBRANCHIATA (FILIBRANCHIA). Lamellibranch or bivalve molluscs with filamentaous gills, as distinct from the Eulamellibranchiata in which the gills form membranes (see *Lamellibranchiata*).

FILIFORM. Thread-like: a term used to describe certain worms, the antennae of insects or crustacea etc.

FILOPLUMES. Minute hair-like feathers.

FILOPODIA. Fine thread-like pseudopodia such as those of Radiolaria.

FILUM TERMINALE. The terminal filament of the spinal cord of a vertebrate.

FIMBRIA. A fringe-like border of an opening, *e.g.* round the mouth of the mammalian oviduct.

FIN-RAYS. Skeletal structures, cartilaginous, bony, horny or fibrous, supporting and giving shape to fins (see *Fin*).

FIN-RAY BOXES. Box-like skeletal structures supporting the dorsal and ventral fins of *Amphioxus*.

FINS. Flattened limbs of a fish or other aquatic animal, used for locomotion, balancing, turning etc. In fish there are usually a number of median dorsal and ventral fins, a caudal or tail fin and paired pectoral and pelvic fins. They are supported and given shape by a number of fin-rays which may be cartilaginous (radialia), horny (ceratotrichia), bony and jointed (lepidotrichia) or fibrous and jointed (camptotrichia).

FIRMISTERNAL. A word used to describe two coracoid or epicoracoid bones which meet and fuse in the middle line adjacent to the sternum, as in frogs.

FIRST AND SECOND VENTRICLES. The cavities in the two cerebral hemispheres of the vertebrate brain.

FISH (PISCES). Cold-blooded aquatic vertebrates with gills, a stream-lined body, a powerful muscular tail and usually with paired pectoral and pelvic fins and median dorsal, ventral and caudal fins. The skeleton may be of cartilage or of bone and there is usually an exoskeleton of scales. Under the present scheme of classification the group does not include those fish-like creatures such as the lamprey which have no proper jaws and are known as Cyclostomata. True fish, as well as all land vertebrates, are *Gnathostomata*, *i.e.* they have jaws developed from the enlarged parts of the branchial arch skeleton.

FISSION. The division of a cell or of a unicellular organism to form new cells. There are several ways in which this may occur, most of which can be observed in various protozoa, algae, bacteria, etc. The chief types are: (*a*) Binary fission in which two uninucleate cells of equal size are formed from one parent-cell; (*b*) Plasmotomy in which a multinucleate cell gives rise to two multinucleate offspring; (*c*) Budding in which small cells break off from a larger parent-cell; (*d*) Multiple fission in which the nucleus divides repeatedly, afterwards forming numerous smaller cells. This sometimes takes place within a cyst, as with *Amoeba* when conditions are unfavourable for binary fission.

FISSIPEDIA. A name formerly used for cats, dogs, bears and similar animals: all the land forms of Carnivora as distinct from the *Pinnipedia* or seals and sea-lions. In a more modern system of classification Pinnipedia are ranked as a separate order and the name Carnivora is used only for the Fissipedia.

FISTULARIIDAE. Tropical and sub-tropical fish having a small mouth at the end of a snout. They include pipe-fish, flute-mouths and gigantic marine sticklebacks.

FIXATION DISC (ECHINODERMS). A disc near the mouth of an echinoderm larva by which it fixes itself to a substratum during metamorphosis from a bilateral larva to a radially symmetrical adult.

FIXATION PAPILLAE (TUNICATES). Three small papillae near the mouth of an ascidian tadpole by which it attaches itself to some solid object before undergoing metamorphosis.

FLABELLUM. (1) Of Crustacea: the flattened and enlarged distal exite of a phyllopodium.

(2) Of Insects: A spoon-like lobe at the tip of the proboscis of a bee.

FLAGELLATA. Protozoa possessing one or more flagella (see *Mastigophora*).

FLAGELLATED CHAMBERS (OF SPONGES). Round or cup-shaped cavities lined with flagellated collar-cells (choanocytes) in the more complex types of sponge.

FLAGELLULAE (FLAGELLISPORES). Spores having flagella, for example the gametes of many protozoa.

FLAGELLUM. A thread-like outgrowth from a cell which, by an undulating movement, can propel a small organism or attract particles to it. Flagella are larger than cilia and fewer in number but otherwise similar. Down each flagellum runs an axial filament attached to a basal granule. The latter is often connected to the nucleus by threads known as *rhizoplasts*. These may also connect it with other structures, notably a *parabasal body* or *blepharoplast*, the function of which is not known. Sometimes, as in *Trypanosoma*, a flagellum runs for some distance parallel to the surface of the cell and is connected to it by a film of protoplasm forming an undulating membrane.

FLAME-CELLS. Large hollow cells containing a number of flagella whose flickering movement gives a flame-like appearance. They are found at the commencement of the excretory tubules of Turbellarians, flukes etc., as well as in *Amphioxus*. Excretory products diffuse into the cavity of the flame-cell and the flagella drive these down the intra-cellular duct leading to the main excretory tube.

FLAT FISH (PLEURONECTIDAE). Fish such as plaice and sole which start life normally symmetrical but later undergo a metamorphosis and come to lie on one side. The head becomes twisted so that the eye of the under side moves over to the upper side. This is quite different from fish such as the skate which are flattened dorso-ventrally and remain symmetrical.

FLAT WORMS. See *Platyhelminthes*.

FLEXOR MUSCLES. Muscles which bring about the bending of a limb.

FLIGHT. In the animal world the ability to fly has been evolved only in insects, reptiles, birds and bats. Limited powers of gliding have been evolved in flying fish (using the enlarged pectoral fins) and in a few mammals other than bats. The wings of insects are formed from thin outgrowths of skin from the thorax. Those of vertebrates are formed by three main modifications of the fore-limb. In birds the wing is made of plumes attached to the fore-arm and to the three digits of the 'hand'; in bats a membranous wing or *patagium* extends between the fore and hind limbs and is supported by four elongated fingers; pterodactyls had a similar membranous wing supported by the enormously lengthened fourth finger.

FLOATING RIBS. The last two or three pairs of ribs in mammals; these are short and do not extend round the chest to the sternum as the anterior ribs do.

FLOCCULUS. The convoluted side portion of the cerebellum of a bird or mammal.

FOETAL MEMBRANES. Extra-embryonic membranes of a vertebrate: the *amnion*, *chorion* and *allantois* which are derived from the embryo but lie outside it and are folded in such a way as to enclose and protect it. Such membranes are only present in reptiles, birds and mammals and are concerned with nutrition and respiration.

FOETUS. A mammalian embryo in its later stages when the main features are clearly recognizable. In man this is after about two months of gestation.

FOLLICLE. (1) Hair follicle: a pit of epidermis having at its base a papilla or group of active cells from which the hair grows.

(2) Ovarian follicle: a sac of cells in the ovary enclosing a developing ovum.

(3) Graafian follicle: the ovarian follicle of a mammal differing from those of other vertebrates by having a large fluid-filled cavity.

FONTANELLE. A hole or gap in the roof of the cranium where the brain is only covered by skin, *e.g.* that between the frontal and parietal bones of a new-born baby.

FORAMEN. Latin: a small hole, from *Forare*, to bore.

FORAMEN LACERUM ANTERIUS. A foramen or opening in the mammalian cranium between the *orbito-sphenoid* and the *alisphenoid* bones making a passage for the oculomotor, trochlear and abducens nerves and the ophthalmic branch of the trigeminal nerve.

FORAMEN LACERUM MEDIUM. An opening in the mammalian skull for the passage of the internal carotid artery between the *basisphenoid bone* and the *tympanic bulla.*

FORAMEN LACERUM POSTERIUS. An opening in the mammalian skull between the *periotic* and the *exoccipital* bones for the passage of the glosso-pharyngeal nerve, the vagus nerve, the spinal accessory nerve and the internal jugular vein.

FORAMEN OF MAGENDIE. An opening in the roof of the fourth ventricle of the mammalian brain, through which cerebro-spinal fluid can pass from the brain cavities to the space between the meningeal membranes.

FORAMEN MAGNUM. The large opening at the hind end of a vertebrate skull through which the spinal cord passes to the brain.

FORAMEN OF MONRO. A passage linking each cerebral cavity with the third ventricle in the brain of a vertebrate.

FORAMEN OVALE. (1) An opening for the passage of the mandibular nerve between the alisphenoid and the periotic bones of a mammalian skull.

(2) An embryonic opening in the septum separating the two auricles of the mammalian heart.

FORAMEN OF PANIZZA. A small opening connecting the two systemic aortae of a crocodile or similar reptile at the point just beyond the semilunar valves where the two aortic arches cross one another.

FORAMEN ROTUNDUM. An opening in the mammalian skull through which the maxillary branch of the trigeminal nerve emerges from the brain.

FORAMEN TRIOSSEUM. An opening enclosed by the scapula, clavicle and coracoid of a bird at the point where these three bones meet. The tendon of the minor pectoral muscle, which raises the wing, passes through this opening.

FORAMEN OF WINSLOW. An opening connecting the peritoneal cavity with the space enclosed by the great omentum on the right of the stomach in reptiles, birds and mammals.

FORAMINIFERA. Amoeboid protozoa, usually marine, having perforated shells of chitinous, calcareous or siliceous matter through the pores of which a network of pseudopodia extends. The shell is usually many-chambered and often arranged in a spiral. Some live on the sea bottom; others are planktonic. The shells of millions of foraminifera form the chief constituent of chalk.

FORCEPS. Claspers for copulation on the hind end of the abdomen of certain insects.

FORCIPULATE OSSICLES. Small pincer-like *pedicellariae* covering the surface of a star-fish or other echinoderm. They usually consist of a basal ossicle and two 'jaws' which may be crossed (see *Pedicellariae*).

FORE-BRAIN (PROSENCEPHALON). The front part of the vertebrate brain consisting of the olfactory lobes, the cerebral hemispheres and the thalamencephalon, the latter bearing the pineal organ and the pituitary body.

FORE-GUT. See *Stomodaeum*.

FORESKIN. See *Prepuce*.

FORMICOIDEA. Ants: Hymenopterous insects with elbowed antennae and with a well marked abdominal constriction or *petiole*. They usually live in highly organized colonies containing many types or castes. The queen is the only ant capable of laying eggs; mating takes place during a nuptial flight in which several colonies take part together. The females then cast off their wings and start colonies in the ground. The majority of ants in a colony are workers or incomplete females without wings and with large heads. In some species, however, there are many different castes each with its own specialized features. The social activities of ants include such varied occupations as storing food, tending and 'milking' aphids, cultivating fungi, enslaving other ants, defending the colony, etc. Some types of 'soldier ants' use their extra large jaws for defence; others shoot out formic acid from glands at the hind end of the body.

FORNIX. An arched longitudinal tract of white fibres running beneath the cerebral region of the mammalian brain. It is formed from the union of two strands of fibres known as the *taeniae hippocampi* running along the edges of the hippocampal region.

FORNIX OF GOTTSCHE (VALVULA CEREBELLI). A projection from the roof of the brain at the junction of the optic lobes and the cerebellum in some vertebrates.

FOSSA OVALIS. A depression in the inter-auricular septum of the mammalian heart marking the position of the embryonic *foramen ovale* connecting the two auricles.

FOURTH VENTRICLE. The cavity in the *medulla oblongata* of the vertebrate brain.

FOVEA CENTRALIS. The thinnest part of the retina of the eye, usually in the centre of the yellow spot in certain vertebrates including man. It contains numerous cones and is a region of very acute vision and colour perception.

FREE-MARTIN. A mammal having male external features but having the internal sex-organs of a female although usually sterile. Such a condition is sometimes found in twins of unlike sex (*e.g.* in cattle) where the development of one is influenced by the sex hormones of the other owing to fusion of the placental circulation.

FRENULUM. A strong bristle or group of bristles projecting forwards from the base of the hind wing in some moths and other insects. It interlocks with a hook-like retinaculum on the underside of the fore-wing and enables the two pairs of wings to act as one.

FRILLED ORGAN. An organ of attachment at the anterior end of certain parasitic tapeworms.

FRINGILLIDAE. Finches: small sparrow-like birds with short thick swollen beak without a notch and with a basal swelling. They include the goldfinch, bullfinch, canary, house-sparrow and many others.

FRONS. The frontal sclerite of the head of an insect, bounded dorsally by the *epicranium* and laterally by the *genae*.

FRONTAL BONES. Dermal bones covering the front part of the brain of a vertebrate.

FRONTAL GANGLION. A median ganglion in the head of an insect just in front of the brain and connected with the ventricular ganglion on the hind part of the oesophagus. It forms part of the visceral or sympathetic system.

FRONTAL GLAND. A gland in the front of the head from which poison is discharged from certain species of termite.

FRONTAL LOBE. The anterior part of each cerebral hemisphere in the mammalian brain, separated from the lateral lobe by the *Sylvian fissure*.

FRONTAL PORE. The pore of the frontal gland of a termite or other insect (see *Frontal gland*).

FRONTAL SINUSES. Spaces extending from the nasal cavity into the frontal bones of mammals.

FRONTO-PARIETAL BONES. Bones formed from fusion of the frontals and parietals in the skulls of some vertebrates.

FULGORIDAE. Lantern-flies: large brilliantly coloured tropical insects of the group Homoptera (plant bugs).

FUNDATRIX. A term used in entomology with particular reference to Aphides and other plant-lice to denote the first wingless female of the season who produces numerous offspring similar to herself by parthenogenesis and often viviparously. The latter are termed *fundatrigeniae*.

FUNICULUS. FUNICLE. A strand or cord of connective tissue holding any organ in place.

FURCA. (1) Caudal: a fork-like structure formed by two appendages on the last abdominal segment of certain insects and crustacea.

(2) Labellar: a fork-like skeletal structure giving support to the two spongy pads on the proboscis of a blow-fly or similar insect.

FURCULA. (1) The united clavicles of a bird.

(2) The forked tail of certain insects, notably the 'spring-tails' or Collembola. In these insects the furcula is normally bent forwards and retained beneath the abdomen by a 'catch' or retinaculum known as the *hamula*. When the catch is released the furcula springs back propelling the insect forwards.

(3) The names furca and furcula may be used in a general sense for any fork-like organ or appendage.

FUSULAE. Cylindrical projections on the spinnerets of spiders, each containing numerous tiny tubes from which the silk is secreted. When very large they are termed spigots.

G

GADIDAE. See *Gadiformes* below.

GADIFORMES (ANACANTHINI). Cod-like fish usually covered with small smooth scales; having fins without spinous rays and having the pelvic fins forward in the jugular or thoracic position. The air bladder if present has no duct.

GAERTNER'S CANAL. The atrophied mesonephric duct in female reptiles, birds and mammals. In these animals the mesonephros or middle kidney is replaced by a metanephros or hind kidney, but its duct may remain as a vestige.

GALAGINAE. A group of lemurs of Madagascar and Africa having large ears and long tail. They sleep during the dry season and consume fat which has been deposited at the base of the tail.

GALEA. A hood-like outer lobe on the maxilla of an insect.

GALEO-. Prefix from Latin *Galea*: a helmet.

GALEOIDEA. A family of sharks having two dorsal fins without spines; having an anal fin and five pairs of gill-slits. The group includes Basking sharks, Tiger-sharks, Blue sharks, Hammer-heads and dogfish.

GALEOPITHECIDAE. See *Dermoptera*.

GALL BLADDER. A bladder or sac which receives and stores the bile secreted by the liver in a vertebrate.

GALLI. Fowls, pheasants, turkeys etc. A more restricted term than Galliformes.

GALLIFORMES. Fowls, pheasants, turkeys and similar birds including also grouse, partridges and quails. They are usually good runners which feed and nest on the ground. Many are polygamous; the young are praecoces.

GALLINACEOUS. See *Galliformes* above.

GALLS. Abnormal growths of plant tissues caused by various organisms which irritate the plant and possibly lead to the production of some type of growth hormone. The principal gall-causing insects or *gallicolae* belong to three different orders: (1) Gall-wasps and some saw-flies belonging to the Hymenoptera; (2) Gall-midges belonging to the order Diptera; (3) Certain Aphids and their allies among the Hemiptera. Galls are also sometimes formed by mites, eel-worms and fungi.

GALTON'S LAW. A person or animal derives $\frac{1}{2}$ of his inherited characters from the two parents; $\frac{1}{4}$ from his grandparents; $\frac{1}{8}$ from his great-grandparents and so on. This is merely a statistical approximation independent of any biological theory of genetics.

GAMETE. A male or female reproductive cell, *i.e.* a cell with the haploid number of chromosomes which, when it combines with another haploid cell at fertilization, gives rise to a diploid cell or zygote from which the offspring develops. Male gametes are usually small and motile (spermatozoa); female are larger and contain yolk (ova). In some organisms, however, the gametes may be alike (isogametes).

GAMETO-, GAMO-. Prefix from Greek *Gamos*: marriage.

GAMETOCYST. A cyst within which gametes are produced, as for instance in some protozoa.

GAMETOCYTE. A spermatocyte or an oocyte: the parent cell of a spermatozoon or an ovum.

GAMETOGENESIS. The formation of gametes (see *Oogenesis* and *Spermatogenesis*).

GAMOCYST. A cyst in which union of gametes takes place, as in some Sporozoa.

GAMOGONY (GAMETOGENESIS). The formation of gametes; the term *gamogony* is frequently used with reference to the life-cycles of protozoa to denote the formation of gametes from the previous stage which is called the *gamont* (*cf. Sporogony, Schizogony*).

GAMONT. The stage which gives rise to gametes in the life-cycle of a protozoon.

GANGLIA HABENULAE. A pair of thickenings in the roof of the brain near the base of the pineal stalk in some vertebrates.

GANGLION. A small compact mass of nerve cells and connective tissue. In a vertebrate the chief ganglia are: the dorsal root ganglia of the spinal nerves; the sympathetic ganglia forming chains on each side of the spinal cord; the collateral ganglia in the mesenteries. In many invertebrates (*e.g.* annelids and arthropods) there is a ventral chain of ganglia arranged segmentally.

GANOIDEI. GANOID FISH. In the older system of classification the group included a miscellaneous collection of fossil and present day fish all having thick dermal scales but otherwise not having much in common. More recently this system has been shown to be unsound and the various fish hitherto classed as 'ganoids' have been put into several different groups.

GANOID SCALES. GANOIN. GANOINE. Fish-scales in which there is a hard shiny layer of *ganoin* on the upper surface and a bony layer of *isopedine* below. This type of scale is found in a thin form on many modern bony fish and in a thicker form in such fish as the sturgeon and the gar-pike which used to be termed 'ganoids'. In the thinner scales, however, the layer of ganoin may be lost.

GAS-GLAND. Any gas-producing organ, especially the oxygen-secreting part of a fish's swim-bladder.

GASSERIAN GANGLION. The dorsal root ganglion of the trigeminal nerve of a vertebrate.

GASTERO-, GASTRO-. Prefixes from Greek *Gaster, gasteros*: stomach.

GASTEROPODA (GASTROPODA). Molluscs having a large flattened 'foot', a distinct head with tentacles and eyes, and usually having a visceral hump covered by a spiral shell.

GASTEROSTEIDAE. Sticklebacks: small fresh-water fish, usually carnivorous, having the pelvic fins well back and having a number of erectile spines in front of the soft dorsal fin.

GASTRALIA. The so-called 'abdominal ribs': splint-like membrane bones in the ventral abdominal wall of the crocodile and in many extinct reptiles.

GASTRIC. Appertaining to the stomach.

GASTRIC JUICE. The fluid secreted by the glands of the stomach; in mammals this contains pepsin, rennin and dilute hydrochloric acid.

99

GASTRIC MILL. The elaborate arrangement of cuticular teeth which help to break up food in the proventriculus or 'stomach' of some crustacea.

GASTRIN. A hormone produced by the stomach wall of a vertebrate. Its formation is probably stimulated by the presence of certain food substances. It then travels in the blood-stream and when it again reaches the stomach it stimulates the latter to secrete digestive juices.

GASTROLITH. A mass of calcareous material sometimes found in the proventriculus of a crustacean. It is probably formed from calcium withdrawn from the exoskeleton before moulting.

GASTROSTEIFORMES. See *Catosteomi*.

GASTROTRICHA. Minute unsegmented worms which swim by means of their ciliated epidermis. They are hermaphrodite and appear to have some affinities with the Rotifera and with the Nematoda. They are, however, sufficiently distinctive to be classed in a separate phylum.

GASTROZOOID. The feeding type of polyp in a coelenterate colony, having a mouth which may or may not be surrounded by tentacles. In some of the larger Siphonophora such as the Portuguese Man-of-war there are numerous other zooids specialized for stinging and catching prey. In this case the gastrozooids are purely for ingestion and digestion.

GASTRULA. The stage in an embryo in which the primary germ layers have been laid down (see *Gastrulation*).

GASTRULATION (EMBRYOLOGY). The process following cleavage in which, by a complex series of cell-movements, the primitive germ layers, *ectoderm*, *mesoderm* and *endoderm* are laid down in definite positions ready to give rise to the principal organs of the later embryo.

GAVIALIDAE. Crocodiles with long slender snouts; harmless to humans and usually feeding on fish. Present day species living in India, Burma, Borneo and North America attain a length of up to 20 ft. but some fossil forms were 50 ft. long.

GECKONIDAE (ASCALABOTA). Geckos: harmless nocturnal lizards found in most of the warmer parts of the world. Many have digits which can exert a suction enabling them to climb smooth vertical surfaces when hunting for insects. The tongue is protrusible and the eyes are covered with a transparent nictitating membrane.

GEMMA. GEMMULE. A bud or outgrowth which can form a new individual, as in *Hydra*.

GEMMATION. Reproduction by budding as in *Hydra*.

GEMPYLIDAE. Mackerel-like fish of the open sea, widespread, descending to considerable depths and usually breeding near rocky islands.

GENAE. The lateral cuticular plates of an insect's head, below and behind the eyes.

GENE. A hereditary factor or unit of inherited material which carries the power of transmitting some particular quality from one generation to the next. There is reason to believe that the genes are made of some type of nucleo-protein (derivatives of ribonucleic acid) and are arranged in linear series along the chromosomes.

GENICULATE ANTENNAE. Antennae which have an 'elbow' or right-angled bend as in a wasp.

GENICULATE GANGLION. A ganglion on the Facial nerve (seventh cranial nerve) of a vertebrate, corresponding to the dorsal root-ganglion on a spinal nerve.

GENICULUM. See *Genu.*

GENITAL. Appertaining to the reproductive organs, *e.g.* genital artery and vein, genital ducts, apertures etc.

GENITAL ATRIUM. A vestibule or cavity into which the ducts from the male or female genital organs open; in the case of hermaphrosite organisms both ducts may lead into the same atrium.

GENITAL BURSA. A sac formed by enlargement of a genital duct. Sometimes, as for instance in certain Echinoderms, it may have a respiratory function.

GENITALIA. Reproductive organs and their accessory parts. It is usual to distinguish between the internal genitalia (testis, ovary and their ducts and glands) and external genitalia (penis, vagina, claspers etc.)

GENITAL POUCH. Any recess acting as an adjunct to the genital aperture.

GENITAL RACHIS (ECHINODERMS). A ring of tissue near the aboral or upper surface of a star-fish or similar echinoderm, bearing branches on which the gonads are borne.

GENOTYPES. Individuals having the same genetical constitution (*cf. Phenotypes*).

GENU. GENICULUM. Latin: a knee.

(1) The joint of the thigh or femur with the lower leg; any joint roughly corresponding to this, as for instance in insects; a bent structure resembling a knee.

(2) Genu (of the brain). The bent anterior part of the *corpus callosum* in the mammalian brain.

GENUS. A group of organisms consisting of a number of species closely resembling one another. In the Linnaean or binomial system of nomenclature the first name is that of the genus; the second that of the species (see Appendix on Classification).

GEOMETRIDAE. Geometers or 'earth-measurers': caterpillars which move by a looping action owing to the absence of abdominal feet except at the hind-end. They progress by holding on with the thoracic legs and then drawing up the hind-part of the body. The abdominal feet then grip the support and the whole body moves forwards.

GEOMYIDAE. Pouched rats: burrowing rodents of North America with large cheek pouches.

GEPHYREA. Large marine annelids with little or no segmentation. They include the *Echiuroidea* which have the prostomium elongated into a proboscis, and the *Sipunculoidea* in which the prostomium is absent in the adult.

GEPHYRO-. Prefix from Greek *Gephura*: a bridge.

GEPHYROCERCAL. A type of fish-tail like that of the eel, symmetrical and pointed resembling the diphycercal type but dervied secondarily by simplification of a more complex type.

GERMARIUM. The formative region of an ovary or testis containing the primordial germ-cells (oogonia or spermatogonia).

GERMINAL DISC (BLASTODISC). A disc of cells on the top of the yolk-sac formed by partial cleavage of a large-yolked egg such as that of a bird. In such an egg the large amount of yolk prevents complete cleavage from taking place.

GERM-LAYER (PRIMARY GERM-LAYER). A layer of distinctive cells in an embryo, each layer always giving rise to certain definite tissues or structures as the organism develops. The majority of multicellular animals are *triploblastic*, i.e. they have three embryonic layers: *ectoderm*, *mesoderm* and *endoderm*. In the phylum Coelenterata, however, there are only two layers, the mesoderm being absent.

GERRHOSAURIDAE. Lizards of Africa and Madagascar having bony scales and reduced limbs.

GESTATION. The period during which a developing mammal is in the uterus of the parent before birth.

GIANT NERVE FIBRES. Extra thick nerve fibres of some invertebrates, specialized for conducting motor impulses at a more rapid speed than the normal fibres.

GILL BARS (PHARYNGEAL BARS). Cartilaginous bars supporting the tissue between the gill-clefts in lower chordates such as *Amphioxus*: precursors of the branchial skeleton of fish.

GILL BOOKS. Respiratory organs of certain aquatic arachnids such as the King Crab (*Limulus*). They consist of piles of leaflets in which blood circulates and a special muscular mechanism for opening and shutting them in the water, thus facilitating gaseous exchange.

GILL COVER. See *Operculum*.

GILL POUCHES. Outgrowths of the wall of the pharynx forming the beginning of the gills in fish and in the larvae of amphibia. In land animals they do not usually perforate but may disappear or take on a different function, *e.g.* the Eustachian Tube.

GILLS. Respiratory organs by which aquatic animals can obtain dissolved oxygen from the surrounding water. The essential feature of all such organs is a membrane or an outgrowth of the body-wall having a large surface and a plentiful blood supply. External gills are found on many annelids, molluscs and other invertebrates as well as on the young larvae of amphibia. Crustacea usually have gills attached to the upper parts of the legs and enclosed in a branchial chamber formed from the overhanging carapace. Fish usually have internal gills within gill-slits. Water entering the mouth passes out through these slits and in doing so passes over the gills.

GILL-SLITS. See *Gills* (above) and also under *Visceral Clefts*.

GINGLYMODI. An order of fish represented by many fossil types, but at the present day only by the gar-pike (*Lepidosteus*). The fins are of the modern or *actinopterygian* type supported entirely by dermal fin-rays; the body is covered with thick ganoid scales and the tail is of an intermediate or hemi-heterocercal type.

GIRAFFIDAE. Giraffes: even-toed ungulates of Africa having long legs and long neck and attaining a height of up to 19 ft. The head bears two horns which are present in both sexes from birth. These have a bony core, are short and covered with skin. There are only two known species of giraffe but the rare Okapi of tropical West Africa is included in the same family.

GIRDLES. See *Pectoral girdle* and *Pelvic girdle*.

GIRDLE BONE (SPHENETHMOID). A single bone replacing the two orbitosphenoids and extending into the ethmoid region in frogs and toads.

GIZZARD. A part of the alimentary canal having thick muscular walls often with hard horny teeth. Its main function is to break up hard food before this reaches the main digestive region. Gizzards are present in birds and in many invertebrates.

GLABELLA. The middle region of the head of a Trilobite, divided by furrows into five somites and separated from the lateral portions by two longitudinal grooves.

GLAND. An organ which secretes specific chemical compounds such as enzymes or hormones.

GLANS PENIS. The expanded part of the corpus spongiosum at the extremity of the mammalian penis.

GLAUCO-. Prefix from Greek *Glaukos*: sea-green.

GLAUCONIDAE. Snakes having teeth in the lower jaw only and having vestiges of a pelvic girdle and femora. About thirty species exist in Mexico, Africa and S.W. Asia.

GLAUCOTHOE. The larval stage of a hermit-crab.

GLENOID FOSSA (GLENOID CAVITY). (1) A cup-like hollow in the scapula into which the head of the humerus fits.
  (2) A shallow depression on the squamosal bone for the articulation of the lower jaw of a mammal.

GLIA. See *Neuroglia*.

GLIRES. A group or cohort comprising the two classes Rodentia (rats, mice, squirrels etc.) and Lagomorpha (rabbits and hares). Both groups were formerly included in the Rodentia but this word now has a more restricted use.

GLIRIDAE. Dormice: small arboreal rodents with long hairy tails. They are found in many parts of the world; the European species hibernates.

GLISSON'S CAPSULES. Envelopes of fibrous connective tissue enclosing the lobules of the vertebrate liver.

GLOBIGERINA. A genus of pelagic Foraminifera with perforated calcareous shells whose chambers are arranged in a spiral bearing long spines. They are found in very great numbers on ocean beds and give their name to the well known *Globigerine Ooze* which eventually gives rise to chalk and limestone.

GLOBULINS. A group of proteins which yield the amino-acid *glycine* on hydrolysis. Like the albumens they are coagulated by heat but unlike them, they are very soluble in salt solution. They are found in many plant and animal cells as well as in blood-plasma.

GLOCHIDIUM. The larva of a lamellibranch mollusc such as a mussel, having a small bivalve shell and a tentacle or sucker for attachment to a fish after being released from the gill-chamber of the parent.

GLOEA. A sticky secretion exuded by some protozoa.

GLOMERULUS (GLOMUS). A small cluster of capillaries such as those in the Malpighian bodies of the vertebrate kidney. The term glomus is used for the more diffuse capillary network in the early stages of development before definite Malpighian bodies have been formed.

GLOMUS. See *Glomerulus* above.

GLOSSAE. A pair of tongue-like lobes in the middle of the labium of an insect (see also *Paraglossae*).

GLOSSO-. Prefix from Greek *Glossa*: a tongue.

GLOSSOPHARYNGEAL NERVE. The ninth cranial nerve of a vertebrate going from the hind-part of the brain to the tongue and pharynx and, in the case of fish, to the first gill also.

GLOTTIS. The opening from the back of the pharynx into the trachea.

GLOW WORM. The wingless female of the beetle *Lampyris noctiluca*, bearing phosphorescent organs on the sides of the abdomen; used apparently to attract the males. The exact nature of the chemical action which produces light is not understood but it is generally believed that a substance named *luciferin* is oxidized with the help of an enzyme known as *luciferase*. The energy so produced forms a continuous spectrum of visible light with very little heat. It depends upon a plentiful supply of oxygen and is under the control of the nervous system.

GLYCOGEN. A polysaccharide sometimes known as 'animal starch', built up from glucose and stored in the livers of many animals.

GLYPTODONTIDAE. Giant extinct armadillos of South and Central America, having a rigid carapace formed of a large number of bony scutes joined together. Some species attained a length of 17 ft.

GNATHO-. Prefix from Greek *Gnathos*: a jaw.

GNATHOBASE. An enlarged basal segment forming a toothed jaw-like structure sometimes present on the legs or other appendages of crustaceans, spiders etc.

GNATHOBDELLIDAE. Leeches having three chitinous jaws; they include the medicinal leech, the horse leech and some tropical land-leeches.

GNATHOCHILARIUM. A mouth-part of a millipede resembling the labium of an insect but bearing a pair of flat tongue-like organs known as *lamellae linguae* and a number of short sensory palps.

GNATHOSTOMATA. All vertebrates having a true pair of jaws formed from the visceral arch skeleton, as distinct from the Cyclostomata (*e.g.* lampreys) which have a circular mouth without jaws. All true fish as well as amphibia, reptiles, birds and mammals are gnathostomata.

GOBIIFORMES (GOBIOMORPHI, GOBIIDAE). Gobies: spiny-rayed fish with four pairs of gills; with the pelvic fins below the pectorals and usually without an air-bladder. They live in fresh and in salt water and are carnivorous.

GOLGI BODIES. Minute coiled threads of protein and lipoid material found in the cytoplasm of most animal cells and a few plant cells. Their function is not known with certainty.

GOMPHODONTIA. Extinct mammal-like reptiles characterized by having broad molar teeth and in some cases large incisors. Most of the fossil remains are very incomplete. One genus *Theriodesmus* had remarkably mammalian fore-limbs and feet. The group is sometimes included in the Theriodontia.

GON-, GONO-. Prefixes from Greek *Gonos*: offspring, seed.

GONAD. An organ in which gametes are produced: a testis or an ovary.

GONAPOPHYSES. Paired abdominal appendages used for copulation and oviposition in insects etc.

GONIOPHOLIDAE. Fossil crocodiles of the Upper Jurassic Period having dorsal bony plates, a moderately long snout and internal nostrils far back between the palatines and the pterygoids.

GONOCOELS. Parts of the coelom within the gonotomes (*q.v.*).

GONODUCT. A duct from a gonad, *i.e.* the *oviduct* from the ovary and the *vas deferens* from the testis.

GONOPHORE. (1) Any structure bearing gonads.

(2) A gonozooid: a specialized polyp on certain colonial coelenterates,

resembling a sessile medusa and bearing gonads. It is bell-shaped and on a short stalk. The gonads are borne on a central manubrium or spadix.

GONOPODS. Claspers: clawed copulatory appendages on the genital segments of insects, myriopods etc.

GONOPORE. The opening of the gonoduct (oviduct or vas deferens). The term is used for small genital apertures such as are found in insects, earthworms etc.

GONOTHECAE. Chitinous cups forming part of the perisarc surrounding the blastostyles or reproductive structures in colonial coelenterates of the order Calyptoblastea. Examples of these may be seen in the well-known genus *Obelia*.

GONOTOMES. Segmented parts of embryonic mesoderm growing down from the myotomes and giving rise to the gonads (*e.g.* in *Amphioxus*).

GONOZOOIDS. Specialized polyps of a colony, bearing gonads (see *Gonophore*).

GONYS. The lower edge of a bird's beak, reaching from the angle of the chin to the tip.

GORGONIANS (GORGONACEA). Sea-fans: Alcyonarian polyps forming upright branching colonies with a fan-like horny exoskeleton.

GRAAFIAN FOLLICLES. Fluid-filled spherical cavities in which the oocytes develop in the ovary of a mammal (see *Follicle*).

GRANULOCYTES. White blood corpuscles having granular cytoplasm. According to the type of stain readily taken up by the granules, these cells are classified as basophil, acidophil or neutrophil.

GRAPTOLITES (GRAPTOLITHINA). Pelagic colonial coelenterates whose horny exoskeletons are found as fossils in the Palaeozoic rocks. The polyps apparently budded off from one another and remained in contact with the parent. In the later Palaeozoic rocks up to the Carboniferous Period more complex polymorphic types known as Dendroid Graptolites were very common.

GRAVIGRADA. Ground sloths: large extinct edentates of South America whose skull and teeth resembled those of sloths, but whose vertebral column, limbs and tail were more like those of ant-eaters. The best known is the *Megatherium* of Pleistocene times, about the size of an elephant but with shorter limbs and a small head.

GREEN GLANDS. Excretory organs of crustacea situated in front of the cephalothorax and having openings at the bases of the antennae. Each gland consists of an end-sac, a spongy 'labyrinth', a tubular duct and a dilatable bladder.

GREGARINIDEA (GREGARINIDA). Parasitic sporozoa, included in the subclass Telosporidia, living in the intestines and coelom of arthropods, annelids and other invertebrates. They undergo a complex life-cycle usually involving multiple fission in the sexual and in the asexual stages.

GREY MATTER. The part of the brain or of the spinal cord where most of the nerve-cells are congregated: the internal part of the spinal cord and the external part of the cerebrum and cerebellum.

GREY NERVE FIBRES. Nerve fibres of the non-medullated type (*i.e.* without a fatty sheath), connecting the sympathetic ganglia with various internal organs.

GRUB. An apodous insect larva (*q.v.*).

GRUIDAE. Cranes: long-necked, long-legged wading birds which are also powerful flyers. They are found in most parts of the world except New Zealand and the Pacific Islands.

GRUIFORMES. Marsh and water birds with schizognathous jaw (cleft palate) and without a crop. The young are covered with down. They include cranes, coots, water hens, moorhens, corncrakes and bustards.

GRYLLIDAE. Crickets: orthopterous insects similar to grasshoppers but distinguished from them by their brown colour, by the long unsegmented anal cerci and by their less predominantly vegetarian habits.

GRYLLOBLATTOIDEA (GRYLLOBLATTIDAE). The most primitive of all orthopterous insects, combining the features of stick-insects and of crickets. The few known species are wingless and occur in the mountains of North America and Japan.

GRYLLOTALPIDAE. Mole-crickets: orthopterous insects with large flattened fore-legs modified for burrowing.

GUBERNACULUM. An elastic cord connecting the epididymis of the testis with the wall of the scrotal sac in mammals. It is long at first and its subsequent shortening causes the descent of the testes from the abdomen into the scrotum.

GULA. A median ventral sclerite on the head of an insect, situated between the genae and separating the labium from the 'occipital' foramen through which the nerves, oesophagus etc. pass into the thorax.

GUT. The alimentary canal.

GYMNO-. Prefix from Greek *Gumnos*: naked.

GYMNOBLASTEA (ANTHOMEDUSAE). Hydrozoa (colonial coelenterates) covered by a horny perisarc which stops short at the bases of the polyps. They thus differ from the Calyptoblastea in which the perisarc forms cups or thecae round the polyps.

GYMNOCERATA. Insects of the group Hemiptera-heteroptera comprising most of the terrestrial forms of bug. They are distinguished from the aquatic forms (Cryptocerata) by their conspicuous antennae. Notable among the Gymnocerata are the bed-bugs (*Cimicidae*) and the plant-eating 'shield-bugs' (*Pentatomidae*).

GYMNODONTES. Globe-fish, sun-fish and porcupine-fish of tropical and sub-tropical seas. They have rough scales, scutes or spines and the teeth are fused to form a beak. In some species the body is inflatable.

GYMNOLAEMATA. Polyzoa with a ring of oral tentacles (*lophophore*) and without an epistome or ridge overhanging the mouth.

GYMNOMERA. Water-fleas (Cladocera) in which the carapace is reduced and consists only of that part forming the brood-pouch. The rest of the trunk is uncovered and the limbs are usually long and prehensile (*cf. Calyptomera*).

GYMNOPHIONA (APODA). Worm-like amphibia without limbs or girdles and usually with small calcified scales embedded in the skin in transverse rows. The eyes are small and functionless and on each side of the head is a small pit containing a protrusible tentacle. They are found in India, S.E. Asia, Central Africa and Central America.

GYMNOSPORES. Naked spores of protozoa, *i.e.* without spore-cases. They may be amoeboid, flagellate or ciliate.

GYMNOSTOMATA. A suborder of ciliate protozoa possessing a mega- and micronucleus and having a ciliated surface but no ciliary mechanism round the mouth and gullet.

GYMNOTIDAE. Electric eels and similar fish of South America, having an elongated body, a pointed tail and no dorsal, caudal or pelvic fins. The electric organs are along each side of the tail.

GYRI. The convolutions of the brain in higher vertebrates.

GYRINIDAE. Whirligig beetles: aquatic beetles characterized by a rapid gyratory movement. The middle and hind-legs are flattened and oar-like with a fringe of long stiff bristles. A peculiar feature of these beetles is that the compound eyes have two definite regions with large and small facets respectively. One theory is that the upper portion serves for vision in air and the lower portion for vision under water.

H

HABENULAR COMMISSURE. A transverse nerve tract running across the roof of the third ventricle of the vertebrate brain just beneath the pineal organ.

HAECKEL'S BIOGENETIC LAW. An organism approximately recapitulates its evolutionary history during its embryonic development.

HAEM-, HAEMO-, HAEMATO-. Prefixes from Greek *Haima, haimatos*: blood.

HAEMAL ARCH. A ventral arch on each of the caudal vertebrae of a fish, formed by fusion of the two lateral processes down the mid-ventral line. The caudal artery and vein run through the passage formed by these arches.

HAEMATIN. An iron-containing pigment resulting from decomposition of haemoglobin.

HAEMATOBIUM. An organism which lives in blood.

HAEMATOBLAST. A parent-cell of red blood corpuscles.

HAEMATOCHROME. A red pigment in the eye-spots of flagellated protozoa and some unicellular algae.

HAEMATOCRYAL. Cold-blooded. Poikilothermous.

HAEMATOCYTE. HAEMACYTE. HAEMOCYTE. A blood corpuscle.

HAEMATOCYTOZOON. A parasite living inside a blood corpuscle, *e.g.* th malaria parasite.

HAEMATOGENESIS. Blood development.

HAEMATOPHAGOUS. Feeding on blood.

HAEMATOPOIESIS. HAEMOPOIESIS. Blood formation in bone-marrow etc.

HAEMATOTHERMAL. Warm-blooded. Homothermous.

HAEMATOZOON. An animal parasite living in blood.

HAEMOCHORIALIS (EMBRYOLOGY). A type of placenta found in certain mammals, including Man, in which the epithelium, the connective tissue and the capillaries of the uterus are broken down so that maternal blood comes to circulate in lacunae close to the capillaries of the foetal membranes. This greatly helps the exchange of substances between foetus and mother.

HAEMOCOEL. HAEMOCOELE. A large blood-filled cavity or sinus surrounding most of the organs in arthropods, molluscs and many other invertebrates. Blood from the arteries does not go into capillaries but goes directly into the haemocoels, so that every organ is bathed in it. Finally it reaches the pericardial sinus from which it enters the heart through small openings called *ostia*.

Embryologically the haemocoel is quite distinct from the coelom. The former is the primary body-cavity or blastocoel and never communicates with the exterior, whereas the latter is a secondary body-cavity formed within the mesoderm and always communicates with the exterior by gonoducts, excretory ducts etc.

HAEMOCYANIN. A blue-green, copper-containing respiratory pigment in the blood of some arthropods and mollusca.

HAEMOCYTES. A general name for blood corpuscles, *viz.* erythrocytes (red) and leucocytes (white).

HAEMO-ERYTHRIN. HAEMERYTHRIN. A red pigment in the blood of certain invertebrates.

HAEMOGLOBIN. The respiratory pigment in the blood of vertebrates and a few invertebrates. It consists of a complex and variable protein (*globin*) combined with a substance called *haem* or *haematin* containing iron. The oxygenated form is scarlet and the de-oxygenated form bluish red.

HAEMOSPORIDIA. Protozoa which are parasites in the blood corpuscles of vertebrates. They belong to the order *Coccidiomorpha*; have naked sporozoites and motile zygotes (ookinetes). They are transmitted from one vertebrate host to another by blood-sucking insects, mites, leeches etc. The best-known example is the malaria parasite transmitted by the Anopheline mosquito.

HAEMULIDAE. Carnivorous spiny-rayed fish of warm seas.

HAIR. (1) Of Mammals: a filament of cornified epidermal cells growing from a papilla at the base of a socket or follicle.

(2) Of Insects etc. (See *Microtrichia* and *Macrotrichia*.)

HALF VERTEBRAE. The two embryonic halves of a vertebra known as the *Hypocentrum* and the *Pleurocentrum* (*q.v.*).

HALLUX. The first digit on the hind-foot of a tetrapod: the big toe of a human.

HALO-. Prefix from Greek *Hals, halos*: salt.

HALOLIMNIC. Living in salt marshes.

HALOPHILIC. Living in salt water.

HALTERES. Balancers: the reduced hind-wings of dipterous insects, shaped like short drumsticks. It has been shown in some species that if they are removed the insect soon loses its balance and falls to the ground.

HAMULA (OF INSECTS). A structure which, together with the furcula, enables insects of the order Collembola (spring-tails) to leap forwards into the air. The hamula is formed from a pair of appendages on the ventral side of the abdomen. Before leaping, the caudal furcula is bent forwards and locked to the hamula. When this 'catch' is released, the furcula springs back propelling the insect violently forwards.

HAMULI. (1) Hooks by which the anterior and posterior wings of a bee or similar insect interlock when in flight.

(2) Small hooks by which the barbules of a feather interlock.

HAPALIDAE. Marmosets: small monkeys of Central America, having a large brain and a rounded skull. The fore-limbs are shorter than the hind-limbs and the thumbs are not opposable. The tail is bushy and longer than the whole body but is not prehensile.

HAPLO-. Prefix from Greek *Haplos*: single, simple.

HAPLODACTYLINA. Carnivorous sea-breams of the South Pacific.

HAPLOID. Having half the full number of chromosomes in a cell. Spermatozoa and ova exhibit this phenomenon so that when they combine to form a zygote, the full chromosome number is restored.

HAPLOMI. Pike-like fish (see *Esociformes* and *Esocidae*).

HAPLOPODA. Fresh-water Cladocera (water-fleas) belonging to the group *Gymnomera* (*q.v.*), having a very elongated head and body with six slender legs.

HAPLOSPORIDIA. Parasitic protozoa belonging to the subclass Neosporidia, in which the adult stage is a syncytium and whose spores possess cases with a lid, but have no pole-capsules. Most of them infest aquatic invertebrates such as annelids.

HARDERIAN GLANDS. Glands on the inner side of the eye in many reptiles, birds and mammals. They lubricate the nictitating membrane (third eyelid).

HARPAGO. A claspette: one of a small pair of claspers on the hind end of the abdomen of certain insects, *e.g.* mosquitoes. They do not form the principal tail forceps but are smaller and nearer to the median line.

HATSCHEK'S NEPHRIDIUM. An unpaired excretory organ of *Amphioxus* consisting of numerous flame-cells and fine tubules whose main collecting duct opens into the pharynx just behind the mouth.

HATSCHEK'S PIT (PRE-ORAL PIT). A mucus gland in a small depression slightly to the right of the middle line in the oral hood of *Amphioxus*: the remains of an embryonic anterior gut diverticulum.

HAUSTELLUM. The large distal portion of the proboscis of a blow-fly or similar insect: probably a specialized development of the prementum and part of the mentum. It bears at its extremities two large oral lobes or *labella* containing numerous tubes or *pseudotracheae* through which food is sucked.

HAVERSIAN CANALS. See *Haversian Systems* below.

HAVERSIAN SYSTEMS. Concentric rings of bone with *lacunae* or spaces containing bone-cells and *canaliculi* containing fine protoplasmic threads linking the cells with one another. Each series of rings surrounds a Haversian Canal in which is a blood vessel. Fully formed bone contains numerous such systems.

HEAD. The specialized anterior part of an animal on which are the mouth and the chief sense organs. It is often enlarged to accommodate the brain or the principal ganglia.

HEAD FOLD (EMBRYOLOGY). A fold of blastoderm which bends under and limits the anterior end of an amniote embryo separating it from the yolk sac. By the formation of this and a corresponding tail fold, the embryo is lifted up from the underlying yolk and becomes 'pinched off' from the surrounding tissues. The space enclosed by the head fold forms the beginning of the fore-gut.

HEAD-FOOT. The combined head and foot of a snail or similar gasteropod mollusc.

HEAD-KIDNEY. See *Pronephros*.

HEART. The chief pumping organ whose pulsations drive blood round the circulatory system. Usually it consists of an enlarged muscular part of a blood vessel and contains at least two valves to prevent back flow. It may contain more than one chamber; that which receives blood being an auricle and that which drives out blood being a ventricle. Fish have one auricle and one ventricle; amphibians and most reptiles two auricles and one ventricle; birds and mammals two of each. In insects and other arthropods the heart is an enlarged dorsal vessel with a number of ostia or openings through which blood enters from the pericardial sinus.

HELICO-. Prefix from Greek *Helix, helikos*: spiral.

HELICOTREMA. A narrow canal at the tip of the spiral cochlea in the mammalian ear. It connects the upper cavity or *Scala vestibuli* with the lower cavity or *Scala tympani* (see also *Scala vestibuli*).

HELIO-. Prefix from Greek *Helios*: the sun.

HELIORNITHIDAE. Fin-foots: small crane-like birds of the tropics having a small head and a thin neck and having toes with broad flaps and pointed claws.

HELIOZOA. Spherical fresh-water protozoa of the class Rhizopoda, sometimes enclosed in a perforated siliceous or chitinous sphere. The outer layer of protoplasm is highly vacuolated giving increased buoyancy. Radiating pseudopodia with axial filaments can, by their successive contractions, cause a rolling form of locomotion.

HELMINTHOLOGY. The study of worms.

HEMEROBIIDAE. Brown lacewing-flies: predatory insects of the order Neuroptera, smaller and more hairy than the closely related green lacewing-flies. They are to be found on the bark of pines and other conifers.

HEMI-. Prefix from Greek *Hemi*: half.

HEMI-AZYGOS VEIN. A name sometimes given to the left azygos vein of a mammal. It takes blood from the wall of the thorax into the right azygos vein which opens into the right anterior vena cava. It is considered to be homologous with a part of the left posterior cardinal vein of the lower vertebrates.

HEMIBRANCH. A half-gill, *i.e.* one with only one vascular surface. This is normally the case of the last gill on each side in a fish. The others, each situated between two gill-clefts, have vascular lamellae on both sides and are called *holobranchs*.

HEMIBRANCHII. Fish of the stickleback type (Gastrosteiformes), having a small mouth at the end of a snout and having some of the pharyngeal and branchial bones reduced or absent. The group includes the small present-day sticklebacks, a few large marine types, some extinct varieties and also snipe-fish, trumpet-fish and pipe-fish.

HEMICEPHALOUS. A term used to denote insect larvae in which the head is incomplete posteriorly and is partly embedded in the prothorax.

HEMICHORDATA (HEMICHORDA). Primitive worm-like animals having certain vertebrate features such as gill-clefts and a small notochord in the anterior region of the body only. An example is the Acorn-worm *Balanoglossus*.

HEMIELYTRON. HEMYLYTRON. An insect's wing in which the basal half is hardened and the remainder membranous. Such wings are found in plant bugs of the order Hemiptera-heteroptera.

HEMI-HETEROCERCAL. A type of fish-tail intermediate between the heterocercal and the homocercal types. The tail appears to be symmetrical but the hind-end of the vertebral column is bent upwards extending some way into the dorsal lobe of the tail-fin. Such a state is found in *Amia*.

HEMIMETABOLA. Insects which do not undergo a complete metamorphosis (see *Exopterygota*).

HEMIMYARIA (SALPIDA). Tunicates in which there is no larval stage and in which the adult has no tail. The nervous system is degenerate and the gill clefts are not divided by longitudinal bars.

HEMIPHRACTINAE. Frogs of South America having teeth in both jaws, separate Eustachian tubes and distinct tympana.

HEMIPTERA (RHYNCHOTA). A large order of insects including bugs, cicadas, aphides etc. The mouth-parts are adapted for piercing and sucking, with mandibles and maxillae in the form of long stylets lying in a trough-like labium; palps are reduced or absent. For convenience this large order of 40,000 species is divided into two suborders named respectively *Homoptera* and *Heteroptera*. The former have fore-wings of uniform consistency, either leathery or membranous; the latter have them half membranous and half thickened. There are also many wingless forms.

HEMIPTERYGOID. The first part of the pterygoid bone in certain birds such as the penguin. In these the pterygoids grow forwards and the anterior parts later lose their connection with the parent bones. They may disappear or fuse indistinguishably with the palatine bones.

HENLE'S LAYER. A layer of horny cells forming part of the inner root-sheath of a mammalian hair (*cf. Huxley's layer*).

HENSEN'S DISCS. Lightly staining discs running transversely across a fibril of striped muscle when the latter is in a relaxed state. Each Hensen's disc divides the 'A-disc' (Dobie's dot) into two parallel halves, but when the muscle contracts these two halves become one (see also *Dobie's Dots*).

HEPATIC. Appertaining to the liver.

HEPATIC CAECUM (MESENTERIC CAECUM. PYLORIC CAECUM). A branch or diverticulum from the mid-gut, in which digestion and absorption may take place.

HEPATIC PORTAL SYSTEM. Veins carrying blood with dissolved food substances from capillaries in the wall of the intestine to the liver in vertebrates.

HEPATIC SINUS. A sinus or enlarged vein taking blood from the liver to the heart in a fish.

HEPATO-. Prefix from Greek *Hepar, hepatos*: liver.

HEPATO-PANCREAS. A digestive gland or 'liver' which may combine the functions of digestion and absorption, *e.g.* in the Mollusca.

HEPATO-PANCREATIC DUCT. A common duct leading into the duodenum from the gall bladder and the pancreas, as for instance in the frog.

HEPIALIDAE. Ghost Swift Moths: primitive Lepidoptera with elongated narrow wings which interlock by means of a jugum or process arising from the base of the fore-wing. The caterpillars damage young trees by tunnelling into the pith of the roots or young stems. The adult moth has no functional mouth-parts and does not feed.

HERMAPHRODITE. Having both male and female reproductive organs on the same individual, as for instance in the earthworm.

HESPERIIDAE. Skipper butterflies which fly with swift jerky movements and rest with wings flat; the larvae usually feed on grass.

111

HETERACANTH (HETERACANTHOUS). A type of dorsal fin in spiny-rayed fish, whose spines can be raised or depressed at will and in the depressed position are turned slightly to one side or the other alternately. Sometimes the spines are asymmetrical (*cf. Homacanth*).

HETERO-. Prefix from Greek *Heteros*: other.

HETEROCERA. Moths: a name used in the old method of classification which divided all Lepidoptera into butterflies and moths. The name is no longer used and the order is more rationally classified according to the wing-venation.

HETEROCERCAL TAIL. An unsymmetrical fish-tail in which the vertebral column is bent slightly upwards and the dorsal lobe of the caudal fin is smaller than the ventral lobe. Such a tail is found in dogfish, sharks etc.

HETEROCOELA. An order of calcareous sponges in which the flagellated chambers are simple thimble-shaped extensions of the main cavity.

HETEROCOELOUS. A term used to denote a vertebra whose centrum has one face convex and the other concave (*cf. Amphicoelous*).

HETEROCOTYLEA. Ectoparasitic Trematoda (flukes) having a large posterior sucker often subdivided into numerous smaller ones.

HETERODACTYLOUS. A term used to describe the feet of birds in which the first and second toes are directed backwards and the third and fourth forwards (*cf. Zygodactylous*).

HETERO-DIPHYCERCAL TAIL. A fish-tail intermediate between the heterocercal and the diphycercal types, usually having the dorsal lobe much less developed than the ventral lobe.

HETERODONT. The condition occurring in most mammals and a few extinct reptiles of having several different types of tooth, *viz.* incisors, canines, premolars and molars.

HETEROGAMETES (ANISOGAMETES). Gametes which show a clear distinction between male and female, *e.g.* the spermatozoon and the ovum.

HETEROLEPIDOTIDAE (HEXAGRAMMIDAE). Small spiny-rayed shorefish of the North Pacific, sometimes having several lateral lines.

HETEROMETABOLA. Insects which do not undergo a complete metamorphosis: an alternative name for *Hemimetabola* or *Exopterygota* (*q.v.*).

HETEROMI (DERCETIFORMES). Eel-shaped fish having an air bladder without an open duct and usually having pelvic fins in the abdominal position. In this they differ from true eels which have no pelvic fins.

HETEROMYIDAE. Burrowing rodents of North America with long hind-limbs and tail.

HETERONEMERTINI. Nemertine worms without stylets on the proboscis (see also *Metanemertini*).

HETERONEREIS. A modified form of the marine worm *Nereis* in which the sexual organs and the parapodia become greatly enlarged during the breeding season and the whole worm leads a very much more active existence.

HETERONEURA. Butterflies and moths which have the veins differently arranged in the fore and hind wings; the group includes nearly all the common Lepidoptera (*cf. Homoneura*).

HETEROPHIL. A type of white blood corpuscle whose granules stain differently according to the species of animal from which they come, *i.e.* they may be basophil, acidophil or neutrophil.

112

HETEROPTERA. A suborder of bugs (Hemiptera) having the fore-wings in the form of *hemelytra*, *i.e.* having the proximal half of the wing horny and the distal half membranous. Included in this group are most plant-bugs, water-boatmen, water-scorpions and the common bed-bug. The latter is placed in the Heteroptera on account of the general similarity of its body and mouth-parts to those of the rest of the group. The absence of wings is considered to be a degenerate or secondary feature which has arisen in connection with its parasitic habits.

HETEROSTRACI. Fossil ostracoderm fish having the head and anterior part of the body covered by a dorsal shield composed of several plates and the hind-part covered by smaller rhombic plates.

HETEROTRICHA (HETEROTRICHIDA). Protozoa having a permanently open 'gullet' with an undulating membrane and an adoral wreath and having the rest of the body covered completely with cilia. They include the well known fresh-water *Stentor*.

HETEROZYGOUS. HETEROZYGOTE. Hybrid: an individual which has inherited opposite allelomorphic characters from each of its parents (*cf.* *Homozygous*).

HEX-, HEXA-. Prefix from Greek *Hex*: six.

HEXACANTH EMBRYO. A spherical embryo bearing six hooks, characteristic of certain tapeworms.

HEXACTINELLIDA. Sponges with a siliceous skeleton composed of six-rayed spicules often forming an elaborate lattice structure as in the 'Venus Flower Basket' (*Euplectella*). They are all marine and are generally only found at very great depths.

HEXACTINIAN. A type of Zoantharian sea-anemone in which the membranes or mesenteries in the body-cavity consist either of six pairs or a multiple of six.

HEXAGRAMMIDAE. See *Heterolepidotidae*.

HEXAPODA. An alternative name for insects, so-called on account of the six legs.

HIBERNATION. The winter sleep characteristic of many animals in temperate climates. In complete hibernation, such as that of the frog, no food is taken; metabolism slows down and the temperature drops. The animal does not wake up until the spring. In such animals as bats, however, hibernation is only partial and the animal has to wake up from time to time in order to feed.

HIND-BRAIN (METENCEPHALON). The posterior part of the vertebrate brain consisting chiefly of the cerebellum and the medulla oblongata.

HIND-GUT. See *Proctodaeum*.

HIPPO-. Prefix from Greek *Hippos*: a horse.

HIPPOBOSCIDAE. Two-winged insects belonging to the Cyclorrhapha and frequently living as blood-sucking parasites on mammals or birds. A peculiarity of the group is their mode of birth. Eggs are not laid but the young larvae develop to an advanced state inside the body of the parent. Almost as soon as they are born they commence to pupate and the fully formed fly soon hatches out. The Sheep-ked, *Melophagus* is an example.

HIPPOCAMPUS (HIPPOCAMPAL LOBES, HIPPOCAMPAL CORTEX). The parts of the cerebral cortex nearest to the middle line in higher vertebrates. In reptiles this region is comparatively thin and is apparently associated with the sense of smell. In mammals, however, it is greatly thickened to form two lobes, one in each cerebral hemisphere, connected together by the Hippocampal commissure. It forms part of the *Archipallium* in reptiles and of the *Neopallium* in mammals. The latter is considered to be the seat of the intelligence.

HIPPOPOTAMIDAE. Hippopotami: even-toed ungulates (Artiodactyla) with thick almost hairless skin, a huge head and a large heavy body. The legs are short and each foot has four toes with nail-like hoofs all resting on the ground. They are herbivorous semi-aquatic animals confined to Africa.

HIPPOTRAGINAE. African ruminants with long horns in both sexes. They include the Sable Antelope, the Blaubok, the Equine Antelope etc.

HIRUDINEA. Leeches: hermaphrodite Annelids without chaetae or parapodia; with a short body having a small fixed number of segments each divided externally by grooves giving the appearance of a greater number. There is a sucker at each end and the mouth usually has three horny jaws. Most species suck blood and store it in numerous distensible pouches leading from the mid-gut. The blood is kept in a liquid state by an anticoagulant substance (hirudin) produced by the salivary glands.

HIRUNDINIDAE. Swallows and martins: migratory insectivorous birds with broad triangular beak, small weak feet and a long forked tail.

HISTIOCYTES. See *Clasmatocytes*.

HISTO-. Prefix from Greek *Histos*: a web, tissue.

HISTOGENESIS. The formation of new tissues.

HISTOLYSIS. The breaking down of tissues.

HOLO-. Prefix from Greek *Holos*: whole.

HOLOBLASTIC CLEAVAGE. Complete cleavage of a fertilized ovum in which the planes of cleavage go right across the ovum forming a definite number of new cells known as *blastomeres*. Eggs with little yolk undergo complete cleavage of this type, but those with a larger amount of yolk (*e.g.* birds' eggs) only undergo partial or *meroblastic* cleavage in which a disc of cells develops on one side of the yolk-sac.

HOLOBRANCH. A complete gill: one with vascular lamellae on both sides (*cf. Hemibranch*).

HOLOCEPHALI. A group of cartilaginous fish represented at the present day only by the *Chimaeridae*, but well represented in Jurassic times. They resemble sharks but have a compressed head, a small mouth and an operculum over the gill clefts; the pectoral fins are greatly enlarged and in some species the tail is long and whip-like. The present day genus *Chimaera*, the so-called 'King of the herrings' is found off the coasts of most continents.

HOLOCEPHALOUS. A term used to describe insect larvae in which the head is fully developed (*cf. Hemicephalous* and *Acephalous*).

HOLOCEPHALOUS RIBS. Ribs which articulate with the thoracic vertebrae by a single broad head touching both the centrum and the neural arch of each vertebra. Such ribs are found in primitive reptiles and amphibia.

HOLOCRINE. A method of production of digestive juices or other fluids in which the cells of the glands disintegrate to form part of the fluid. Such a mode of secretion occurs in the sebaceous glands of mammals and in the intestines of certain insects.

HOLOGAMY. A form of sexual union occurring in some protozoa in which two full-sized ordinary individuals act as gametes and unite to form a zygote.

HOLOMASTIGINA. Fresh-water or marine protozoa, usually spherical in shape and having forty or fifty flagella scattered evenly over the whole surface. Feeding is holozoic by amoeboid action.

HOLOMETABOLA. See *Endopterygota*.

HOLOPTYCHIIDAE. Fossil fish of the Devonian Period having the body covered with thick overlapping cycloid scales; having the tail slightly upturned (hetero-diphycercal) and having pectoral fins with a long central axis.

HOLORHINAL. A term used to describe a bird's skull in which the external nasal openings are oval-shaped and have their posterior borders in front of the posterior ends of the premaxillae (see also *Schizorhinal*).

HOLOSTEI. A group of bony fish represented as fossils from the Mesozoic Period onwards and at present containing the primitive types *Amia* (the 'Bow-fin') and *Lepidosteus* (the *Gar-pike*) from the fresh waters of North America. The body of the former is covered with thin cycloid scales; that of the latter with thick rhombic ganoid scales. The tails of both fish are primitive; *Amia* nearly diphycercal; *Lepidosteus* heterocercal. The pectoral and pelvic fins of both, however, are of the modern or *actinopterygian* type supported by dermal fin-rays.

HOLOTHUROIDEA. Sea cucumbers: elongated echinoderms without arms but with tentacles round the mouth and usually with suctorial tube-feet. The body wall is muscular and contains numerous minute ossicles but no spines. The animal progresses horizontally by means of its tube-feet, the mouth and tentacles being in front.

HOLOTRICHA (HOLOTRICHIDA). Protozoa having cilia uniformly distributed over the whole surface of the body, but without an adoral wreath.

HOLOZOIC. Feeding entirely on organic substances and unable to do photo-synthesis. This is the case with all animals and with those plants which do not contain chlorophyll. Sometimes this is the only means of deciding whether a unicellular organism should be classed as an animal or a plant.

HOM-, HOMO-. Prefix from Greek *Homos*: the same.

HOMACANTH. The type of dorsal fin in spiny-rayed fish in which the spines are symmetrical and when depressed cover one another completely.

HOMALOPSINAE. Viviparous water snakes inhabiting the rivers and estuaries of S.E. Asia and Northern Australia.

HOMALOPTERINA. Carp-like fish without an air-bladder, from the hill streams of East India.

HOMINIDAE. The family of Primates which includes present day Man (*Homo sapiens*) and the various species and varieties from the Pleistocene Period onwards, *viz. Pithecanthropus* of Java, *Homo rhodesiensis, sinensis, heidelbergensis, neanderthalensis* etc.

HOMOCERCAL. The type of tail found on most present day bony fish; symmetrical and supported only by dermal fin-rays. The symmetry, however, is a secondary feature, as the tail has developed from an unsymmetrical or *heterocercal* type by a lengthening of some of the fin-rays. This is shown by the fact that the tip of the vertebral column is turned slightly upwards.

HOMODONT. Having teeth all alike or nearly so, as in most lower vertebrates.

HOMOGAMETES. See *Isogametes*.

HOMOIOTHERMIC (HOMOTHERMOUS). Having blood of a uniform temperature usually higher than that of the surroundings, as in birds and mammals.

HOMOLOGOUS ORGANS. Organs in different species which show similarity of origin and development in the embryo although they may become very much modified and take on a different function in the adult. The small bones in the ear of a mammal, for instance, have been shown to be homologous with certain bones of the jaw in fish and reptiles (*cf. Analogous*).

HOMONEURA. Moths having almost identical venation in the two pairs of wings. The group is a small one and is generally considered to be primitive. It includes the minute *Micropterygidae* with biting mouth-parts and an undeveloped proboscis, as well as the much larger *Hepialidae* or Swift-moths whose larvae do much damage to plants.

HOMOPTERA. A suborder of Hemiptera or bugs, distinguished from the Heteroptera by having the fore-wings of uniform consistency. All the common plant-bugs, aphides, frog-hoppers, cicadas etc. are included in the Homoptera. Many do great damage to plants not only by sucking the sap, but by discharging a large amount of sugary waste or 'honeydew' which blocks the stomata of leaves and harbours micro-organisms.

HOMOTHERMOUS. See *Homoiothermic*.

HOMOZYGOUS (HOMOZYGOTE). A term applied to an individual containing two genes for the same character, *i.e.* it has inherited the same quality from each of its parents (*cf. Heterozygous*).

HONEYCOMB. The reticulum or second chamber of a ruminant's stomach, so-called because its wall has a structure similar to that of the honeycomb from a beehive.

HONEYDEW. A sugary waste substance produced by aphides and other plant-bugs. These insects feed almost continuously on the juices of plants and thereby take in far more sugar and water than they need. The excess is passed out through two small tubes near the tip of the abdomen. It is much sought after by ants.

HOOF. A horny casing of the toe formed from *keratin*, a hardened form of epidermal cuticle. Hoofs are peculiarly characteristic of the Ungulate group of mammals (see *Ungulata*).

HOPLO-. Prefix from Greek *Hoplon*: a weapon.

HOPLOCARIDA (STOMATOPODA). Marine burrowing crustacea belonging to the class Malacostraca, having a small flattened carapace, stalked eyes and five pairs of thoracic limbs bearing *subchelae* or inverted pincers. The second pair of these is greatly enlarged for catching prey.

HORMONE. A chemical substance produced by an endocrine or ductless gland, able to pass in the blood-stream to all parts of the body and to exercise a controlling effect on various metabolic functions. Examples in the vertebrate body are *thyroxin* from the thyroid gland and *adrenalin* from the adrenal glands.

HORNS. The horns of cattle etc. are formed of a layer of epidermal keratin overlying a central bony core which is not deciduous.

The antlers of deer consist of bony processes of the frontal bones covered during growth by a soft velvety skin. Except in reindeer they are on the males only. They are shed annually after mating and grow again with great rapidity.

The 'horn' of a rhinoceros contains no bony process but is an aggregation of fused hairs.

HUMERUS. The upper arm-bone or proximal bone of the fore-limb of a tetrapod vertebrate.

HUXLEY'S LAYER. The innermost layer of cells in the sheath or follicle of a mammalian hair.

HYAENIDAE. Hyaenas: a family of Carnivora, mostly carrion feeders, which, although somewhat dog-like in appearance, are apparently more nearly related to the civets (Viverridae) and to the cats (Felidae). These three families are therefore grouped together in the super-family Feloidea. Hyaenas have four toes on each foot and the hind legs are shorter than the fore. They are confined to Africa and parts of Asia.

HYAL-, HYALO-. Prefix from Greek *Hualos*: glass.

HYALINE CARTILAGE. Clear translucent cartilage consisting of a matrix of *chondrin* with small groups of cells lying in fluid-filled spaces.

HYALOID CANAL. A narrow canal passing through the vitreous humour of the vertebrate eye from the lens to the point of origin of the optic nerve (blind spot).

HYALOID MEMBRANE. A membrane investing the vitreous humour of the vertebrate eye and also forming the lining of the hyaloid canal.

HYALOPLASM. The clear liquid constituent of protoplasm.

HYBRID. A heterozygous individual: the result of crossing two parents with opposite allelomorphic characters (see also *Heterozygous*).

HYDATID CYST. The large bladder-like larva of certain tapeworms, notably of the genus *Echinococcus*. Unlike the small bladder-worms which are the larvae of most tapeworms, this organism is an enormous cyst holding several pints of fluid and capable of budding off numerous daughter-cysts forming a large number of heads or proscolices. The adult *Echinococcus* is a small relatively harmless parasite sometimes found in dogs; the hydatid cyst, however, may develop in the brains of sheep or other domestic animals and occasionally in man causing symptoms of epilepsy.

HYDR-, HYDRO-. (1) Prefix from Greek *Hudor, hudros*: water.
(2) Prefix from Greek *Hudra*: a mythical many-headed water serpent.

HYDRANTH. The normal type of polyp with mouth and oral tentacles found in colonial Coelenterates such as corals.

HYDRATUBA. The hydra-like larva of a jelly-fish which by strobilation forms a segmented larva called a *scyphistoma*; from the latter the young jelly-fish or *ephyrae* break off (see *Scyphistoma* and *Ephyra*).

HYDRIDA. Solitary coelenterate polyps of sedentary habit, usually anchored at the base and having a mouth and oral tentacles at the other end. They are considered to be derived from the more complex colonial types by loss or degeneration of the free-swimming medusoid stage which in most other coelenterates alternates with the hydroid stage.

HYDROCAULUS. That part of the coenosarc which resembles the stem of a plant in colonial coelenterates.

HYDROCOEL. The water vascular system of an echinoderm. In star-fish it consists of a ring-vessel round the mouth with radial vessels leading along the five arms. Numerous branches from these lead to the tube-feet which are normally distended with water enabling them to be used for locomotion. Water enters the star-fish by means of a porous plate or filter known as the *madreporite* on the dorsal surface. Beneath this is the *stone canal*, a calcareous tube leading to the ring vessel.

HYDROCORALLINAE. Colonial coelenterates of the class Hydrozoa which give rise to coral by deposition of calcium compounds in a matrix round the much branching hydrorhiza. From pits in the surface of the coral there arise two types of polyp, *viz. gastrozooids* and *dactylozooids*. The former are normal polyps with mouth and tentacles; the latter are long and slender for catching prey.

Hydrocorallinae are found in tropical seas and differ from the majority of reef corals in that they produce a sexual generation of free-swimming medusae, whereas other corals consist only of polyps.

HYDROID. Similar to *Hydra*: having a sac-like body with a mouth and tentacles at the top and a body-wall made of only two layers of cells (see *Coelenterate* and *Diploblastic*).

HYDROID GENERATION. The asexual phase in the life of a colonial coelenterate of the class Hydrozoa, consisting of a number of polyps whose body-cavities are linked by a common tube or *coenosarc*. Alternating with this generation is the *medusoid* or sexual generation consisting of free-swimming jelly-fish which bud off from the parent colony.

HYDROMEDUSAE. See *Hydrozoa*.

HYDROMETRIDAE. Pond-skaters: agile insects which move rapidly on the surface of water owing to the fact that their skin is unwetted and they do not break the surface-tension film of the water. This is due to the fact that the cuticle is of a waxy chitinous substance and is covered with fine hairs which form an insulating layer between the insect and the water. Pond-skaters belong to the order Hemiptera and to the suborder Heteroptera.

HYDROPHINAE. Sea-snakes of the Indian and Pacific Oceans, having a laterally compressed body and small eyes. They are viviparous and extremely poisonous.

HYDROPHYLLIUM. A leaf-like polyp with no mouth, whose function is to cover or protect other polyps in a colony. Coelenterates of the order Siphonophora, for instance, consist of free-swimming *cormidia* or colonies usually having a large air bladder or *pneumatophore* at the top to keep them afloat; beneath this are numerous polyps of various kinds hanging beneath a covering of hydrophyllia.

HYDRORHIZA. The root-like portion of a colonial coelenterate such as *Obelia*, consisting of a much branched tube by which the colony is attached to the surface of sea-weeds or rocks. From this root-like part there arise stems or *hydrocauls* bearing numerous sedentary polyps whose digestive cavities are all linked by a common tube.

HYDROSPIRES. Long internal pouches radiating out from the centres of certain fossil echinoderms. They opened to the exterior by means of spiracles near the mouth and were probably used for respiration.

HYDROTHECAE. Cup-like portions of perisarc partly enclosing the polyps in colonial coelenterates of the group Calyptoblastea.

HYDROZOA. Coelenterates which usually show an alternation of sessile polyp colonies with free-swimming *medusae* or jelly-fish. The class is large and varied and includes the *Calyptoblastea* and *Gymnoblastea* in which the colony is enclosed by a horny case or *perisarc*; the *Hydrocorallinae* in which the peri-sarc is replaced by a calcareous matrix of coral; the *Siphonophora* consisting of floating colonies of polymorphic zooids and the *Hydrida* or solitary polyps in which the medusa stage is lost.

HYLESININAE. A group of bark-beetles, the best known being the Ash-bark beetle which does much damage to timber by its complex system of tunnels.

HYLIDAE. Tree-frogs whose claw-like digits end in adhesive pads. They are found in many parts of the world and include the smallest known frog which is less than three quarters of an inch long and is found in Peru. Some of them show complex patterns of behaviour. They may, for instance, make pools with mud walls in which they lay their eggs; others lay them on leaves overhanging water so that when the tadpoles hatch they fall in.

HYMEN. (1) A fold of mucous membrane stretching across the entrance to the vagina in some mammals. It is perforated by a small opening and is normally ruptured during coitus.

(2) A sheet of tissue closing the aperture leading from the oviduct into the cloaca in immature females of certain fish.

HYMENO-. Prefix from Greek *Humen*: a membrane.

HYMENOPTERA. Bees, wasps, ants, sawflies etc., insects having two pairs of membranous wings which can be linked together by small hooks when flying. The mouth-parts are of various types; in the case of the bee they are very much lengthened to form a proboscis through which nectar is sucked. The ovipositor may be adapted for sawing, piercing or stinging. Most of the Hymenoptera lead a social life in well organized colonies. The order is divided into two sub-orders: *Apocrita* with a narrow waist and *Symphyta* without.

HYMENOSTOMATA (VESTIBULATA). Ciliated protozoa with a small micronucleus and a larger meganucleus and having a permanently open 'gullet' with an undulating membrane. The best known example is the common *Paramecium*.

HYO-, HYOID. Prefix and adjective from Greek *Huoeides*: Y-shaped.

HYOID ARCH. The second visceral skeletal arch of a vertebrate, immediately behind the mandibular arch and supporting the floor of the mouth. In cartilaginous fish it separates the spiracle from the first gill-slit and consists of a *basihyal* cartilage, two *ceratohyals* and two *hyomandibulae*. In bony fish there are also *epihyals* and *hypohyals*. In land animals in which the spiracle is converted into the Eustachian tube there is usually a flat plate at the root of the tongue and a series of small bones passing upwards towards the cranium.

HYOID CORNUA. See *Cornua*.

HYOID SCLERITE. A small U-shaped sclerite which keeps the pharynx distended in some insects.

HYOMANDIBULA. One of a pair of small cartilages or bones forming the dorsal ends of the hyoid arch of a vertebrate. In fish they are at the junction of the upper and lower jaws and help to support both. In land animals, with the changed mode of articulation of the lower jaw, the hyomandibula becomes modified to form an auditory ossicle (the *columella auris* of amphibia, reptiles and birds and the *stapes* of mammals).

HYOSTYLIC. A type of skull in which the upper jaw nowhere touches the auditory capsule and is suspended only by the hyomandibula and by ligaments. The skull of the dogfish is of this type.

HYPANDRIUM. A sub-genital plate formed from the ninth abdominal sternum in the males of certain insects. It may bear two short stylets, as in the common cockroach.

HYPAPOPHYSIS. A median ventral process on the centrum of a vertebra: usually present on the first two or three lumbar vertebrae of mammals.

HYPAXONIC MUSCLES. In fish and higher vertebrates, the segmental muscles or myotomes are each divided into two by a horizontal septum within which the ribs are formed. The ventral half of each myotome is called the *hypaxonic* muscle and the dorsal half the *epaxonic*.

119

HYPER-. Prefix from Greek *Huper*: above, beyond, greater than.

HYPERDACTYLY. A state in which the number of digits in a limb is more than the normal five, as for instance in the broad paddles of the *Ichthyosaurus*.

HYPERMASTIGINA. Holozoic protozoa possessing a very large number of flagella; some species are symbiotic in the alimentary canals of termites where they help to digest wood.

HYPERMETAMORPHOSIS. The development of an insect in which there are several distinctly different types of larvae having different modes of life.

HYPEROARTIA. A name formerly used for the Petromyzontidae (lampreys).

HYPEROTRETA. A name formerly used for the Myxinidae (hag-fish).

HYPERPARASITE. A parasite which lives on or in another parasite.

HYPERPHALANGY. A state in which the number of phalanges in each digit of a limb is increased, as for example in the paddle-like limbs or 'flippers' of a whale (*cf. Hyperdactyly*).

HYPERPHARYNGEAL BAND. A wide band of tissue running along the roof of the pharynx in Tunicates, from which a row of processes known as *languets* hangs down into the branchial cavity.

HYPERPHARYNGEAL GROOVE (EPIPHARYNGEAL GROOVE). A ciliated groove running along the roof of the pharynx of *Amphioxus* and similar organisms, by which a stream of mucus and food particles is passed back to the oesophagus.

HYPO-. Prefix from Greek *Hupo*: under, below.

HYPOBLAST. An alternative name for the *endoderm* (*q.v.*).

HYPOBRANCHIAL. Beneath the gills.
    (1) Hypobranchial bones or cartilages. Skeletal elements adjacent to the basibranchial and beneath the ceratobranchials in the gill-arch skeleton of a fish.
    (2) Hypobranchial space (crustacea). The part of the branchial chamber below the gills in a crab or similar crustacean.

HYPOCENTRUM. See under *Pleurocentrum*.

HYPOCEREBRAL. Beneath the brain.

HYPOGLOSSAL NERVE. The twelfth cranial nerve of a reptile, bird or mammal: a motor nerve innervating the floor of the mouth and the tongue. In amphibia it is the first spinal nerve coming out immediately behind the skull; in fish it is formed from several spinal nerves joined together.

HYPOHYALS. Bones or cartilages forming part of the hyoid arch (floor of the mouth) in fish; they lie at the base of the arch on each side of the median *basihyal*.

HYPO-ISCHIUM (OS CLOACAE). A cartilaginous or bony extension backwards from each ischium to support the cloaca in many reptiles.

HYPOMERE. The lateral plate mesoderm (*q.v.*).

HYPONOMEUTIDAE. A family of moths whose larvae are very destructive to trees. They include the Larch shoot borer, the Ash-bud moth and the Ermine moth.

HYPOPHARYNX. A chitinous sclerite arising from the floor of the mouth of an insect. Normally it bears the salivary apertures; in blood sucking insects like mosquitoes it is enormously elongated and forms part of the proboscis down which saliva flows.

HYPOPHYSIAL FENESTRA. An opening in the floor of the vertebrate cranium accommodating the pituitary body and also serving for the admission of the internal carotid artery. It is formed in the embryo as a widening of the gap between the two trabeculae (*q.v.*).

HYPOPHYSIAL SAC. A cavity beneath the brain of some vertebrates, representing a duct which originally led from the hypophysis of the pituitary body into the mouth. In cyclostomes it is large and spreading and has lost contact with the pituitary body. In higher vertebrates it is reduced to a cleft between the anterior and intermediate lobes of the pituitary body.

HYPOPHYSIS. That part of the pituitary body which is formed from an up-growth of the ectoderm in the region of the mouth. The whole pituitary body is a combination of this with the *infundibulum* or down-growth from the third ventricle of the brain. In the fully developed organ the hypophysis gives rise to the anterior lobe and the infundibulum to the posterior lobe.

HYPOPYGIUM. The terminal segments of the abdomen of an insect, often modified for copulation.

HYPORHACHIS (HYPOPTILUM). The aftershaft of a feather (*q.v.*).

HYPOSTOME (HYPOSTOMA). Any structure below or immediately behind the mouth, as for instance the oral cone of a hydra.

HYPOSTOMIDES. Small flying fish with large horizontal pectoral fins and with the body entirely covered by bony plates. They inhabit sandy shallows of the Indian and Pacific Oceans.

HYPOTREMATA. Rays: cartilaginous fish with the body flattened dorso-ventrally; with large spreading pectoral fins and with the gill clefts on the ventral side.

HYPOTRICHA. Flattened ciliated protozoa with a few stiff bristles on the dorsal side and with movable cirri on the ventral side. Round the mouth is an adoral wreath of normal cilia and undulating membranes. Many species are ectoparasites on other fresh-water organisms.

HYPSODONT TEETH. Teeth having open roots and persistent pulps so that as fast as they are worn down they continue to grow. As the surface wears away a complicated pattern of enamel, dentine and cement is formed. The changing of this pattern may be used to give an indication of the age of the animal. Such teeth are found in ungulates, rodents and most herbivorous animals.

HYPURAL BONES. Enlarged haemal arches supporting the ventral lobe of the tail-fin of a fish.

HYRACOIDEA. Small plantigrade, squirrel-like mammals, usually arboreal, having four toes on the front foot and three on the hind; having short ears and a short tail. They are usually regarded as a separate order, but apparently have affinities both with the rodents and with the ungulates.

HYRACOTHERIIDAE. Extinct ungulate mammals including some of the probable ancestors of the horse, *viz. Eohippus* and *Hyracotherium* of Eocene times followed by *Pliohippus*, *Miohippus* and others leading to the Equidae or present-day horses. Throughout the series there has been a gradual increase in the size of the animal and a gradual decrease in the number of toes.

HYSTRICOMORPHA. Rodents having thick cheek-bones, one pair of upper incisors, complete clavicles and free fibulae. The group includes Porcupines, Chinchillas, Agoutis, Guinea pigs and Coypu rats, all of which live in South America, and also the Jumping hare of South Africa.

I

IBALIINAE. A group of parasitic gall-wasps which lay their eggs in the larvae of the wood-wasp *Sirex*.

ICHNEUMONOIDEA. A large group of parasitic insects belonging to the *Apocrita*, *i.e.* Hymenoptera having a narrow waist. By means of a sharp piercing ovipositor they lay their eggs inside the body of another insect. The larvae hatch, live for some time in the body of the victim, eventually killing it and emerging to pupate in a silky cocoon.

ICHTHY-. Prefix from Greek *Ichthus*: a fish.

ICHTHYOBORINA. Carp-like fish of Tropical Africa having a rounded scaly body with 12–17 dorsal fin-rays.

ICHTHYODORULITES. Fin-spines of Elasmobranch fish found as fossils in Palaeozoic rocks. They are sometimes the only part of a cartilaginous fish hard enough to become fossilized and are therefore of little use to identify the fish. They have, however, a microscopic structure similar to that of the fin-spines of modern sharks and rays.

ICHTHYOID. Pertaining to or resembling a fish.

ICHTHYOPSIDA. A name originated by Huxley and formerly used for the Amphibia and fish to denote their close relationship to each other.

ICHTHYOPTERYGIUM. A paddle-like limb or fin adapted for swimming, *e.g.* the pectoral and pelvic fins of fish.

ICHTHYORNITHES. Extinct toothed birds from the Cretaceous rocks of North America. The wings were better developed than those of the Jurassic *Archaeopteryx* and the tail was shorter.

ICHTHYOSAURIA. Fish-lizards: large extinct fish-like reptiles whose skeletons have been found from the Triassic to the Cretaceous periods. They had four paddle-shaped limbs with many digits and many phalanges. The tail was fish-like and the head was large with a long snout and many teeth.

ICHTHYOTOMI. Extinct cartilaginous fish of the Carboniferous and Permian periods, having leaf-like pectoral fins (*archipterygia*), a continuous dorsal fin, a long body and a slender diphycercal tail. *Pleuracanthus* is the best known example.

ICTERIDAE. American Orioles or 'starlings'.

IDIO-. Prefix from Greek *Idios*: peculiar to oneself.

IDIOCHROMATIN. Generative chromatin. Chromatin contained in the reproductive nucleus of a cell or in the micronucleus of those protozoa which have a mega- and a micronucleus (*cf. Trophochromatin*).

IDIOGASTRA. A name sometimes given to the *Orussoidea*, a small group of parasitic Hymenoptera intermediate between the Apocrita and the Symphyta.

IGUANIDAE. Iguanas: a family of lizards of North and South America, some Pacific islands and Madagascar. The body is without bony plates but these may be present on the head, sometimes in the form of horn-like tubercles. Teeth are sometimes on the palatine bones as well as on the jaws.

ILEUM. The hind-part of the small intestine of a vertebrate. In the mammalian alimentary canal it comes after the jejunum or after the duodenum and ends at the *sacculus rotundus* where it meets the colon and the caecum. In some vertebrates, however, it is less distinct and the name may be used vaguely for the hind part or sometimes for the whole of the small intestine; it is also used to denote any analogous part in some invertebrates.

ILIAC ARTERIES. Blood vessels leading from the dorsal aorta to the hind limbs of a tetrapod vertebrate.

ILIAC VEINS. Blood vessels returning blood from the hind-limbs to the posterior vena cava in a tetrapod vertebrate.

ILIUM. The anterior dorsal bone on each side of the pelvic girdle in a tetrapod vertebrate, fused to the ischium and the pubis and also to the transverse processes of one or more sacral vertebrae. This fusion of bones forms a strong support for the body in the region of the hind legs.

ILYSIIDAE. Snakes of the Boa type having the cranial bones more or less solidly united; having teeth on the mandibles, pterygoids, palatines and maxillae, and having a vestigial pelvic girdle. They are found in South America and S.E. Asia.

IMAGINAL DISCS (IMAGINAL BUDS). Clusters of cells which undergo rapid division to form the rudiments of future organs during the metamorphosis of an insect.

IMAGO. The adult form of an insect after metamorphosis.

IMPLANTATION. The attachment of the developing mammalian embryo to the uterus of its parent. In the first place this is by *trophoblastic villi* which penetrate corresponding crypts in the uterine wall. Later attachment is by a placenta formed from the foetal membranes.

INCISOR PROCESS. A sharp biting part on the mandible of an arthropod or other invertebrate (see also *Molar process*).

INCISORS. The front teeth of a mammal, usually sharp and chisel-like for biting.

INCUBATION PERIOD. (1) The time between the laying of an egg and the hatching of the young bird, reptile etc. if kept at the necessary temperature. (2) The interval between an infection and the onset of a disease.

INCUS. The Anvil: the middle one of the three auditory ossicles of a mammal, between the *malleus* and the *stapes*. The incus is homologous with the *quadrate bone* of a lower vertebrate as can be shown by its embryological development.

INDETERMINATE CLEAVAGE. Cleavage of a fertilized egg in which there are at first no specific recognizable organ-forming areas (*cf. Determinate cleavage*).

INDRISINAE. Herbivorous short-tailed lemurs of Madagascar having long hind legs on which they walk with the arms held above the head.

INFERIOR OBLIQUE MUSCLES. See *Eye muscles*.

INFERIOR RECTUS. See *Eye muscles*.

INFERIOR VENA CAVA (POSTERIOR VENA CAVA, POSTERIOR CAVAL VEIN). The main vein returning blood from the body to the right auricle of the heart in higher vertebrates.

INFRA-. Latin prefix: below, beneath.

INFRA-CLAVICLE. One of three membrane-bones forming part of the pectoral girdle of certain primitive fish. The other two are the *clavicle* and the *supra-clavicle* but in a more recent system of nomenclature the latter are called respectively the *cleithrum* and *supra-cleithrum*, whilst what was formerly named the infra-clavicle is now called the clavicle. The change of name brings the system more into line with recent opinions regarding the homologies of the various bones with those of other vertebrates.

INFRA-ORBITAL GLANDS. A pair of salivary glands below the orbits in some mammals but not in man.

INFRA-TEMPORAL ARCADE. A bar of bone in the skulls of certain reptiles, joining the jugal and the quadrate bones and forming the lower boundary of the *inferior temporal fossa* (see *Diapsida* and *Synapsida*).

INFUNDIBULAR GLAND. See *Saccus vasculosus*.

INFUNDIBULUM. That part of the pituitary body which is formed from a down-growth of the Thalamencephalon in the vertebrate brain (see also *Hypophysis*).

INFUSORIA. See *Ciliophora*.

INGESTION. The act of taking in food. Eating.

INGUINAL CANAL. The passage leading from the abdominal cavity into the scrotal sac, through which the testes descend in a mammal (see *Scrotum*).

INHALENT CANAL. Any passage or canal through which water enters an organism, *e.g.* those leading into the flagellated chambers of sponges.

INIOMI. Lantern-fishes: deep sea fish with luminous spots along the sides of the body.

INK-SAC. A sac of dark brown fluid on the ventral side of a cuttle-fish or similar cephalopod mollusc. It opens into the rectum and when the animal is attacked a dense 'smoke-screen' is ejected.

INNOMINATE ARTERY. A name given to various short arteries which bifurcate into two principal arteries, *e.g.* that which leads from the aorta in some vertebrates and branches to form the subclavian and the carotid arteries.

INNOMINATE BONE. Half of the vertebrate pelvic girdle consisting of the three bones *ilium*, *ischium* and *pubis* which may become fused together in the adult.

INNOMINATE VEIN. (1) A name given to various short veins formed by the union of two other veins.

(2) A vein present in some mammals including humans by which the left anterior vena cava is connected with the right and so loses its own opening into the right auricle.

INQUILINE. Commensal: applied to an animal which lives in the home of another and shares its food.

INSECTA. Insects: Arthropods which breathe by spiracles and tracheae and have the body divided into a distinct head, thorax and abdomen, the former bearing antennae and usually compound eyes. The thorax consists of three segments with three pairs of legs and usually two pairs of wings; the abdomen is segmented and without legs. The mouth-parts are complex and variably adapted for different modes of feeding.

Insects form the largest known group of animals consisting of about 700,000 species.

INSECTIVORA. An order of small mammals, usually terrestrial, characterized by having the teeth nearly all similar and roughly conical. The limbs are normally pentadactyl and plantigrade. Shrews, hedgehogs and moles are included in this group.

INSECT LARVAE. (1) *Protopod larvae*. Primitive parasitic larvae with barely incipient limb-buds and with no segmentation of the abdomen; similar to an early embryo which is usually in the egg.

(2) *Polypod* or *Eruciform larvae*. Typical caterpillars with six legs on the thorax and a number of 'cushion feet' on the abdomen.

(3) *Oligopod* or *Campodeiform larvae*. These are usually predatory and therefore have efficient sense organs and long legs but no 'cushion feet'; they are common among beetles.

(4) *Apodous* or *legless larvae*. The body is segmented and has a minute head with few sense organs and no legs or limbs of any kind. The absence of these is probably a secondary feature arising from the fact that they are either fed by other members of the colony, as in bees, or the eggs are laid in suitable food such as meat or dung, as in house-flies etc.

INSEMINATION. Any device by which sperm cells are introduced into the body of a female prior to fertilization of the ovum.

INSESSORIAL. Feet or claws specially adapted for perching.

INSPIRATION. The taking of air into the lungs or other breathing organs.

INSTAR. A stage in the development of an insect between two moults. There are usually a number of larval instars, varying from three to about thirty, followed by the pupa and then by the adult or imago. The pupal stage may be omitted.

INTEGUMENT. An outer protective covering: epidermis, cuticle, feathers, scales etc.

INTER-. Latin prefix: between, among.

INTER-AURICULAR SEPTUM. A septum separating the right from the left auricle in the hearts of amphibia, reptiles, birds and mammals.

INTERBRANCHIAL SEPTUM. A septum or partition separating each gill-slit from the next in a fish. It is supported by a bony or cartilaginous gill-arch and bears on each side the lamellae or filaments which constitute the gill proper.

INTERCALARY VEINS (OF INSECTS). See *Cross-veins*.

INTERCALARY DISCS (INTERCALATED DISCS). Transverse markings, formerly regarded as septa, across the fibres of cardiac muscle; their function is obscure.

INTERCENTRA. Small bones between the vertebrae in certain reptiles; they represent the detached and reduced *hypocentra*, the main part of the vertebra being a *pleurocentrum* (*q.v.*). In the tails of some reptiles the intercentra bear V-shaped haemal arches and are known as *chevron bones*.

INTERCLAVICLE. A membrane-bone, usually Y-shaped, between the two clavicles in some amphibia, reptiles and monotremes.

INTERCOSTAL. Between the ribs, as of arteries, veins, muscles etc.

INTERDORSAL. That part of a vertebra which is formed embryonically from the anterior dorsal part of each sclerotome, the corresponding ventral part being known as the *interventral*. The interdorsal and interventral parts together form the *pleurocentrum* (*q.v.*) which is the main part of the centrum in reptiles, birds and mammals (see also *Basidorsal, Basiventral, Pleurocentrum, Intercentrum* etc.).

INTERHYAL BONES. Small rod-like bones formed by the ossification of ligaments connecting the epihyal with the symplectic bone in the jaw of a bony fish.

INTERLOBULAR VEINS. Branches of the hepatic portal vein running between the lobules of the vertebrate liver.

INTERMEDIUM. The median proximal carpal or tarsal bone of the pentadactyl limb. In the fore-limb it is between the radiale and the ulnare; in the hind-limb between the tibiale and the fibulare.

INTERNAL CAROTID ARTERY. That branch of the common carotid artery which supplies blood to the brain and internal parts of the head.

INTERNAL JUGULAR VEINS. Veins bringing blood from the brain and internal parts of the head to the anterior venae cavae and so to the right auricle of the heart.

INTERNAL NARES. Openings from the nasal cavity to the inside of the mouth; only present in land vertebrates, lung-fish and osteolepids.

INTERNAL RECTUS (ANTERIOR RECTUS). One of the six muscles (four rectus and two oblique) which move the eyeball in a vertebrate.

INTERNASAL SEPTUM. A partition dividing the right and left nasal passages in many terrestrial vertebrates.

INTEROCEPTORS. Sensory receptors within the body of an animal, *e.g.* in the walls of the stomach and intestines.

INTER-OPERCULAR BONE. One of three small bones supporting the operculum of a bony fish. It is between the opercular and the sub-opercular bones and is connected by a ligament with the angle of the mandible.

INTER-PARIETAL BONE. A small bone partially separating the parietals in the skulls of many mammals.

INTERPHASE. A resting phase sometimes occurring between the first and second meiotic divisions of a cell. Normally each cell during meiosis gives rise to two haploid cells and these immediately divide again giving four cells in all. If the second process is delayed, however, an interphase may occur during which the chromosomes disappear.

INTER-RADIAL POSITIONS (OF COELENTERATES). Positions half-way between the per-radial positions on the disc of a circular jelly-fish or medusa. The names per-radial and inter-radial are convenient for designating the positions of tentacles, gonads, canals etc. (see *Radial canals*).

INTER-RENAL BODIES. Endocrine organs between the kidneys in a fish, homologous with the adrenal cortex in mammals (see *Suprarenal* and *Adrenal*).

INTERSPINOUS BONES. An alternative name for the *somactids* or basal skeletal elements of the median fins of a bony fish.

INTERSTITIAL CELLS. Cells occupying the interstices between other cells, as for instance those in the testis or ovary which secrete secondary sex hormones; those in the body of a Hydra or similar animal which are able to develop into nematoblasts or other types of cell etc.

INTERVENTRAL. See under *Interdorsal*.

INTERVERTEBRAL DISCS. Pads of fibro-cartilage between the centra of adjacent vertebrae in many animals. In the centre of each disc is a soft area, the *nucleus pulposus*, which represents the vestige of the embryonic notochord.

INTESTINES. The parts of the alimentary canal which follow the stomach. In mammals these are the duodenum, jejunum and ileum which comprise the small intestine, and the colon, caecum and rectum which comprise the large intestine.

INTRA-. Latin prefix: within.

INTRACELLULAR DIGESTION. Digestion of small particles of food in vacuoles within a cell after having been engulfed by means of pseudopodia. Such digestion takes place not only in protozoa such as *Amoeba*, but also in certain amoeboid cells in the endoderm lining the digestive cavity in Coelenterates and Platyhelminthes.

INTRALOBULAR VEINS. Veins running down the centre of each lobule in the vertebrate liver. They collect up blood which has passed through numerous fine radial vessels in the lobules and convey it to the hepatic vein.

INVAGINATION. An in-pushing of the outer layer of an organism. At the mouth and anus, for instance, the epidermal layer is often folded in to form a covering for part of the oesophagus and the rectum. These parts are known as the *stomodaeum* and the *proctodaeum* respectively. A similar process takes place in the early development of many embryos when a blastula (hollow ball of cells) is converted into a gastrula (sac-like structure) by invagination of one side to form the endoderm.

INVERTEBRATA. All animals which do not have a vertebral column (see *Vertebrata*).

INVERTED RETINA. The arrangement in the eyes of vertebrates whereby the nervous cells are in front of the sensory cells and the light has to pass through the former before stimulating the latter. This arrangement comes about owing to the rolling up of the embryonic neural plate to form a neural tube, followed by the outgrowth of this tube to form the primary optic vesicles.

INVOLUNTARY MUSCLE (UNSTRIPED MUSCLE). Muscular tissue consisting of narrow spindle-shaped cells each with a centrally placed nucleus and with no transverse striations. Such muscles have the power of slow contraction carried out automatically and unconsciously; the movements of the alimentary canal are an example.

IPINAE. Bark beetles and Ambrosia beetles: a subfamily of the Scolytidae characterized by having the head concealed by the lengthened prothorax. They damage trees by burrowing between the wood and the bark; the Ambrosia beetles apparently cultivate fungi on the walls of their tunnels in order to provide food for the larvae.

IRIDOCYTES. Modified cells, *e.g.* those in a squid, which cause iridescence or a play of colour due to diffraction and interference.

IRIS. A pigmented disc between the cornea and the lens in the eyes of vertebrates and of certain mollusca such as the squid and cuttle-fish. A central hole, the pupil, allows light to pass through to the retina. Radial and circular muscles in the iris enable the size of the opening to be varied in order to regulate the amount of light entering the eye.

IRIS CELLS (OF INSECTS, CRUSTACEA ETC.). Pigmented cells surrounding each ommatidium or optical unit of a compound eye. In bright light they spread round each crystalline cone, isolating it from its neighbours, so that vision is *mosaic*; in dim light they may retract so that the light entering each ommatidium merges with that of its neighbours causing continuous but rather ill-defined vision.

ISCHIOPODITE. The first of the five segments of the leg in such crustaceans as the crayfish, crab or lobster. Named in order from the base outwards the segments are: the *ischiopodite, meropodite, carpopodite, propodite* and *dactylopodite*.

ISCHIO-PUBIC BAR. A bar of cartilage forming a simplified pelvic girdle in many cartilaginous fish.

ISCHIO-PUBIC FORAMEN (OBTURATOR FORAMEN). An opening separating the ischium from the pubis on each side of the pelvic girdle in the higher vertebrates.

ISCHIUM. The posterior ventral bone on each side of the pelvic girdle of a tetrapod vertebrate.

ISLETS OF LANGERHANS. Microscopic endocrine organs in the pancreas, secreting the hormone *insulin* which regulates the amount of sugar in the blood.

ISO-. Prefix from Greek *Isos*: equal.

ISOCERCAL (DIPHYCERCAL). Having a symmetrical tail-fin with the vertebral column extending along the median axis, as in many primitive fish.

ISOGAMETES (HOMOGAMETES). Gametes in which there is no obvious distinction between male and female.

ISOGAMY. (1) The ability to form isogametes.
(2) The union of two isogametes.

ISOPODA. Crustaceans belonging to the class *Peracarida* with flattened body and no carapace and with limbs nearly all alike. They include the common wood-louse, the sand-slaters and pond-slaters (*Asellus*) and some species which are parasitic on other crustaceans or on fish.

ISOPTERA. Termites or white ants: social insects which do much damage to woodwork and other materials. They live in large communities in huge nests tunnelled in wood or built of earth and wood cemented together. As with true ants there are numerous castes: queen, workers, soldiers etc. The queens sometimes grow up to four inches or more in length.

ISOSPONDYLI. Bony fish with soft flexible fin-rays; having the air-bladder connected to the throat by an open duct and having the pelvic fins well back on the body. Examples are herring, salmon and many others.

ISOSPORES. Spores which are all alike: asexual spores (*cf. Anisospores*).

ITER. The full name in its Latin form is *Iter a tertio ad quartum ventriculum*: a passage connecting the third and the fourth ventricles of the vertebrate brain. It is also called the *Aqueduct of Sylvius*.

ITONIDAE. Gall-midges: an alternative name for the *Cecidomyidae* (*q.v.*).

J

JACOBSON'S ANASTOMOSIS. A network of nerves connecting the seventh cranial (facial) nerve with the ninth (glosso-pharyngeal) in many vertebrates.

JACOBSON'S ORGANS. A pair of pouches which communicate with the nose and with the mouth in many vertebrates; their function is doubtful but is probably connected with the smelling of food in the mouth. They are very highly developed in snakes.

JAPYGIDAE. A family of Diplura or bristle-tailed insects distinguished from other members of the order by being large and darkly pigmented. Like all the

Diplura they are somewhat primitive and may be considered to be a link between the insects and the Myriopoda.

JASSIDAE. An alternative name for Cicadellidae or leaf-hoppers, a family of agile plant-bugs which infest trees and leap from leaf to leaf with great speed.

JAWS. See *Mandible* and *Maxilla*.

J-DISC. A light coloured region running transversely across the fibres of striated muscles between the darker *A-discs* or *Dobie's Dots*. Each J-disc is itself crossed by a thin dark disc known as the *Z-disc* or *Krause's membrane*.

JEJUNUM. The part of the small intestine between the duodenum and the ileum in some mammals.

JOHNSTON'S ORGAN. A collection of chordotonal receptors near the base of the antenna of an insect. Their function is probably to enable the insect to perceive movements of its antennae (see *Chordotonal receptor*).

JORDAN'S ORGAN (CHAETOSOMA). A sense-organ of uncertain function on the heads of certain lepidoptera.

JUGAL BONE. A membrane-bone forming part of the anterior border of the orbit in the vertebrate skull. In mammals part of the jugal extends backwards and fuses with the forward directed process of the squamosal bone to form the zygomatic arch.

JUGULAR ANASTOMOSIS. A transverse vein connecting the right and left external jugular veins in some mammals.

JUGULAR FINS. Pelvic fins which in some fish are situated far forwards in front of the pectoral fins (*e.g.* in Blennies).

JUGULAR VEINS (INTERNAL AND EXTERNAL JUGULAR). Veins taking blood back from the head to the anterior venae cavae and so to the heart in a vertebrate.

JUGULARES (BLENNIIFORMES). Fish whose pelvic fins are forward in the jugular position.

JUGUM (JUGAL LOBE). The small posterior lobe or process at the base of the fore-wing in certain insects by means of which, in some cases, the fore- and hind-wings are locked together during flight.

K

KAMPTOTRICHIA (CAMPTOTRICHIA). Fibrous, jointed fin-rays such a those of the lung-fish.

KARY-, KARYO-. Prefix from Greek *Karuon*: a nut (nucleus).

KARYASTER. A stellate collection of chromosomes.

KARYOGAMY. The union of two nuclei in sexual reproduction; the union of cytoplasm is known as *plasmogamy* and the whole process involving the two types of union is *syngamy*.

KARYOKINESIS. An alternative name for *Mitosis* (*q.v.*).

KARYOLYMPH (KARYENCHYMA, KARYOCHYLEMA). The fluid part of a nucleus.

KARYOSOME. An aggregation of chromatin in a resting nucleus.

KAT-, KATA-, KATH-. Various forms of the Greek prefix: down.

KATABOLISM. The breaking down of complex substances into simpler ones in a living organism with liberation of energy.

KATADROMOUS. A term used to denote fish such as the eel which make their way from rivers down to the sea to breed: the opposite of *anadromous*.

KERATIN. A hard horny protein-like substance from which claws, nails, hoofs etc. are formed.

KERATOSA. Sponges of the class Demospongiae having a skeleton or mtarix made of the horny substance *spongin*; the common bath-sponge belongs to this group.

KIDNEY. The principal excretory organ of a vertebrate or of certain invertebrates. The vertebrate kidney consists of a large number of *Malpighian bodies* each consisting of a *glomerulus* or minute coiled blood vessel, invested by a funnel-shaped capsule (*Bowman's capsule*). Waste liquid passes from the blood of the glomeruli into the capsules and thence along numerous uriniferous tubules to the ureter from which it passes to the bladder and so to the exterior. During the embryonic development of a vertebrate the first excretory organ is a 'head-kidney' or *pronephros*. This is later replaced by the 'middle kidney' or *mesonephros* and finally by a 'hind-kidney' or *metanephros*. In fish and amphibia no metanephros develops, the mesonephros remaining as the functional kidney.

KINETO-. Prefix from Greek *Kinein*: to move.

KINETONUCLEUS. An alternative name for the *Parabasal body* (*q.v.*), a small structure resembling a nucleus in some protozoa, notably in the Flagellates. It is near the basal granule of the flagellum and is usually connected with it and with the nucleus by protoplasmic threads or *rhizoplasts*. The functions of both basal and parabasal bodies are obscure and sometimes the two are apparently combined in one structure which has also been named the kinetonucleus.

KINETOPLAST. An alternative name for the kinetonucleus (see above).

KIONOCRANIA. A name sometimes given to those lizards which possess a rod-shaped bone (the epipterygoid) extending from the parietal to the pterygoid on each side in close contact with the membranous or cartilaginous wall of the skull.

KNERIIDAE. Small loach-like fish from the fresh water of tropical Africa.

KÖLLIKER'S PIT. Sometimes called the 'olfactory pit': a small depression on the left side of the oral hood in *Amphioxus*. It represents the embryonic neuropore, the spot where the cavity of the neural tube opened to the exterior.

KRAUSE'S CORPUSCLES. Sense organs of touch in the mammalian skin.

KRAUSE'S MEMBRANE. See *Z-disc*.

KURTIFORMES (KURTIDAE). Fish of the Indian and Pacific Oceans having a compressed body, a short snout, a single dorsal fin and a very long spiny anal fin.

L

LABELLA. Oral lobes or pads sometimes present at the distal end of the proboscis of an insect. They may be purely sensory, as in the mosquito, but in the blowfly and similar insects they are greatly enlarged and partly covered by a membrane containing a series of channels or *pseudotracheae* up which liquid food is sucked.

LABI-. Prefix and root-word from Latin *Labium*: a lip.

LABIAL CARTILAGES. Small cartilaginous skeletal elements at the corners of the mouth in some fish.

LABIAL GLANDS. Salivary glands arranged in a row in the upper and lower jaws of some reptiles. In poisonous snakes those of the upper jaw are modified to form the poison-glands which usually have ducts leading to the bases of the fangs; the latter are grooved or perforated making channels down which the poison flows.

LABIAL HOOKS. Hooks or spines on the labial palps of the nymphs of dragonflies (Odonata). In these insects the whole labium forms an enlarged prehensile organ known as the *mask* which is capable of being shot out rapidly to catch tadpoles or other creatures.

LABIAL PALPS. (1) Palps or feelers borne on the labium of an insect.
(2) A pair of ciliated palp-like organs near the mouth of a lamellibranch mollusc.

LABIA MAJORA. Two large folds of skin covering the entrance to the vulva in a female mammal; they are said to be homologous with the two halves of the scrotum of the male.

LABIA MINORA. Two small internal folds of skin behind the labia majora of some mammals.

LABIDOPHOROUS. Having pincer-like organs.

LABIUM. The lower lip of an insect formed by the union of a pair of appendages homologous with the second maxillae of a crustacean. It consists of the fixed *mentum* and the movable *prementum* to which are attached laterally the two labial palps and distally the *ligula*. The latter, in a generalized insect such as the cockroach, consists of a pair of inner lobes or *glossae* and outer lobes or *paraglossae*. In some specialized mouth-parts such as those of the bee, the ligula may be lengthened to form a tubular proboscis.

LABRIDAE. Wrasses: marine fish, often brilliantly coloured, having the anterior teeth strong and canine-like; having the body covered with cycloid scales and having the pelvic fins well forward in the thoracic region.

LABRUM. The upper lip of an insect or crustacean.

LABYRINTH. The parts of the inner ear of a vertebrate including the three semicircular canals, the utriculus and sacculus, the endolymphatic sac and the spiral cochlea.

LABYRINTHODONTIA. Large extinct amphibia of the Carboniferous, Permian and Triassic periods having an armour-plated skin and having complex teeth in which the enamel and dentine were much folded.

LABYRINTHODONT TEETH. Teeth in which the enamel and dentine are thrown into complex folds.

LACERTIDAE. Lizards of the Old World without osteoderms on the body; having a bifid tongue, well developed pentadactyl limbs and a long brittle tail.

LACERTILIA. Lizards: reptiles with horny epidermal scales and usually with four limbs but these may sometimes be reduced or absent giving a snake-like appearance. The skull is *streptostylic*, a condition in which the quadrate bones and the upper jaws are movable relatively to the brain-case. Breathing is by lungs and is effected by movements of the ribs, but there is no diaphragm. Other features include a cloaca, a transverse anal opening and paired copulatory organs.

LACERTILIA VERA. All lizards with the exception of chamaeleons. The latter differ from normal lizards in having a long tongue, in various modifications of the skull and in the absence of clavicles.

LACHRYMAL BONE. A membrane-bone in the anterior part of the orbit in most vertebrates. In land animals the duct from the lachrymal gland runs through a foramen in or adjacent to this bone.

LACHRYMAL DUCT. See *Lachrymal gland*.

LACHRYMAL GLAND. The tear-gland beneath the upper eyelid of many vertebrates. The fluid from this gland continually washes the front of the eye and drains through the lachrymal duct into the nose.

LACINIA. The inner spiny lobe of the maxilla of an insect, sheathed by the outer hood-like lobe or *galea*.

LACINIA MOBILIS. A movable, toothed structure on the mandibles of certain crustaceans, similar to, but not homologous with the lacinia of an insect, since the latter occurs not on the mandibles but on the maxillae. The lacinia mobilis is a characteristic feature of the Peracarida, a group which includes the sand-slaters, wood-lice and numerous fresh water and salt water shrimps.

LACTEALS. The central lymphatic cavities in the villi of the small intestine, into which digested fats are absorbed. The fluid in them is a milky emulsion containing fat globules suspended in lymph. From the lacteals fats are carried to all parts of the body.

LACUNA. Any space between cells or a space containing cells within a matrix such as bone.

LACUNAE OF TROPHOBLAST. Spaces or cavities in the trophoblastic part of a mammalian placenta which become filled with maternal blood diffusing through from the wall of the uterus.

LACUNAR SYSTEM OF ECHINODERMS. This corresponds to the blood vascular system of higher animals but in echinoderms, such as the star-fish, it consists of strands of *lacunar tissue* containing small spaces and channels through which the watery blood flows. The strands lie in cavities forming part of the *perihaemal coelom*.

LAGENA. An outgrowth from the sacculus of the ear in fish and amphibia: a forerunner of the cochlea of mammals but smaller and not coiled.

LAGO-. Prefix from Greek *Lagos*: a hare.

LAGOMORPHA. Rabbits and hares: formerly classed with the rodents but now regarded as a separate order (see *Duplicidentata*).

LAGOMYIDAE. Tailless short-eared hares of mountainous parts of North America, Northern Asia and S.E. Europe.

LAGOPODOUS. Having hairy or feathery feet.

LAMBDOIDAL CREST. A crest of bone between the occipital and parietal regions in the skulls of some vertebrates.

LAMELLI-. Prefix from Latin *Lamella*, diminutive of *Lamina*: a thin plate.

LAMELLIBRANCHIATA (LAMELLIBRANCHIA). Bivalves: molluscs such as mussels, oysters etc., with a symmetrical body and with a shell consisting of two halves or valves. These are joined dorsally by a ligament forming a hinge; ventrally they can be opened to allow the emergence of a large retractile foot. Respiration is by gills in the form of two pairs of perforated lamellae hanging down into the mantle cavity. Through these water is kept circulating by way of an inhalent and an exhalant opening.

LAMELLICORNIA. Beetles such as the stag-beetle and the chafers which have the distal end of each antenna expanded into a flattened comb-like club or lamella.

LAMINA. Latin. A thin plate.

LAMINA CRIBRIFORMA (LAMINA CRIBROSA). See *Cribriform plate*.

LAMINA PAPYRACEA. The lateral part of the ethmoid bone forming the inner wall of the orbit in Primates.

LAMINA PERPENDICULARIS. The median plate of the ethmoid bone continuous with the cartilaginous nasal septum in a mammal.

LAMINIPLANTAR. A word used to describe the metatarsus of a bird which, in some cases such as the thrush, is scaly in front and smooth behind.

LAMNIDAE. Mackerel-sharks: large pelagic sharks with a ventral crescent-shaped mouth and with large lanceolate teeth. They include several man-eating species such as the Blue shark and the Basking shark. Some species are viviparous.

LAMPRIDIDAE. Bony fish with a terminal toothless mouth and with minute scales over the body; the *Opah* or King-fish (4 ft. long) of the North Atlantic is an example.

LANGERHANS. See *Islets of Langerhans*.

LANGUETS. Small tongue-like processes such as those hanging down into the pharynx of a Tunicate.

LANIIDAE. Shrikes or Butcher-birds: large powerful birds with hooked serrated beaks and long claws. They prey on small mammals and birds, impaling them on thorns to be devoured at leisure.

LANUGO. Very fine hair covering a human foetus and disappearing before birth.

LARI. Gulls, skuas and auks: fish-eating birds which are typically good flyers and swimmers with anterior toes webbed. The young are covered with down when hatched but remain for a time helpless in the nest.

LARVA. A free-living embryo usually distinctly different from the adult and incapable of sexual reproduction. In many animals there is a rapid change or metamorphosis from the larval to the adult form, as for instance the tadpole changing into the frog or the caterpillar into the butterfly. Sometimes the same organism passes through a succession of different larval forms.

For types of larva, see under separate headings: *Actinula, Cypris, Dipleurula, Glaucothoe, Glochidium, Insect-larvae, Metazoea, Müller's larva, Mysis, Nauplius, Planula, Pluteus, Protaspis, Rhabditoid, Tornaria, Trochosphere, Veliger, Zoea.*

LARVACEA. Tunicates in which the mature form resembles the larva; this is now generally regarded as a form of *Neoteny* (*q.v.*). In this group the *test* or outer skin, instead of remaining attached to the animal as in other tunicates, can be cast off and a new one formed.

LARYNX. The voice-box of a vertebrate: a widening of the trachea, stiffened and protected by cartilages and usually containing the vocal cords.

LASSO CELL. See *Colloblast*.

LATA TYPE. A mutant with more chromosomes than its parents.

LATER-, LATERO-. Prefix from Latin *Latus, lateris*: a side.

LATERAL LINE SYSTEM (LATERALIS SYSTEM, ACOUSTICO-LATER-ALIS SYSTEM).
A collection of sense-organs known as *neuromasts* by which fish and aquatic amphibia can detect pressure changes, vibrations and low pitched sounds in the water. They consist of clusters of sensory cells in small pits situated along a line down each side of the body and also forming a complicated pattern on the head. The receptor cells in the membranous labyrinth of the ear are similar in origin and function and are usually regarded as part of the same system.

LATERAL PLATE MESODERM. Unsegmented mesoderm in the ventro-lateral region of a vertebrate embryo.

LATERAL VESICLES (LATERAL VENTRICLES). The cavities in the two cerebral hemispheres of the vertebrate brain: the first and second ventricles.

LATEROSPHENOIDS. Paired bones beneath the frontal bones and in front of the pro-otic bones, forming part of the sides of the cranium in a bony fish.

LAURER'S CANAL (LAURER-STIEDA CANAL). A short tube leading from the oviduct to the exterior in endoparasitic flukes such as the liver fluke and from the oviduct to the alimentary canal in ectoparasitic flukes such as *Polystomum*. Its function is uncertain but it has been suggested that fertilization may take place through here in the liver fluke.

LECITHIN. A phospho-lipoid substance in the yolk of most eggs.

LEMNISCI. Glandular organs on the eversible pharynx of the *Acanthocephala* (spiny-headed worms).

LEMNOPHAGI. Mud-eating pike-like fish of tropical America.

LEMUROIDEA (PROSIMIAE). Lemurs: arboreal and usually nocturnal mammals found chiefly in Madagascar and Africa. They are more primitive than true Primates but resemble them in many respects and are usually included in the same group. The large toe and the thumb are both opposable enabling them to grasp tree branches but the tail is not prehensile. The eyes face the front giving them stereoscopic vision and the orbit is almost closed behind by a bar formed from the frontal and jugal bones.

LENTICULAR GANGLION. Also called the *Ophthalmic ganglion* or the *Oculomotor ganglion*: a small ganglion in the orbit of some vertebrates where the ophthalmic branch of the trigeminal nerve forms an anastomosis with the oculomotor nerve.

LEPIDO-. Prefix from Greek *Lepis, lepidos*: a scale.

LEPIDOPIDAE. Small mackerel-like fish with many vertebrae, with a spinous dorsal fin and with the pelvic fins well forward.

LEPIDOPTERA. Butterflies and moths: a large order of insects with over 140,000 species characterized by having pigmented scales all over the wings and body. The mouth-parts are in the form of a long coiled tubular proboscis through which nectar can be sucked. The larvae or caterpillars are usually herbivorous and are of the *polypod* type, *i.e.* they have pro-legs or cushion feet on the abdomen in addition to the three pairs of true legs on the thorax. Pupae are of the *obtect* type and are sometimes contained in silky cocoons.

In systematic classification the old division into butterflies and moths, based on differences in the antennae, has been superseded by a system based on the venation of the wings (see *Homoneura* and *Heteroneura*).

LEPIDOSAURIA. An alternative name for the *Squamata*, a group of reptiles comprising all lizards and snakes as well as the extinct *Dolichosauria* and *Mosasauria*.

LEPIDOSIREN. Lung-fish of South America, differing from the Australian lung-fish in having two lungs.

LEPIDOSTEIDAE (LEPIDOSTEI). Gar-pikes: primitive fish covered with thickened rhomboid scales arranged in oblique rows and articulated together; having a slightly asymmetric tail (hemi-heterocercal) but having paired fins of the modern or *actinopterygian* type (*i.e.* of dermal fin-rays only). They are voracious fish inhabiting fresh water of North and Central America.

LEPIDOSTEOID SCALES. The type of ganoid scales found in the gar-pike *Lepidosteus* consisting of a thick layer of the hard shiny substance *ganoin* covering a layer of *isopedin* or laminated bone. (See also under *Cosmoid scales*, *Ganoid scales* and *Palaeoniscoid scales*.)

LEPIDOTRICHIA. Bony, jointed fin-rays such as are found in most present-day fish other than cartilaginous fish and lung-fish.

LEPISMIDAE. A family of insects belonging to the *Thysanura* or bristle-tails, apparently a link between the lower Apterygota and the Orthoptera.

LEPORIDAE. Rabbits and hares: duplicident rodents with long ears, short tails and long hind-legs (see also *Lagomorpha*).

LEPTO-. Prefix from Greek *Leptos*: fine, small, thin.

LEPTOCEPHALI. Transparent pelagic larvae of eels: given as a generic name before their true nature was realized.

LEPTOLEPIDAE. Extinct salmon-like fish of the Cretaceous period having cycloid scales.

LEPTOMEDUSAE. See *Calyptoblastea*.

LEPTOSTRACA. Small marine crustaceans having a large carapace which is not fused with the sides of the thorax but is held by adductor muscles. The appendages are simple, flattened and biramous; the eyes are on stalks.

LEPTOTENE STAGE. An early stage in meiosis in which each chromosome has a granular appearance due to the formation of chromomeres or particles arranged like a string of beads.

LEUCO-. Prefix from Greek *Leukos*: white.

LEUCISCINA. Roach, bream and similar fresh-water fish having a short un-spined dorsal fin, a pectoral girdle attached to the skull, the pelvic girdle well back, and having a chain of Weberian Ossicles linking the air bladder with the ear.

LEUCOCYTE. A white blood corpuscle, *i.e.* a blood cell containing no res-piratory pigment. There are a great number of types which are usually classified as granular and non-granular and as basophil or acidophil according to their staining reaction. In human blood the chief kinds are *polymorphs*, *lymphocytes* and *monocytes* (see also under separate headings).

LEUCON GRADE. The most complex of the three types of sponge. In these the water enters by narrow inhalent canals leading into flagellated chambers. From these it passes through exhalent canals into the *paragaster* or central cavity and finally out of the sponge by means of *oscula* at the ends of numerous branches.

LEYDIG'S DUCT. An alternative name for the *Wolffian duct* or *Mesonephric duct (q.v.)*.

LIEBERKÜHN'S GLANDS. See *Crypts of Lieberkühn*.

LIENO-GASTRIC ARTERY AND VEIN. Blood vessels supplying the lower part of the stomach and the spleen in some vertebrates.

LIGAMENT. A strip of elastic tissue holding two bones together at a joint.

LIGHT-COMPASS REACTION. Ants and other insects are able to move in a straight line by keeping the sun in a definite direction in relation to their body. If they are kept in the dark for a few hours and then released, they will start off in a direction which makes an angle with the original route equal to that through which the sun has moved in the meantime.

LIGULA. The distal part of the labium of an insect consisting typically of a pair of inner lobes or *glossae* and a pair of outer lobes or *paraglossae*. In some insects the latter are reduced or absent and in bees the glossae are greatly lengthened to form a tubular proboscis.

LIMACINE (LIMACIFORM). Resembling a slug.

LIMICOLAE. Plovers, curlews, redshanks and similar birds: typical waders but not as a rule habitual swimmers.

LIMNO-. Prefix from Greek *Limne*: a marsh.

LIMNETIC (LIMNOPHILOUS, LIMNOBIOTIC). Inhabiting marshes.

LIMNO-PLANKTON. The plankton of marshes and swamps.

LINGUAL ARTERY AND VEIN. Vessels supplying blood to the tongue; in tetrapoda the artery is a branch of the carotid and the vein returns to the external jugular.

LININ. A viscoid substance forming an apparent mesh throughout the nucleus of a cell.

LINKAGE. The carrying of several different genes on one chromosome so that the qualities associated with these genes are generally inherited together as if they formed a single unit. Occasionally such linkages break down as a result of the phenomenon of *chiasma* or *crossing over (q.v.)*.

LIP-, LIPO-. Prefix and root-word from Greek *Lipos*: fat.

LIPASE. Any enzyme which helps to break down fats into fatty acids and glycerine; *steapsin* in mammalian pancreatic juice is an example.

LIPOGENOUS. Fat-producing.

LIPOGENYIDAE. Deep-sea eel-like fish.

LIPOIDS (LIPIDS). Substances with fat-like properties, including true fats (made from fatty acids and glycerol) as well as steroids and other complex substances which may contain phosphorus, nitrogen and other elements.

LIPOLYTIC. Able to decompose fats.

LIPOSTRACA. A suborder of the *Anostraca* or 'fairy-shrimps' including certain minute blind forms found as fossils from the Middle Devonian period.

LIPOTERNA. Extinct American ungulates whose affinities are doubtful as they have certain features resembling those of horses and others resembling those of camels.

LITHITE. An otolith or statolith: a granule of calcium carbonate or other substance inside the cavity of an organ of balance, as for instance in the semi-circular canals of the vertebrate ear.

LITORAL FAUNA AND FLORA. Animals and plants inhabiting the sea-shore and the shallow sea near the shore.

LIVER. A digestive gland whose duct opens into the intestine; in a vertebrate it is the largest gland of the body, producing bile, making and storing glycogen and helping to break down proteins and fats.

LOBI INFERIORES. Lateral branches and thickenings of the pituitary in-fundibulum generally present in fish.

LOBOPODIA. The blunt type of pseudopodia such as those possessed by *Amoeba*.

LONGICORN BEETLES. See *Cerambycidae*.

LOOPERS. See *Geometridae*.

LOPH-, LOPHO-. Prefix from Greek *Lophos*: a ridge or crest.

LOPHIIFORMES, LOPHIIDAE (PEDICULATI). Angler fish: remarkable fish which spend their lives lurking in the mud of tropical shores, with mouth wide open ready to snap up any unwary prey. The head and pectoral fins are very large and the dorsal fin has a long extension hanging down near the mouth like a fishing line. The tip is like a wriggling worm and is used as a bait; if bitten off it can rapidly be regenerated.

LOPHOBRANCHIATE. Having gills composed of small rounded lobes attached to the branchial arches as in sea-horses etc. (see below).

LOPHOBRANCHII. Pipe-fish, sea-horses and similar fish having the skin covered with bony plates and having a prolonged snout bearing a small terminal toothless mouth.

LOPHODONT. Molar and premolar teeth with transverse ridges of enamel running across the crown: the typical grinding teeth of ruminants and many other herbivores.

LOPHOPHORE. A circle or horse-shoe of ciliated tentacles surrounding the mouth in Polyzoa and also in Brachiopoda.

LOPHOTIFORMES. Another name for the 'Ribbon-fish' or *Taeniosomi* (*q.v.*).

LORE. The surface of the head between the eyes and the beak of a bird.

LORENZINI'S AMPULLAE. See *Ampullae of Lorenzini*.

LORICATA. Another name for the armadilloes (see *Dasypodidae* and *Xenarthra*).

LORISINAE. Slow lemurs: sluggish nocturnal animals of Africa and Asia having a rounded head, large eyes and short ears.

LORUM. A V-shaped cuticular plate supporting the elongated labium of a bee or similar insect.

LOXODONT. Having molar teeth with alternate grooves and ridges.

LUCANIDAE. Stag-beetles: large wood-boring beetles belonging to the *Lamellicornia* all of which have the distal parts of the antennae enlarged and flattened. The common stag-beetle *Lucanus cervus*, in which the male is distinguished by its very large antler-like mandibles, is one of the largest British insects.

LUCI-. Prefix and root word from Latin *Lux, lucis*: light.

LUCIAE (PYROSOMATIDAE). Phosphorescent Tunicates which by repeated budding give rise to cylindrical colonies.

LUCIFERASE. An oxidizing enzyme which, acting on the protein *luciferin*, causes luminosity in glow-worms and similar insects.

LUCIFUGOUS. Shunning light.

LUMBAR VERTEBRAE. Vertebrae of the abdominal region to which are attached the muscles of the back.

LUMBRICIDAE. Earthworms: annelids of the order *Oligochaeta*, having a large number of segments each bearing eight chaetae or bristles and without any other appendages. They are hermaphrodite and reproduction is by mutual cross-fertilization.

    The Lumbricidae include most of the common earthworms but not the giant worms of Australia which belong to the *Megascolecidae* (*q.v.*).

LUMINESCENCE (BIOLUMINESCENCE). The production of a cold light by certain insects, fish and other organisms. Little is known of the process but it is believed to involve the oxidation of a substance *luciferin* with the help of a specific enzyme known as *luciferase*.

LUNAR BONE. The *Intermedium* or central bone of the proximal row of carpals in some vertebrates.

LUNG. The principal breathing organ of most land animals. There are many types of lung which are not necessarily homologous in different animals. The following are the chief kinds:

    (1) In Tetrapod vertebrates: a pair of sacs either simple or composed of a large number of alveoli as in birds and mammals. Embryologically they are formed from a branch of the gut.

    (2) In lung-fish: an air-sac acting as an accessory breathing organ. A vascular swim-bladder connected to the back of the pharynx.

    (3) In slugs and snails: a vascular part of the mantle enclosing an air cavity which opens to the exterior by a small spiracle.

LUNG-BOOK. A type of respiratory organ found in scorpions and spiders consisting of a large number of parallel leaflets between which air can circulate; the whole organ is sunk in a pit with a narrow opening or *pneumostome*. Lung-books are believed to have evolved from the gill-books which characterize aquatic arachnids such as the king-crab (*Limulus*).

LUTEIN. A yellow pigment formed in the *Corpus luteum* (*q.v.*).

LUTRINAE. Otters: aquatic carnivorous mammals closely related to badgers, weasels, stoats etc. but having webbed feet. They are usually fresh-water animals and feed on fish but there is also a sea-otter of the North Pacific which feeds on molluscs.

LYCTIDAE. 'Powder-post beetles': small wood-boring beetles which reduce oak and other woods to a fine powder. They differ from the common furniture beetle in having the body flattened dorso-ventrally and in not having the head covered by the prothorax.

LYMANTRIDAE. A family of moths including the Vapourer Moth whose larvae do considerable damage by defoliating forest trees both deciduous and coniferous.

LYMPH. Colourless fluid consisting largely of blood plasma which has filtered through the walls of the capillaries and fills the intercellular spaces of an animal. It serves as an intermediary for conveying food substances and oxygen between the blood and the tissues.

LYMPHATIC VESSELS. Thin walled vessels which convey lymph (plasma and white corpuscles) to various parts of the body and eventually back into the veins.

LYMPH GLANDS (LYMPH NODES). Organs in the vertebrate body which produce and store lymphocytes and destroy bacteria in the lymphatic fluid. The spleen, the tonsils and Peyer's Patches of the intestine are examples of lymph glands.

LYMPH HEARTS. Enlargements of the lymphatic vessels able to pump lymph along by their muscular contractions.

LYMPHOCYTE. A type of white blood corpuscle made in the lymphatic tissues of vertebrates. They have a large spherical nucleus, are amoeboid but apparently not phagocytic. Various functions have been attributed to them such as the production of enzymes, antitoxins etc.

LYTTA (LYSSA). A worm-like mass of muscle, fat and cartilage beneath the tongue in carnivorous mammals.

M

MACANER. A large male ant.

MACERGATE. A large worker ant.

MACHILIDAE. A primitive group of Thysanura or bristle-tailed insects having a pair of long abdominal cerci, a median tail filament and stylets on nearly all the abdominal segments. They show a convergent resemblance to certain higher crustacea (*cf. Lepismidae*).

MACR-, MACRO-. Prefix from Greek *Makros*: large, long.

MACRAUCHENIDAE. Extinct ungulates from the Eocene of Patagonia closely resembling the horse but having three toes on each foot; they also showed certain resemblances to both the camel and the rhinoceros.

MACROCYTE. Any very large cell: particularly applied to *monocytes*, a large type of white blood corpuscle.

MACROCEPHALIC. Large headed. A human head with cranial capacity above 1,450 c.c.

MACROGAMETE (MEGAGAMETE). The larger of two gametes: the female gamete.

MACROLEPIDOPTERA. The larger butterflies and moths: a name formerly used in the classification of Lepidoptera but now obsolete although occasionally used for convenience.

MACROMERE. See *Megamere*.

MACROPODIDAE. Kangaroos, wallabies and kangaroo-rats: marsupials of the *diprotodont* group which have few incisors, have the canines small or absent and have two toes of the hind feet joined together or *syndactylous*. They are mostly jumping animals with long hind legs and a large tail for balancing.

MACROPTEROUS. Large-winged: a term used with particular reference to the various castes of insects such as ants and termites (*cf. Micropterous*).

MACRORHAMPHOSIDAE. Snipe-fish and Bellow-fish having the bones of the skull much prolonged to form a tubular snout with short jaws at the end.

MACRORHYNCHIDAE. Extinct fresh-water crocodiles of the early Cretaceous Period having a long slender snout and a dermal armour of dorsal and ventral plates.

MACROSCELIDIDAE. Small nocturnal jumping shrews.

MACROTRICHIA. Large hairs or *setae* of insects etc. consisting of extended epidermal cells surrounded by cuticle. They may be of various types including the following:

(1) Simple or branching hairs covering most of the body.

(2) Sensory setae with nervous connections which respond to touch, taste, smell etc.

(3) Setae which exude irritant secretions, as on certain caterpillars.

(4) Pigmented scales such as those on butterflies' wings and on the bodies of mosquitoes.

MACRURIDAE. Deep-sea fish which have long dorsal and anal fins extending back to behind the long filamentous tail. The body is covered with spiny scales.

MACRUROUS (MACRURA). Large-tailed: a term used to denote crustaceans such as the lobster in which the body is long, ending in a tail-fan, and which do not have the abdomen doubled up underneath as is the case with the brachyurous crabs.

MACULA. Latin: a spot.

MACULA LUTEA. A yellow spot in the centre of the retina in Primates, birds and a few other vertebrates; it is the point of clearest colour vision containing a high concentration of sensory cones.

MACULAE ACUSTICAE. Sensory receptors in the utriculus and sacculus of the vertebrate ear.

MADREPORITE. A porous opening or filter through which water enters the hydrocoel of an echinoderm. In the common star-fish it is a small perforated plate on the upper surface slightly to one side of the centre. Beneath it is the *stone canal*, a calcareous tube leading to the main circular water-vessel.

MAENIDAE. Carnivorous spiny-rayed shore fish related to the perch.

MAGNUM. One of the distal carpal bones in certain mammals.

MALAC-, MALACO-. Prefix from Greek *Malakos*: soft.

MALACOCOTYLEA. Liver flukes: Trematoda which are usually endoparasites of vertebrates in the adult stage and of molluscs in the larval stages.

MALACOPTERYGII (SALMONI-CLUPEIFORMES). A large group of bony fish including herrings, salmon, trout etc. having soft fin-rays and usually thin cycloid scales. The pectoral arch is suspended from the skull and there are four pairs of gills. The air-bladder if present has a pneumatic duct.

MALACOSTRACA. A very large group of crustaceans including crabs, lobsters, shrimps etc. Normally they have twenty segments with various types of appendages on all except the first. Six segments form the head, eight the thorax and six the abdomen. A carapace normally covers the thorax (see *Caridoid Facies*).

MALAR BONE. An alternative name for the cheek bone or zygomatic arch of a mammal; more particularly used of humans.

MALAXATION. Softening by chewing with the mandibles, as for instance by a wasp.

MALE. Sexually the opposite of female and able to fertilize it prior to reproduction.

Male gamete. A spermatozoon capable of fertilizing an ovum or egg-cell.

Male gonad. The testis or male reproductive organ which produces spermatozoa.

MALLEOLAR BONE. The reduced fibula of certain ungulates.

MALLEUS. The hammer: the first of the three small bones in the middle ear of a mammal, the others being the *incus* or anvil and the *stapes* or stirrup. These three, usually known as auditory ossicles, convey sound vibrations from the tympanic membrane to the inner ear. Anatomically the malleus is homologous with the *articular* bone of the jaw in reptiles.

MALLOPHAGA. Biting lice: ectoparasites of birds and mammals sometimes regarded as a distinct order of insects and sometimes included as a suborder of the Anoplura. Unlike the sucking lice they do not pierce the skin but use their mouth-parts to bite off small particles of feathers, hair etc.

MALPIGHIAN BODY (MALPIGHIAN CAPSULE). One of thousands of small filtering organs which make up the kidney of a vertebrate. It consists of a *glomerulus* or coiled capillary tube closely adjacent to a funnel-shaped sac known as *Bowman's Capsule*. Waste liquid passes out through the wall of the glomerulus into the capsule and thence along a uriniferous tubule to be excreted. Usually a large amount of water is re-absorbed into the blood as the urine passes down the tubule thus enabling the animal to excrete soluble waste substances with the minimum possible loss of water.

MALPIGHIAN LAYER. A layer of actively growing cells at the base of the epidermis of a vertebrate.

MALPIGHIAN TUBULES. The chief excretory organs of insects, arachnida and myriopoda: narrow glandular tubes opening into the alimentary canal near the commencement of the hind-gut.

MAMMALIA. Mammals: warm-blooded hairy quadrupeds having lungs, a four-chambered heart, a left aortic arch, non-nucleated red blood corpuscles, mammary glands for suckling the young and many other features which distinguish them from the reptiles which are believed to have been their ancestors. The lower jaw consists of a single bone on each side, the *dentary* which articulates with the squamosal bone; this contrasts with the jaws of other vertebrates which articulate with the quadrate bone. The teeth of mammals are heterodont and are in sockets. The majority of present-day mammals have a placenta but the two primitive groups, Monotremes and Marsupials have not.

MAMMARY GLANDS. The milk glands of mammals: probably an enlarged and modified form of sweat glands under the control of the sex hormones.

MANDIBLE. (1) *Of vertebrates:* the lower jaw.

(2) *Of arthropods:* one of a pair of laterally moving horny jaws anterior to the maxillae and usually stronger than them.

(3) *Of other invertebrates:* any strong horny jaws.

MANDIBULAR ARCH. The first visceral arch enlarged to form the skeleton of the lower jaw in all *Gnathostomes, i.e.* all vertebrates from fish upwards but not including the lampreys or *Cyclostomes* which have no true jaws.

MANDIBULAR PALPS. Palps or feelers borne on the mandibles of some crustacea and other arthropods.

MANIDAE. Pangolins: scaly ant-eaters of Africa and Asia. Terrestrial or burrowing animals 2 to 5 ft. long and able to roll themselves into a ball. They were formerly classed with the Edentates but are now put in a separate order, the *Pholidota*.

MANTIDAE. Praying mantids: predatory orthopterous insects named on account of their habit of extending the fore-legs forwards in the attitude of prayer when waiting to catch insects upon which they feed. These fore-legs are armed with spines upon which insects can be impaled.

MANTISPIDAE. Predatory insects having a superficial resemblance to Mantids but belonging to the order Neuroptera. The larvae are parasitic on the egg cocoons of certain spiders.

MANTLE. That part of the epidermis of a mollusc which usually covers the dorsal and lateral surfaces and the visceral hump and which typically secretes the shell.

MANUBRIUM. (1) The anterior segment of the sternum of a mammal.
(2) The elongated tubular mouth which hangs down from the centre of the under surface of a medusa (jelly-fish).

MANUS. The hand or fore-foot of a tetrapod vertebrate.

MANYPLIES. The *psalterium* or *omasum*: the third part of the stomach of a ruminant.

MARGINAL ANCHORS. Short adhesive tentacles by which certain primitive types of jelly-fish of the group *Stauromedusae* attach themselves to sea weeds etc.

MARGINAL VESICLES. A name sometimes given to the marginal statocysts of a medusa or jelly-fish.

MARICOLA. Small marine Turbellarians or free-swimming ciliated flatworms belonging to the order *Tricladida* which are characterized by having the gut divided into three main branches.

MARROW (BONE MARROW). Yellow or red pulp in the central cavities of the larger bones. In this tissue the red blood corpuscles and some types of white ones originate and develop.

MARSIPOBRANCHII. An alternative name for the *Cyclostomata*.

MARSUPIALS (MARSUPIALIA). Mammals usually having a pouch or *marsupium* in which they carry the young; they are found all over Australasia and a few types are also found in America. Their anatomy is primitive and resembles in many ways that of reptiles. The young are without a placenta and are born in an extremely small and undeveloped state. Owing to lack of competition with higher animals in Australia they show a considerable degree of divergent evolution, some being herbivorous and some carnivorous; some jumping, others climbing or burrowing.

MARSUPIUM. See *Marsupials* above.

MASK. The enlarged prehensile labium of a dragonfly nymph (see also *Labial hooks*).

MASTACEMBELIDAE. Spiny-rayed eels from Lake Tanganyika and other fresh waters of Africa and India.

MASTAX. The muscular pharynx of a Rotifer containing movable chitinous teeth.

MAST CELLS. Large basophil amoeboid cells found in vertebrate connective tissue and possibly connected with the deposition of fat.

MASTIGO-. Prefix from Greek *Mastix*, *mastigos*: a whip.

MASTIGOBRANCHIAE. Small gill-like processes below the true gills in crabs etc. Their function is to clean the gills.

MASTIGOPHORA (FLAGELLATA). Protozoa, sometimes parasitic, which possess one or more flagella, have no meganucleus and usually reproduce by equal longitudinal fission. The class is divided into *Zoomastigina* which feed holozoically and *Phytomastigina* most of which possess chloroplasts and are holophytic. Some of the latter such as *Chlamydomonas* and *Volvox* have been classed by botanists among the lower algae.

MASTODON. A large extinct elephant-like mammal of the Miocene and Pliocene periods, having a short trunk and long tusks.

MASTOID. That part of the periotic bone which, in some mammals, appears on the surface of the skull between the exoccipital and squamosal bones. In humans it forms a projecting portion behind the ear.

MATURATION (MATURATION DIVISION). The development of mature sperm cells or egg cells involving the process of *meiosis* or reduction of chromosomes.

MAXILLA. (1) One of a pair of large bones of the upper jaw of a vertebrate, between the premaxillae and the palatines. In mammals they are the chief tooth-bearing bones with sockets for the molars and premolars.

(2) The term is sometimes loosely used for the whole upper jaw of a vertebrate.

(3) One of a pair of laterally moving mouth-parts behind the mandibles on most arthropods. In crustacea there may be a first and a second pair of maxillae, the former sometimes being called *maxillules*. In insects the second pair are fused together to form the *labium*.

MAXILLIPEDE. A limb modified for feeding as in crabs, lobsters etc. which have three pairs of maxillipedes immediately behind the maxillae and in front of the legs.

MAXILLULE. The first or smaller maxilla of a crustacean.

MECHANO-RECEPTORS. Sensory receptors which perceive sound, pressure, tension, movement etc., as for instance on the bodies of insects and in the acoustico-lateralis system of vertebrates.

MECKEL'S CARTILAGES. A pair of large cartilages forming the skeleton of the lower jaw in cartilaginous fish such as sharks, skates and dogfish. In bony fish, reptiles and birds, Meckel's cartilage is ossified to form the *articular bone*; in mammals it is reduced to form the *malleus* of the ear.

MECOPTERA. Scorpion-flies: small carnivorous insects recognizable by the large biting mouth-parts and by the upturned abdomen of the male giving it the appearance of a scorpion.

MEDIAN FISSURE OF BRAIN. A deep fissure separating the right and left cerebral hemispheres of the mammalian brain; in lower vertebrates it may only be a shallow furrow.

MEDIASTINUM. A median cavity between the two pleural sacs of a mammal; it contains the heart and other centrally placed organs and blood vessels of the chest.

MEDULLA. The central part of an organ, as for instance of the kidney or adrenal gland.

MEDULLA OBLONGATA. The most posterior part of the brain of a vertebrate containing a cavity, the fourth ventricle, roofed over by the *posterior choroid plexus*. Its chief functions are the co-ordination of sensory impulses and the regulation of the autonomic system.

MEDULLARY CANAL. (1) The marrow cavity of a long bone.
(2) The cerebro-spinal canal.

MEDULLARY PLATE. An alternative name for the *Neural plate* (*q.v.*) of a vertebrate embryo.

MEDULLATED NERVE FIBRES. Nerve fibres whose axis-cylinder is surrounded by a white tubular sheath of *myelin*, a lipo-protein, round which is a thinner membranous sheath known as the *neurolemma*. At intervals the myelin sheath is interrupted by constrictions known as *Nodes of Ranvier*. Most vertebrate nerve fibres are of the medullated type, the chief exceptions being certain of the sympathetic nerves.

MEDUSA. The free-swimming or sexual generation in coelenterates, having the form of a disc-shaped or bell-shaped jelly-fish with tentacles round the rim and with a mouth underneath in the centre. It is derived from the normal polyp or hydroid structure by a flattening process. In the group *Hydrozoa*, medusae alternate with polyp colonies from which they are budded off, but in the *Scyphozoa*, which include the common jelly-fish, the medusa is the dominant type and the polyp stage is reduced or absent.

MEGA-, MEGALO-. Prefix from Greek *Megas* (Fem. *Megale*): large.

MEGACHILIDAE. Leaf-cutting bees: insects which build their nests from neatly cut oval or round pieces of leaves or flower petals. The female bee holds the leaf with her legs while she cuts with her mandibles.

MEGACHIROPTERA. Large fruit-eating bats found in tropical and subtropical regions of the Old World.

MEGACHROMOSOMES. (1) Large chromosomes forming a ring round the normal or microchromosomes in the nuclei of certain protozoa such as those of the groups *Opalinidae* and *Chonotricha*. They are thought to correspond with the meganucleus of other ciliated protozoa.

(2) Abnormally large chromosomes, sometimes a hundred times the normal size, found in the salivary glands of many dipterous insects.

MEGADRILI. An order of Oligochaeta including most common earthworms and a few fresh-water species; they are distinguished from the *Microdrili* by having more segments and by the absence of eyes.

MEGAGAMETE. The larger of two gametes: the female gamete or ovum.

MEGAGAMETOCYTE. The mother-cell of a megagamete.

MEGALOPTERA. A suborder of Neuropterous insects having primitive wing-venation but otherwise being somewhat varied and including alder-flies, snake-flies and the giant American *Corydalis* with its six-inch wing-span.

MEGALOSPHERIC FORM. One of two forms of shell produced by some Foraminifera, the other being the *microspheric*. They represent an alternation of generations in the life-cycle, the microspheric form reproducing asexually; the megalospheric form sexually.

MEGAMERES. The larger cells or blastomeres formed when a fertilized egg undergoes unequal cleavage. These larger cells are in the neighbourhood of the *vegetative pole*; they are filled with yolk and give rise to the endoderm of the future embryo (see also *Micromeres*).

MEGANEPHRIDIUM. The normal type of nephridium or excretory organ of an earthworm consisting of a coiled tube with a nephrostome or ciliated funnel at one end and a nephridiopore leading to the exterior at the other. The name *meganephridium* is given to distinguish them from the *micronephridia* which are found in some of the larger worms. These are much smaller structures and there may be a very large number in each segment.

MEGANUCLEUS. The larger of two nuclei found in some protozoa, particularly the Ciliophora. It is made of *trophochromatin* as distinct from the *idiochromatin* of the micronucleus. Apparently it regulates the normal metabolic functions of the cell but disappears during or just before cell division (see also *Micronucelus*).

MEGAPODIIDAE. Large-footed fowl-like birds of Australia and S.E. Asia having short wings and short beaks. The eggs are laid in holes in the ground and develop without incubation; the young can fly almost immediately.

MEGASCOLECIDAE. Giant earthworms of Australia sometimes up to 11 ft. in length and having more than five hundred segments.

MEGASPHERIC FORM. See *Megalospheric form*.

MEGATHERIIDAE. Large extinct ground sloths of South America. The body was as large as an elephant but with shorter limbs; the skull and teeth were primitive like those of the small present-day sloths. Evidence from caves in Patagonia shows that the giant sloths were contemporary with primitive man.

MEIBOMIAN GLANDS. Modified sebaceous glands in the eyelids of mammals.

MEIOSIS (REDUCTION DIVISION). A process of cell division whereby the number of chromosomes is halved. The process usually consists of two successive divisions during the first of which the chromosomes arrange themselves in homologous pairs round the 'equator' of the cell. They then move apart to opposite poles and the cell divides in such a way that half the chromosomes go into each new cell. In the second division, which usually follows immediately, there is no further reduction of chromosomes; thus an original diploid cell produces four haploid cells. Division of this kind usually takes place in the formation of gametes prior to sexual reproduction.

MEISSNER'S PLEXUS. A network of nerves lying within the muscular coats of the intestines of a vertebrate.

MELANIN. A dark brown or black pigment in skin or hair.

MELEAGRINAE. Turkeys: a subfamily of the *Phasianidae* originating in North America. They are terrestrial birds which may roost in trees but make nests on the ground and usually lay numerous eggs. They are frequently polygamous, the males being larger and more brightly coloured than the females.

MELINAE. Badgers: mammals of the family *Mustelidae* related to weasels and otters. Although belonging to the order Carnivora, many of them are insectivorous or omnivorous. Locomotion is plantigrade and the long feet have large non-retractile claws which are of great use for burrowing.

MELIPHAGIDAE. Honey-eaters: small brightly coloured birds of Australia having a long curved beak and a long tail.

MELOIDAE. A family which includes Oil beetles and Blister beetles. The larvae of the former are parasitic in the nests of certain bees; the latter are of economic importance as the source of the drug *cantharidin*.

MEMBRACIDAE. Tree-hoppers: plant-bugs of the group Homoptera found mostly in the tropics and characterized by a large hood-like extension of the prothorax.

MEMBRANA BASILARIS (BASILAR MEMBRANE). A membrane separating the *scala media* from the *scala tympani* in the cochlea of the mammalian ear. When stimulated by sound, its vibrations are communicated to the sensory receptors in the *Organ of Corti* and hence to the auditory nerves.

MEMBRANA ELASTICA. See *Elastica*.

MEMBRANA SEMILUNARIS. A transverse semilunar fold of skin across the trachea at the junction of the bronchi, forming part of the syrinx or singing organ of a bird.

MEMBRANE BONES (DERMAL BONES). Bones which are formed in the dermis and not from an initial cartilaginous skeleton. In mammals they include the clavicles and some of the bones of the skull, but in bony fish and in reptiles there are many more such bones. Primitive fish of the Devonian and Silurian periods often had a thick 'armour plating' of dermal bones over the head and sometimes over the whole body; throughout the course of evolution there has been a tendency towards the reduction of their number.

MEMBRANELLA (MEMBRANULA). A membrane formed by the union of cilia in certain protozoa of the class Ciliophora. They are usually found in the neighbourhood of the mouth and by their undulations help to gather food particles.

MEMBRANE OF REISSNER. See *Reissner's Membrane*.

MEMBRANOUS LABYRINTH. The inner part of the vertebrate ear consisting chiefly of the three semicircular canals which form the organ of balance and the spiral cochlea which contains the organ of hearing.

MENDELISM. The discoveries and theory of Gregor Mendel relating to heredity. The main points of the theory are as follows:

(1) Hereditary factors (genes) are carried by the gametes. Any individual therefore receives factors from both parents.

(2) Such factors exist in opposite or contrasted pairs known as *allelomorphs*, *e.g.* tall or short, round or wrinkled (peas); long wing or vestigial wing (of fruit flies) etc.

(3) Only one of a pair of contrasted characters can be carried in a single gamete (Law of Segregation: Mendel's First Law).

(4) When two gametes combine to form a zygote, each one of a pair of contrasted characters may combine, by pure chance, with either of another pair (Law of Independent Assortment: Mendel's Second Law). If two like factors combine, the resulting progeny is said to be *Homozygous* for that factor and can only produce offspring with the same quality. If two opposite factors combine, the progeny is *Heterozygous* or Hybrid and when this produces gametes there will be, on an average, half of each type formed.

(5) One character may be *dominant* and the other *recessive*, the former showing itself in a hybrid and the latter remaining hidden. If a hybrid is crossed with a similar hybrid (or in the case of a plant self-pollinated), the next generation will, on the average, contain three quarters showing the dominant quality and one quarter showing the recessive quality.

MENINGES. The membranes covering the brain and spinal cord of a vertebrate. These are the tough outer *Dura Mater* and the inner more delicate vascular *Pia Mater*.

MENISCUS. An intervertebral disc: a plate of fibro-cartilage between two vertebrae.

MENSTRUAL CYCLE. See *Oestrous cycle*.

MENSTRUATION. Destruction of the mucous membrane of the uterus and consequent bleeding which occurs periodically in humans, anthropoid apes and a few other mammals.

MENTUM. A part of the base of the labium of an insect between the *submentum* and the *prementum*.

MENURIDAE. Lyre Birds: large brightly coloured birds inhabiting warm forests in many parts of the world. The males are distinguished by having two outer tail quills very much enlarged and curved in the form of a lyre.

MERO-. Prefix from Greek *Meros*: part, portion.

MEROBLASTIC CLEAVAGE. Partial cleavage such as occurs in eggs having a large quantity of yolk. In such eggs cleavage only takes place in the region of the animal pole giving rise to a disc of cells on the side of the yolk-sac while the latter remains undivided.

MEROCRINE. The normal method of secretion by digestive glands, sweat glands etc., in which the cells secrete their fluids repeatedly without undergoing any marked change in their structure or appearance (*cf. Holocrine*).

MEROGAMETES. A term used with particular reference to Protozoa to denote gametes formed by fission of the parent cell as distinct from *hologametes* which are formed from the whole cell.

MEROPIDAE. Bee-eaters: rapid flying birds resembling swallows but with variegated plumage and having a compressed beak which curves gently downwards. They are found in temperate and tropical parts of the Old World.

MEROPODITE. The second of the five segments of the leg in such crustaceans as the crayfish, crab or lobster. The segments named in order from the base are: *ischiopodite, meropodite, carpopodite, propodite* and *dactylopodite*.

MEROZOA. Tapeworms (Cestoda) in which a large number of segments or *proglottides* are budded off behind the head or *scolex*. They are thus distinguished from the *Monozoa* which are unsegmented. Each proglottis in a tapeworm is a complete biological unit containing hermaphrodite reproductive organs able to produce a large number of eggs. Mature segments become detached and are cast out with the faeces of the host.

MEROZOITES (SCHIZOZOITES). Small cells formed by multiple asexual fission of certain protozoa.

MES-, MESO-. Prefix from Greek *Mesos*: middle.

MESADENIA. Accessory glands arising from the mesodermal lining of the seminal vesicles, *e.g.* in some insects.

MESENCEPHALON. The mid-brain of a vertebrate comprising the optic lobes and the *crura cerebri*. Through its centre runs the *Iter* or *Aqueduct of Sylvius* connecting the third and fourth ventricles.

MESENCHYME (MESENCHYMA). Embryonic mesoderm giving rise to connective tissues, skeletal tissues and blood.

MESENTERON. The mid-gut: that part of the alimentary canal which is lined by endodermal tissue and in which digestion and absorption of food normally take place.

MESENTERY. (1) The peritoneum: a thin double membrane in which the principal visceral organs of a vertebrate hang from the dorsal body-wall.
(2) Vertical membranes or partitions in the body-cavity of a sea-anemone or similar organism.

MESETHMOID. A cartilage-bone forming the front of the cranium and separating the brain from the nasal cavity in a vertebrate. It is perforated by numerous holes through which pass branches of the olfactory nerves; the perforated part is known as the *cribriform plate*.

MESOBENTHOS. Life on the sea-bottom at depths between about 100 and 500 fathoms.

MESOBLAST. An alternative name for the *mesoderm* (*q.v.*).

MESOCARDIUM. Dorsal and ventral sheets of mesentery by which the heart of a vertebrate is suspended in the pericardial cavity during its early development. At a later stage these disappear except for small portions of the dorsal mesentery at each end of the heart.

MESOCEREBRUM. The mid-brain of an insect or crustacean (see *Deutocerebrum*).

MESOCOLON. The mesentery or membrane supporting the colon of a vertebrate.

MESODERM. The middle of the three primary embryonic cell-layers, the other two being the *ectoderm* and the *endoderm*. The mesoderm gives rise to muscles, blood and skeletal tissue.

MESODERMAL CRESCENT (EMBRYOLOGY). A crescent of granular cytoplasm visible at an early stage in the *mosaic* type of egg and later giving rise to the mesoderm of the embryo.

MESODERMAL POUCHES. Segmentally arranged pouches of mesoderm on each side of the notochord, afterwards giving rise to the somites in a vertebrate embryo.

MESOGASTER. The mesentery or peritoneum supporting the stomach of a vertebrate.

MESOGLOEA. A gelatinous non-cellular layer between the endoderm and the ectoderm in diploblastic animals. In larger Coelenterates such as some jellyfish this layer may contain cells which have invaded it from the other layers at a late stage of development and are in no way homologous with the middle layer or *mesoderm* of the higher triploblastic animals.

MESOMETRIUM. The mesentery or peritoneum supporting the mammalian uterus and oviducts.

MESONEPHRIC DUCT (WOLFFIAN DUCT). The duct of the mesonephros or middle kidney which, in male fish and amphibia, acts as a urinogenital duct conveying seminal fluid from the testis as well as excretory products from the kidney. In reptiles, birds and mammals, the mesonephric duct becomes the *vas deferens* and is quite distinct from the ureter or *metanephric* duct (see *Mesonephros* below).

MESONEPHROS (WOLFFIAN BODY). The middle kidney: the second stage of three excretory systems in vertebrates. All vertebrate embryos first have a head-kidney or *pronephros*. This is later replaced by a *mesonephros* which remains as the functional kidney in fish and amphibia. In mammals, birds and reptiles, however, it is later replaced by a *metanephros* or hind-kidney. In these animals the mesonephros atrophies in the female, but gives rise, in the male, to the tubules of the testis.

MESOPLANKTON. Plankton below about 100 fathoms (see *Nekton*).

MESOPTERYGIUM. The middle one of three cartilages forming the base of the pectoral fin in a cartilaginous fish. The others are the *propterygium* and the *metapterygium*.

MESORACHIC FINS (RACHIOSTICHOUS FINS). Fins whose skeleton has a central axis with branches each side as in the veins of a pinnate leaf. The paired fins of many fossil fish are of this type and are known as *Archipterygia*. The Australian lung-fish (*Ceratodus*) of the present day also has them, but it is doubtful whether in this case they are primitive or secondary.

MESORCHIUM. The mesentery or peritoneum supporting the testes of vertebrates.

MESOSAURIDAE. Extinct reptiles of the primitive group *Rhynchocephalia* having numerous fine teeth, a long neck and short hatchet-shaped ribs.

MESOSOMA. The anterior part of the *opisthosoma* or 'abdomen' of scorpions and certain other Arachnida.

MESOSTERNUM. The middle part of the sternum next to the xiphisternal cartilage in those vertebrates such as the frog whose sternum is in several parts.

MESOTARSAL JOINT. The point at which the ankle can bend in lizards and similar reptiles. Owing to fusion of the proximal tarsal bones with the tibia and fibula and of the distal tarsals with the metatarsals, there is only one position where bending can take place, *viz.* between the proximal and distal tarsals.

MESOTHELIUM. A layer of flattened or squamous mesodermal cells lining the coelom of a vertebrate.

MESOTHORAX. The middle segment of the thorax of an insect.

MESOVARIUM. The mesentery or peritoneum supporting the ovary in the vertebrate body.

MET-, META-. Prefix from Greek *Meta*: after, behind.

METABASIPODITE (PRE-ISCHIOPODITE). An additional distal segment on the basipodite of some crustacean limbs. Normally the basipodite consists of one segment only, but in some genera of the group Malacostraca this additional segment is found.

METABOLA. Insects which undergo metamorphosis, whether slight (*Hemi-metabola*) or complete (*Holometabola*).

METABOLISM. The sum-total of all chemical changes which go on in a living organism: *anabolism* plus *katabolism*.

METABOLY. Changing of the shape of a protozoon by contraction and expansion of the protoplasm and stretching of the pellicle: Euglenoid movement.

METACARPAL BONES (METACARPUS). The bones forming the skeleton of the flat part of the hand or fore-foot of a tetrapod vertebrate.

METACEREBRUM (TRITOCEREBRUM). The hind part of the brain of a crustacean or an insect, consisting of the fused ganglia of the third somite.

METACESTODE STAGE. A stage in the life of certain tapeworms intermediate between the *cysticercus* or bladder-worm and the adult tapeworm. In this intermediate larval stage some segments are present but little growth takes place until the worm is transferred to another host. In the genus *Dibothrio-cephalus*, for example, the first larval stage is spent in a fresh-water cyclops. When this is eaten by a fish it develops into a metacestode; the adult tapeworm only develops when the fish is eaten by a human.

METACHRONAL RHYTHM. The peculiar movement like the effect of wind on a field of corn, shown by cilia bending one after the other in regular succession.

METACROMION. A small projection on the mammalian scapula at its narrow end, at right angles to the ridge or spine (see also *Acromion process*).

METADISCOIDAL PLACENTA. A type of placenta characteristic of primates, at first diffuse but becoming discoidal by restriction of the villi to a particular area. It thus bears a secondary resemblance to the true discoidal type found in rodents, insectivora and bats.

METAGENESIS. Alternation of sexual and asexual generations implying that sexual reproduction has been postponed. In most coelenterates, for instance, the *medusa* or sexual individual does not come directly from the egg, but buds off from the hydroid or asexual phase which precedes it.

METAMERIC SEGMENTATION (METAMERISM). Division of an animal into a number of segments each containing a repetition of the same organs, as for instance in the earthworm.

METAMORPHOSIS. A rapid and complete transformation from a larval to an adult form, as for instance the change from a chrysalis to a butterfly.

METANEMERTINI. Nemertine worms whose long proboscis is armed with stylets.

METANEPHRIC DUCT. The ureter of a reptile, bird or mammal: the duct leading from the kidney to the urinary bladder (see *Metanephros*).

METANEPHRIDIUM. A nephridium or excretory tubule of the type found in the earthworm, having an internal ciliated funnel into which waste products pass from the coelom.

METANEPHROS. The hind-kidney: the functional kidney in reptiles, birds and mammals which supersedes the *mesonephros* or middle kidney of fish and amphibia.

METAPHASE. A stage in mitosis or meiosis in which the chromosomes are arranged in a ring round the equator of the cell (see *Mitosis* and *Meiosis*).

METAPLASM. Lifeless constituents of protoplasm.

METAPLEURAL FOLDS. Folds of the lateral body-wall which grow downwards and backwards meeting in the mid-ventral line to make and enclose the atrium of *Amphioxus*, the opercular cavity of tadpoles etc.

METAPNEUSTIC. A term used to describe insects which have spiracles only at the posterior end of the abdomen, as for instance in the mosquito larva.

METAPODIA. A general name for the metacarpals and the metatarsals.

METAPOPHYSES. A pair of anterior dorsally directed processes on certain mammalian vertebrae. On the inner border of each metapophysis is a facet, the *pre-zygapophysis* on which articulates the *post-zygapophysis* of the next vertebra in front.

METAPTERYGIUM. The posterior and usually largest of the three basal cartilages in the pectoral fins of a cartilaginous fish (*cf. Mesopterygium* and *Propterygium*).

METAPTERYGOIDS. Membrane bones adjacent to and posterior to the pterygoids in the skull of a bony fish.

METASICULA. The hind part of a *sicula*, the primary cone-shaped exoskeleton of a Graptolite, from which polyps of a colony are budded off (see *Graptolite*).

METASOMA. The posterior part of the opisthosoma or 'abdomen' of certain arachnids. In scorpions, for instance, it forms a flexible tail for wielding the terminal sting.

METASTERNITE. A plate formed of fused sterna between the bases of the hind limbs of scorpions etc.

METASTERNUM. The sternum or ventral plate on the metathorax of an insect.

METASTOMA. Any organ or part behind the mouth; commonly used to denote the lower lip of an arthropod.

METATARSAL BONES (METATARSUS). Bones forming the flat part of the hind foot of a tetrapod vertebrate, between the tarsals and the phalanges.

METATHERIA. An alternative name for the *Marsupialia* (*q.v.*).

METATHORAX. The hind segment of the thorax of an insect.

METATROCH. A secondary ring of cilia behind the first ring or *prototroch* on the trochosphere type of larva which is characteristic of certain annelids and molluscs.

METAZOA. All multicellular animals.

METAZOEA. A late larval stage in the development of crabs and similar crustacea, in which there is a carapace and the rudiments of thoracic limbs.

METENCEPHALON. The hind-brain of a vertebrate including the *cerebellum* and the *medulla oblongata*.

METEPIPODITES. Epipodites or lateral appendages borne on the coxopodite of a crustacean; those borne on the pre-coxa are known as pre-epipodites.

METOTIC. Behind the ear: a term used with reference to the head-somites of a vertebrate embryo. The first three somites are *pro-otic* or in front of the ear-capsule, whilst all those after and including the fourth are *metotic*.

METRIORHYNCHIDAE. Extinct marine crocodiles of the Jurassic period having a fairly long snout, broad nasal bones, a bony ring in the sclerotic coat of the eye, but no bony plates in the skin.

MICELL (MICELLE, MICELLA). Minute threads or elongated particles of protein or other material within the protoplasm of a cell.

MICR-, MICRO-. Prefix from Greek *Mikros*: small.

MICRANER. An abnormally small male ant.

MICRERGATE. A dwarf worker ant.

MICROCEPHALIC. Small headed.

MICROCHIROPTERA. The smaller bats which are mostly insectivorous and comprise the majority of bats in most parts of the world (*cf. Megachiroptera*).

MICROCHROMOSOMES. Normal chromosomes (*cf. Megachromosomes*).

MICRODRILI. An order of small, usually aquatic oligochaete worms with few segments and possessing minute eyes on the prostomium. Reproduction is commonly asexual, but sexual forms exist and may be recognized by the presence of a clitellum.

MICROGAMETE. The smaller and usually more mobile of two gametes: the male gamete.

MICROGAMETOCYTE. The mother-cell of a microgamete.

MICROLECITHAL. Having little yolk in an egg.

MICROLEPIDOPTERA. The smallest moths including the primitive *Micropterygidae* only a few millimetres in length. Butterflies and moths were formerly classified according to size as *Macrolepidoptera* and *Microlepidoptera*, but this division is an artificial one and has now been replaced by a more natural classification based on the type of wing-venation.

MICROMERES. The smaller blastomeres or cells formed when a fertilized egg undergoes cleavage. These smaller cells are in the neighbourhood of the *animal pole* of the egg, the region away from the yolk where the ectoderm of the embryo develops (see also *Megameres*).

MICRONEPHRIDIA. Numerous small excretory tubules found in some of the

151

giant earthworms of Australia (*Megascolecidae*). There may be many hundreds of these in each segment in addition to the normal pair of *meganephridia*. It is thought that they develop by the breaking up of a single coiled tube into a multitude of smaller ones.

MICRONUCLEUS. The smaller of two nuclei found in some protozoa such as those of the group *Ciliophora*. It contains *idiochromatin* and is concerned chiefly with cell division. The *meganucleus* on the other hand, the larger of the two, contains *trophochromatin* and is concerned more with general metabolism of the cell (see *Meganucleus*).

MICROPTEROUS. Small winged (see under *Macropterous*).

MICROPTERYGIDAE. A family of moths seldom more than a few millimetres across and of world-wide distribution. They belong to the *Homoneura*, a suborder in which the venation is almost identical in the two pairs of wings. In many other ways they are primitive, feeding chiefly on pollen by means of biting mouth-parts not elongated into the normal proboscis.

MICROPYLE. In addition to its botanical meaning this word is used to denote small pores in the shell of an insect's egg through which spermatozoa can enter. They are necessitated by the fact that the shell is secreted before fertilization.

MICROSAURIA. Small lizard-like creatures of the Upper Carboniferous and Lower Permian Periods, sometimes classed as reptiles and sometimes put with the Stegocephalian amphibians.

MICROSPHERIC FORM. See under *Megalospheric form*.

MICROSPORIDIA. A suborder of protozoa of the group *Cnidosporidia* characterized by the formation of spores which each bear a single pole-capsule resembling a nematocyst. Many are parasites of silkworms, bees and other insects and it is thought that the thread of the pole capsule helps to anchor them to the gut of the host.

MICROTRICHIA. Minute cuticular hairs such as those on the wings of insects.

MID-BRAIN. See *Mesencephalon*.

MIDDLE EAR. The part of the ear of a terrestrial vertebrate between the *tympanum* or outer ear-drum and the *fenestra ovalis* or inner ear-drum. This space contains air and is connected with the back of the pharynx by the *Eustachian tube*. Sound vibrations are conducted across it by the auditory ossicles, of which there are three in mammals and one in lower vertebrates.

MIDRIFF. An alternative name for the diaphragm which separates the thorax from the abdomen in mammals.

MILK GLANDS. (1) Of Mammals. See *Mammary Glands*.

(2) Of Insects. In certain viviparous species such as the tse-tse fly the young are nourished in a 'uterus' by special nutrient glands.

MILK TEETH. The first or temporary set of teeth in mammals.

MILLEPEDE. See *Diplopoda*.

MIMICRY. Protective resemblance of one species to another. See also under *Batesian mimicry* and *Müllerian mimicry*.

MIRACIDIUM. The first larval form of a liver fluke: a flat ciliated freeswimming organism which, in many species, must find and enter a water snail before developing into later larval stages known as the *Redia* and the *Cercaria*.

MITES. Acarina which are usually small and may be free-living scavengers, predatory or occasionally ectoparasites. They differ from the blood-sucking ticks in having clawed chelicerae and leg-like pedipalps (see *Acarina*).

MITOCHONDRIA (CHONDRIOSOMES). Minute rod-shaped bodies, consisting of protein and lipoid material, found in the cytoplasm of most cells. It has recently been shown that their chief function is the production of respiratory enzymes.

MITOME. A protoplasmic network.

MITOSIS. The normal process of cell division consisting of the following stages:
(1) The *Prophase* in which chromosomes make their appearance and the nuclear membrane disappears.
(2) The *Metaphase* in which chromosomes arrange themselves round the equator of the cell, each chromosome then dividing longitudinally to form two chromatids.
(3) The *Anaphase* in which the newly formed chromatids move towards opposite poles of the cell and become new chromosomes.
(4) The *Telophase* in which the new sets of chromosomes re-form nuclei and two new cells are formed.

MITRAL VALVE. The bicuspid valve between the left auricle and the left ventricle of the mammalian heart.

MIXED NERVES. Nerves containing both afferent and efferent fibres, as for example the spinal nerves of a vertebrate.

MIXIPTERYGIUM (PTERYGOPODIUM). A pelvic fin provided with a clasper or copulatory appendage as in the males of sharks, rays and other cartilaginous fish.

MNIOTILTIDAE. American warblers: small birds similar to, but not in the same family as the European warblers, thrushes, blackbirds etc.

MOA. See *Dinornithidae*.

MOLAR PROCESS. A biting part on the mandible of an arthropod or other invertebrate (see also *Incisor process*).

MOLAR TEETH. Grinding back teeth of mammals, having no predecessors in the milk teeth.

MOLIDAE. Sun-fish and similar types having a very large round body, a beak-like mouth, small fins and a short filamentous tail. They bask on the surface of the open sea but apparently do not inflate the body.

MOLLUSCA. Soft-bodied unsegmented animals usually having a calcareous shell secreted by the loose dorso-lateral skin which is called the mantle. The principal organs are often concentrated into a large *visceral hump* and locomotion is generally by means of the single *foot*, a muscular extension of the ventral part of the body.
The chief classes of mollusca are: *Amphineura* with a shell made of several transverse dorsal plates; *Lamellibranchiata* or bivalves with a shell of two lateral plates known as valves; *Gasteropoda* usually having a single spiral shell like that of a snail; *Cephalopoda* whose shells may be either spiral or reduced or absent. In this group the octopus has no shell; squids and cuttle-fish have reduced internal shells; *Nautilus* and the fossil ammonites have spiral chambered shells.

MOLOSSINAE. Small bats of tropical regions having long thick tails, short strong legs and large feet; unlike most bats they can progress rapidly on the ground.

MOLPADIDA. Sea-cucumbers (*Holothuroidea*) of burrowing habit, without tube feet but with simple tentacles round the mouth.

MONACHINAE. Seals of the family *Phocidae* having the hind limbs stretched out backwards on each side of the tail and united to it by a membrane. They include the *Monk-seal* of the Mediterranean and Atlantic as well as the *Sea-leopard* and others of the Antarctic. All these differ from the common British seals in having no claws on the toes.

MONAXONIDA. Sponges of the class *Demospongiae* having a skeleton composed of monaxons or simple needle-like spicules of silica.

MONIMOSTYLIC (MONIMOSTYLICA). A term used to describe reptilian skulls in which the quadrate bone is fixed and the upper jaw incapable of separate movement. They include *Sphenodon* and the crocodiles.

MONO-. Prefix from Greek *Monos*: alone, single.

MONOCONDYLEA. Birds and reptiles which have a single occipital condyle.

MONOCYTE. The largest form of white corpuscle in vertebrate blood, actively phagocytic and having a spherical nucleus.

MONODELPHIA. An alternative name for *Eutheria* or placental mammals, *i.e.* all mammals except Monotremes and Marsupials.

MONOGENEA. Ectoparasitic flukes whose life-histories do not involve an intermediate host. They are thus distinguished from *Digenea* such as the liver fluke which spend their lives in at least two hosts. Monogenea are usually parasites of fresh-water fish and amphibia, clinging either to the skin or to the gills, rectum or bladder.

MONOGONOPORIC (MONOGONOPOROUS). Having a single genital aperture for male and female elements.

MONOHYBRID INHERITANCE. Inheritance of one pair of allelomorphic genes or hereditary factors.

MONOPHYODONT. Having only one set of teeth.

MONOPNEUMONA. Lung-fish with a single lung, as for instance the Australian *Ceratodus*.

MONOPYLAEA (NASSELLARIA). Radiolarians whose central capsule is perforated by pores concentrated into one 'pore-plate'.

MONOSOME. An unpaired sex chromosome (see *Sex Chromosome*).

MONOTHALAMIA. Fresh-water Foraminifera having a soft single-chambered shell and simple unbranched pseudopodia.

MONOTOCARDIA. Gasteropods such as the whelk and the periwinkle which have only one gill and whose hearts have only one auricle.

MONOTOCOUS. Producing one at a birth.

MONOTREMES (MONOTREMATA, ORNITHODELPHIA, PROTO-THERIA). Mammals which lay eggs and have many primitive or reptilian features. The only living kinds are the spiny ant-eater (*Echidna*) and the Duck-billed platypus (*Ornithorhynchus*).

MONOZOA. Small tapeworms which do not form proglottides or segments; they are mostly parasites of fish.

MONRO, FORAMINA OF. Passages leading from the first and second ventricles to the third ventricle in the vertebrate brain.

MORMOPINAE. Small bats of the West Indies and Central America having expanded leaf-like appendages on the chin and not on the nose as in some other bats.

MORMYROIDEA. African fresh-water fish having an elongated snout.

MORPH-. Prefix from Greek *Morphe*: form, shape.

MORPHALLAXIS. Regeneration: growth of a part of a parent into a new individual.

MORPHOGENESIS. Evolution of form and structure.

MORPHOLOGY. The study of form and structure.

MORULA. A solid ball of cells resulting from the repeated cleavage of a zygote in the early stages of embryonic development.

MOSAIC EGGS. Eggs which, at a very early stage, show definite areas or zones which will later develop into particular tissues and organs.

MOSAIC VISION. In the compound eyes of crustaceans and insects, each cone-shaped optical unit or *ommatidium* is surrounded by a dark pigmented sheath separating it from its neighbours. A narrow pencil of light therefore falls on the retinula at the base of each crystalline cone and the whole eye will produce a *mosaic image* composed of as many points of light as there are cones. In some nocturnal insects, however, the pigmented sheaths can be retracted so that the points of light will merge together to form a *superposition image* which is brighter but less distinct than the mosaic image.

MOSASAURIA. Large extinct marine reptiles of the Upper Cretaceous period, having two pairs of clawless five-toed limbs, a lizard-like head and a long body with over a hundred vertebrae.

MOTHS (HETEROCERA). The division of the Lepidoptera into butterflies and moths is now considered to be somewhat artificial and a more scientific classification based on the venation of the wings has superseded it. The names butterfly and moth are, however, retained for convenience. A butterfly differs from a moth in having club-like swellings on the ends of the antennae and in its habit of folding the wings vertically when at rest.

MOTOR NERVE. A nerve which conveys an impulse directly to a muscle and causes it to contract.

MOULTING. (1) Of birds and mammals. A seasonal loss of feathers or hair sometimes accompanied by a change in colour.
(2) Of Arthropods (see *Ecdysis*).

MOULTING GLANDS. Glands in the skin of an arthropod which, at the time of moulting, produce an enzyme capable of liquefying the endocuticle and so facilitating the shedding of the two outer layers, the exocuticle and the epicuticle.

MOUTH. Strictly the term should be used for the opening through which food enters the alimentary canal. It is frequently used, however, as a synonym for the buccal cavity.

MOUTH-PARTS (OF ARTHROPODA). Modified paired appendages used for feeding. In insects they are normally the *labrum, mandibles, maxillae* and *labium,* some or all of which may be modified to form piercing stylets or a suctorial proboscis. In crustacea the chief mouth parts are the *mandibles, maxillules, maxillae* and several pairs of *maxillipedes.*

MUCIN. See *Mucus.*

MUCOUS CANALS (OF FISH). The lateral line and its branches on the surface of the head. Each line consists of a groove or a subcutaneous canal opening at intervals along its length and containing both mucus-secreting cells and sensory cells (see *Lateral line* and *Lorenzini's ampullae*).

MUCOUS GLANDS. Glands which secrete mucus, as for instance in many parts of the alimentary canal.

MUCOUS GLAND OF SNAIL. See *Pedal gland*.

MUCOUS MEMBRANE. Epithelium which secretes mucus, *e.g.* the lining of the alimentary canal.

MUCUS. A slimy solution of mucin or other viscous substance secreted by a mucous gland or mucous membrane.

MUGILIDAE. Grey mullet: fish which inhabit shores and brackish lagoons feeding on organic debris and using a muscular stomach like the gizzard of a bird. The body is covered with large cycloid scales and the anterior dorsal fin is stiffened with four spines.

MUGILIFORMES. See *Percesoces*.

MÜLLERIAN DUCT (PRONEPHRIC DUCT). The duct from the *pronephros* or head-kidney, the first excretory organ to be developed in a vertebrate embryo. In later development this kidney degenerates and is replaced by a middle kidney (*mesonephros*) or by a hind-kidney (*metanephros*). In adult females the Müllerian duct becomes enlarged to form the oviduct but in males it disappears.

MÜLLERIAN MIMICRY. A form of mimicry common among insects in which two harmful or distasteful species resemble each other. Each benefits from the similarity because would-be predators, tasting one, learn to avoid the other also.

MÜLLER'S LARVA. A planktonic larva characteristic of Polyclads (marine Turbellarians): it is a small ovoid organism with projecting lobes and bands of cilia and a large cilium at each end. The mouth is near the hind-end and there are several sensitive eye-spots situated anteriorly.

MÜLLER'S ORGAN. See *Wheel organ*.

MULLIDAE. Red Mullet: edible fish of the Tropics, whose body is slightly compressed and covered with large thin scales. There are two large erectile barbels beneath the mouth.

MULTIPLE FISSION. Spore formation of Protozoa, especially Sporozoa, in which the nucleus divides several times before division of the cytoplasm. The latter then divides into as many parts as there are nuclei, but may leave some residual protoplasm.

MULTIPOLAR. A term used to describe a neurone or nerve-cell having many dendritic processes or nerve-fibres leading from it.

MULTITUBERCULATA. An alternative name for the *Allotheria* (*q.v.*).

MURAENIDAE. Eels: elongated fish with smooth skin, without pelvic fins and with a symmetrical pointed or *gephyrocercal* tail fin.

MURIDAE. Rats, mice etc. Small rodents of the *simplicident* type, *i.e.* having only one pair of upper incisors; having short or medium length tails either scaly or hairy; having well-developed clavicles and having the tibia and fibula united. There are about 720 species in almost every part of the world.

MUSCICAPIDAE. Fly catchers: small, rapidly flying migrant birds having a thin beak with a hooked curved point.

MUSCLE. Tissue which by its contraction can bring about the movement of an animal. There are three chief types of muscular tissue, *viz.* striated or voluntary muscle; unstriped or involuntary muscle and cardiac or heart muscle (see also under separate headings).

MUSHROOM BODY. A crowd of nerve cells grouped together into a mushroom shape, their axons forming the stalk. Large numbers of such bundles are frequently present in the fore-brains of insects.

MUSHROOM GLAND. A cluster of glandular seminal vesicles at the junction of the two vasa deferentia with the ejaculatory duct of a male insect.

MUSOPHAGIDAE. Touracos or plantain eaters: cuckoo-like birds of Africa, whose feathers have a red pigment (*turacin*) containing copper and a green pigment (*turacoverdin*) containing iron. These pigments are washed out of the living bird by the rain but are rapidly re-formed.

MUSTELINAE. Weasels, stoats, ferrets, polecats etc. Small carnivorous mammals with short toes, sometimes partly webbed and usually with sharp claws. They are found in most parts of the world and may be terrestrial or arboreal. A few kinds such as the ermine stoats change their fur to white in the winter. These and others such as the mink are bred for the fur.

MUTATION. The sudden appearance of an individual with heritable features which are markedly different from the average type. Such a change may be due to an alteration in the number of chromosomes (polyploidy), an alteration in the structure of a chromosome or a chemical or physical change in an individual gene. Mutations form the raw material of evolution by natural selection. They are usually rare but may sometimes be brought about artificially by exposure to X-rays or to radioactive particles and radiations.

MYCETINAE. Howling monkeys of Central and South America. They habitually walk on all four legs and have long prehensile tails.

MYCETO-. Prefix and root-word from Greek *Mukes*: mushroom.

MYCETOCYTES. Specialized cells in the bodies of insects etc. containing symbiotic bacteria or protozoa necessary for the life of the host. Those which help to break down cellulose in termites and cockroaches are an example. Sometimes these cells are grouped together in a special organ called the Mycetome.

MYCETOPHILIDAE. Fungus-gnats: small insects whose larvae, white maggots with dark brown heads, infest mushroom beds. They may stray to other crops and do considerable damage but decaying mushrooms are their principal food.

MYCETOZOA (MYXOMYCETES). Slime-fungi: organisms intermediate between plants and animals, but now usually classed with the protozoa. In the active form they consist of a multinucleate mass of naked protoplasm able to move in an amoeboid manner. Reproduction, however, is by plant-like spores.

MYEL-, MYELO-. Prefix from Greek *Muelos*: marrow.

MYELENCEPHALON. An alternative name for the *medulla oblongata* of the vertebrate brain. It is behind the cerebellum and contains a cavity, the fourth ventricle, covered by a vascular membrane, the *posterior choroid plexus*.

MYELIN SHEATH. A fatty sheath surrounding the axis cylinder of a nerve fibre and contained within a thin tubular membrane, the neurilemma.

MYELOCYTES. Cells in the bone marrow of a vertebrate, giving rise to the polymorph type of white blood corpuscle.

MYLIOBATIDAE. Eagle-rays: very large flat cartilaginous fish, sometimes up to 20 ft. across, having a long whip-like tail and large pectoral fins which spread along the sides of the body. These have two lobes appearing as separate smaller fins on the tip of the snout.

MYLODONTIDAE. Extinct sloth-like mammals of South America about as large as a rhinoceros and having a large number of bony plates embedded in the skin. Their remains have been found together with those of humans in a cave in Patagonia and there is evidence that they were domesticated by primitive man. It is possible that some members of the family still survive.

MYMARIDAE. Fairy-flies: the smallest known insects, sometimes only one fiftieth of an inch long. They are included among the Chalcid wasps and feed largely on the eggs of other insects.

MYO-. (1) Prefix from Greek *Mus, muos*: muscle.
(2) Prefix from Greek *Mus, muos*: mouse.

MYOBLAST. A cell which gives rise to a muscle fibre.

MYOCARDIUM. The muscular wall of the vertebrate heart consisting of *cardiac muscle*. This differs from normal or voluntary muscle by reason of the fact that the fibres form a network and have the power of rhythmic contraction even when isolated from the body.

MYOCOEL. The coelomic cavity in each myotome or muscle segment of a vertebrate.

MYOCOMMA. A septum or partition of connective tissue between two myotomes or blocks of muscle.

MYOCYTE. Any type of muscle cell.

MYODOME. The eye-muscle canal: a cavity beneath the cranium housing some of the rectus muscles in many bony fish.

MYOFIBRILLAE. Muscle fibrils: the term may apply either to the fine parallel fibrils running longitudinally in the cells of unstriated muscle or to those which make up the larger fibres of striated muscle.

MYOGALINAE. Mole-like insectivora differing from true moles in not having the extra or sixth digit on the front feet. The group includes the burrowing mole-shrews of North America and Japan as well as some web-footed aquatic forms of Europe and Asia.

MYOMERE. See *Myotome*.

MYOMORPHA. A large group of mouse-like rodents including rats, mice, dormice, hamsters, voles, lemmings etc., as well as some burrowing forms and the long-legged jumping Jerboas.

MYONEMES. Contractile fibrils in the ectoplasm of many protozoa.

MYOPHRISKS. (1) Sometimes used synonymously with *myoneme*.
(2) Muscle-like rings round the bases of the spines of certain Radiolaria.

MYOPSIDA. Squids: free-swimming cephalopod molluscs whose eyes have a well developed cornea. They are allied to the cuttlefish, having ten tentacles, two of which are longer than the others.

MYOSEPTUM. See *Myocomma*.

MYOTOMES (MYOMERES). Segmentally arranged blocks of muscle, as for instance in the body and tail of a fish. In higher vertebrates much of the embryonic segmentation is lost in the adult owing to specialization of muscles for limb movement.

MYRIOPODA. Centipedes and millepedes: elongated arthropods with numerous legs and having a distinct head bearing clusters of simple eyes and a single pair of antennae. Breathing is by means of tracheae similar to those of insects. The main difference between centipedes and millepedes is that the latter have two pairs of legs on each segment.

MYRMECOPHAGIDAE. South American ant-eaters, sometimes up to five feet in length, belonging to the order Edentata. They have a long toothless snout and a sticky worm-like tongue with which they lick up the ants after scraping down an ant-hill with their enormous claws.

MYRMECOPHILES. Miscellaneous creatures which inhabit the nests of ants either as true guests, fed and tended by the ants, or as scavengers.

MYRMELEONIDAE. Winged insects whose larvae, known as ant-lions, make pit-like snares for capturing their prey. They belong to the order Neuroptera.

MYSIDACEA. Small carnivorous shrimp-like crustacea of the group Pera-carida, usually without gills and breathing through the thin lining of the carapace.

MYSIS LARVA (SCHIZOPOD LARVA). The shrimp-like larva of certain crustacea, having biramous limbs on all the thoracic segments. It is so-called from its resemblance to the genus *Mysis* (see *Mysidacea* above).

MYSTACOCETI (MYSTICETI). Toothless whales (see *Balaenoidea*).

MYX-, MYXO-. Prefix from Greek *Muxa*: slime.

MYXINOIDEA (MYXINOIDEI, MYXINIDAE). Hag-fish: a group of Cyclo-stomata or fish-like creatures without jaws, similar in general structure to the lamprey and living as parasites on other fish. They bore their way into the body of the host by means of horny teeth on their lips and tongue, secreting a large amount of mucus as they do so. They differ from lampreys in having six to fourteen gill-clefts on each side and in being without a dorsal fin.

MYXOMYCETES. See *Mycetozoa*.

MYXOPTERYGIUM. See *Mixipterygium*.

MYXOSPONGIAE. Sponges without the usual skeleton of spicules.

MYXOSPORIDIA. A group of Sporozoa which are usually parasitic on fresh-water fish. They are multinucleate organisms which give rise to spores having two or four *pole capsules* resembling nematocysts. The latter are able to shoot out threads by which the organism can anchor itself to the tissues of the host.

N

NACREOUS LAYER. The innermost or pearly layer of the shell of a mollusc, consisting mostly of calcium carbonate secreted by the epithelium of the mantle.

NAILS. Plate-like thickenings of the horny layer of the epidermis at the tips of the digits. Like claws and hoofs they are formed of *keratin*, a hard protein-like substance.

NANDIDAE. Small carnivorous spiny-rayed fish from fresh waters of West Africa, S.E. Asia and South America.

NANOPLANKTON. Floating sea plankton.

NARCO-. Prefix and root word from Greek *Narke*: torpor.

NARCOBATOIDEA (TORPEDINIDAE). Electric rays: flat cartilaginous fish having a pair of large electric organs between the pectoral fins and the head (see *Electric organs*).

NARCOMEDUSAE. Small stinging jelly-fish of the order Trachylina including some fresh-water forms from African lakes and from North American rivers. Although belonging to the Hydrozoa, which usually show alternation of generations, this group is characterized by dominance of the medusa and reduction or loss of the hydroid phase.

NARES. Nostrils: internal and external openings of the nasal cavity. All vertebrates have a pair of external nares by which the olfactory organs are connected with the surrounding medium; land animals also have internal nares or *choanae* opening into the mouth. These enable the animal to breathe continuously at the same time as it is feeding. Primitively they open into the front of the buccal cavity, but in mammals and some reptiles they are much farther back, the air passage being separated from the buccal cavity by the secondary palate.

NASAL BONES. A pair of membrane-bones roofing over the nasal cavity of a vertebrate. They extend forwards from the frontal bones and are bounded laterally by the maxillae and premaxillae.

NASAL CAPSULES (OLFACTORY CAPSULES). Those parts of the vertebrate skull which enclose the nasal cavities.

NASAL CAVITIES. See *Nasal sacs.*

NASAL PITS. Pits or depressions by which the olfactory organs communicate with the exterior in fish, amphibian larvae etc. In lower forms they are not connected in any way with the mouth but in air-breathing vertebrates they break through to form the internal nostrils.

NASAL SACS (NASAL SINUSES, NASAL CAVITIES). Paired cavities in the vertebrate skull in which are situated the olfactory organs (see also *Nares*).

NASO-PALATINE CANAL (OF CYCLOSTOMATA). A passage connecting the nasal cavity and the pituitary sac with the hind-part of the buccal cavity in the *Myxinidae.*

NASSELLARIA. See *Monopylaea.*

NASUTES. Soldier-termites: specialized forms within the genus *Nasutitermes,* able to defend the colony by discharging an acrid secretion from glands situated at the end of a long snout or rostrum.

NATURAL SELECTION. The theory propounded by Charles Darwin to explain the process of evolution. The principal points of the theory are as follows:
(1) Most organisms produce a very large number of offspring.
(2) A shortage of food causes a struggle for existence.
(3) Variations are constantly occurring and some of these may be inherited.
(4) Variations which have some advantage in the struggle for existence tend to be perpetuated in the race; those which are disadvantageous tend to die out. Changes in climate or other conditions, however, cause different features to be advantageous at different times.
(5) The accumulation of small variations can lead to large differences producing new species and ultimately new genera, families etc.

NAUCORIDAE. Water bugs: Heteropterous insects with flattened body and clawed front-legs. Although they can fly they normally live under water among the vegetation, taking down bubbles of air trapped between the wings and the abdomen.

NAUPLIUS LARVA. The typical larva of a crustacean: a minute creature with rounded unsegmented body, three pairs of simple appendages and a single median eye.

NAUTILOIDEA. Cephalopod molluscs, closely related to the fossil *Ammonites* having numerous tentacles round the mouth and living in a spiral chambered shell.

NAVEL. The umbilicus: the place of attachment of the umbilical cord to a vertebrate embryo.

NAVICULAR (NAVICULARE). Boat-shaped: a name given to certain carpal and tarsal bones, not necessarily homologous in different animals.

NECROPHORIDAE. Burying beetles: brightly coloured insects which bury the corpses of small animals by excavating beneath them till they sink below the soil. The eggs are then laid in the decaying corpse which provides food for the larvae.

NECTARINIIDAE. Sun-birds: brilliant metallic-coloured birds of India, Africa and North Australia.

NECTOCALYCES. Bell-shaped polyps, without tentacles or manubrium, found in some of the floating colonial coelenterates. By their rhythmic contraction they are said to help in the locomotion of the colony.

NECTON (NEKTON). A general name for those forms of life which inhabit middle depths of the sea (*cf. Plankton* and *Benthos*).

NEEDHAM'S SAC. The terminal reservoir of the sperm duct in a cephalopod such as the cuttlefish.

NEMATO-. Prefix from Greek *Nema, nematos*: a thread.

NEMATOBLASTS (CNIDOBLASTS). Cells which contain nematocysts or stinging threads. They are found in the ectoderm of most coelenterates (see *Nematocyst* below).

NEMATOCALYX. See *Nematophore*.

NEMATOCERA. Two-winged flies having long many-jointed antennae. They include crane-flies, mosquitoes and many common forms of gnat.

NEMATOCYST. The active part of a nematoblast consisting of a poison sac and a stinging thread. The latter is shot out when the *cnidocil*, a small projection on the surface of the nematoblast, is touched. Batteries of such cells form the stinging mechanism of jelly-fish, sea-anemones and other coelenterates.

NEMATODA. Thread worms and round worms: one of the largest of all the phyla of invertebrates. They are unsegmented, pointed at both ends, without a coelom and with a smooth skin. The group includes a large number of para-sites, a few free-swimming forms and some soil dwellers.

NEMATOMORPHA. Thread-like aquatic worms resembling nematodes but in many ways more primitive. They may start life as parasites of insects but afterwards become free-swimming. *Gordius*, the hair-worm, is a well known example.

NEMATOPHORE (NEMATOCALYX). A specialized protective polyp of a coelenterate colony, consisting of a long thin amoeboid thread projecting out of a small cup; particularly characteristic of the genus *Plumularia*.

NEMERTEA (NEMERTINI). Proboscis worms: long thread-like worms, usu-ally marine, with unsegmented body and ciliated ectoderm. They are charac-terized by an extremely long eversible proboscis lying in a sheath on the dorsal side of the alimentary canal. This proboscis may have stylets for piercing its prey or may be used for coiling round the prey.

NEMOPTERIDAE. Predatory insects of the order Neuroptera having larvae with very long necks and adults with fine thread-like hind-wings.

NEO-. Prefix from Greek *Neos:* new.

NEO-DARWINISM. The theory that Natural Selection operates, not on acquired characters, but on hereditary genes (see also *Natural Selection*).

NEOGNATHAE. One of the two major groups of living birds, the other being the *Palaeognathae*. The former includes all the *Carinatae* or flying birds with the exception of the Central American *Tinamus*. This is an anomalous genus and is now grouped with the *Ratitae* or flightless birds on account of the structure of the skull. In the Neognathae the prevomer is small and does not touch the pterygoids; in the Palaeognathae it touches them and is larger.

NEOPALLIUM. That part of the cerebral cortex of vertebrates which is specially large in higher mammals and whose chief function is general co-ordination and intelligence. It forms a thickening in the dorsal region near the hippocampal lobes.

NEOPTERYGII. All present-day bony fish with the exception of lung-fish, sturgeons, gar-pikes and a few others. They are characterized chiefly by the fact that the scales are thin or absent; the fins are supported only by dermal fin-rays (lepidotrichia) and the tail is homocercal or secondarily symmetrical.

NEORNITHES. All known birds except the extinct *Archaeopteryx*. They all have a shortened tail with the last few vertebrae fused to form a pygostyle and all except the fossil group *Odontolcae* are toothless.

NEOSPORIDIA. Protozoa of the class Sporozoa, distinguished from others of the class by the fact that the adult in the vegetative stage is a syncytium able to form spores continuously within itself. Most are parasites of invertebrates, but one group, the *Sarcosporidia*, lives in the muscle fibres of mammals including man.

NEOSSOPTILES. The nestling feathers of a young bird, preceding the adult feathers or *teleoptiles*. The various types of adult feather, namely filoplumes, plumulae and pennae are preceded respectively by pre-filoplumes, pre-plumulae and pre-pennae.

NEOTENY. The temporary or permanent retention of larval structures. In extreme cases the animal is able to breed while maintaining its juvenile form. The latter phenomenon, also known as *paedogenesis*, is well exhibited in the case of the *axolotl*. This animal, the larva of a salamander, normally retains its external gills and other juvenile features owing to a deficiency of thyroid secretion and yet is able to produce offspring similar to itself. If given thyroxin, however, metamorphosis rapidly takes place.

NEOTRAGINAE. A subfamily of African antelopes with horns only on the male. They include the Klipspringer, the Grysbock, the Steinbock and the small Royal Antelope which is only ten inches high.

NEPHR-, NEPHRO-. Prefix and root-word from Greek *Nephros*: kidney.

NEPHRIDIOPORE. The external opening of a *nephridium* (*q.v.*).

NEPHRIDIUM. A primitive excretory organ usually consisting of an ecto-dermal ingrowth forming either a single or a branched tubule ending in one or more *Flame cells*. These cells are hollow and contain a number of cilia whose movements suggest a flickering flame. Waste products are excreted into the tubule and the beating of the cilia sends them along to the exterior. The tubule itself often consists of an intracellular duct passing through cylindrical cells known as *drain-pipe cells*. This type of nephridium is found in Platyhelminthes and in some Annelids.

Sometimes a nephridium is combined with a coelomoduct or tubule of mesodermal origin leading from the coelom to the exterior. Such a structure, known as a *nephromixium*, is found in the earthworm.

NEPHROCOEL. The coelomic cavity within a *nephrotome* (*q.v.*).

NEPHROCYTES. Specialized cells in the bodies of insects and other arthropods, usually in strands on each side of the heart. They are believed to be capable of absorbing unwanted nitrogenous waste substances from the blood but little is known of their mode of action.

NEPHROMIXIUM. A composite excretory tubule such as those in many annelids. It consists of a coelomoduct or tube of mesodermal origin leading from the coelomic cavity towards the exterior, combined with a true nephridium or tube of ectodermal origin leading inwards from the epidermis (see also under *Nephridium*).

NEPHRON. A functional unit of the vertebrate kidney consisting of a *Malpighian body* (glomerulus and capsule) together with a uriniferous tubule.

NEPHROS. A kidney (see *Pronephros, mesonephros* and *metanephros*).

NEPHROSTOME. The internal aperture of a nephridium or nephromixium.

NEPHROTOMES. Segmentally arranged parts of the mesoderm of a vertebrate embryo giving rise later to the tubules of the kidney.

NEPIDAE. Water scorpions: aquatic insects of the group Heteroptera, resembling a dark brown leaf or stick crawling about on the bottom of a pond, having a long breathing tube resembling a sting at the hind-end and bent front legs like the pincers of a scorpion. They have wings but are unable to fly.

NERVE. A bundle or a number of bundles of nerve fibres bound together with connective tissue and surrounded by a sheath (see *Nervous System*).

NERVE CORD. The main bundle of nerves of the central nervous system. In many invertebrates it consists of a solid ventral cord with ganglia in each segment; in vertebrates it is a hollow dorsal tube running inside the vertebral column and enlarging at the anterior end to form the brain.

NERVE NET. A network of nerve cells and fibres with no central cord, ganglia or brain. Such a simple nervous system is found in *Hydra* and other coelenterates.

NERVE RING (OF ECHINODERMATA). A ring of nerve fibres surrounding the mouth of a star-fish or similar creature, connecting and co-ordinating the action of the radial nerves in the five arms.

NERVOUS SYSTEM. The chief means of conducting impulses from one part of an animal to another in order to co-ordinate the various senses and activities. It consists of numerous nerve-cells or *neurones* with branching processes whose endings are in close proximity to one another. The small gaps between them are known as *synapses*. Normally the stimulation of a receptor such as a sense organ sends an electrical impulse along a nerve fibre to a synapse where it produces a chemical substance (*e.g. sympathin* or *acetyl-choline*). This starts another impulse in the neighbouring nerve fibre so that the message is relayed from one part of the body to another. In higher animals the cell-bodies of the neurones are usually concentrated in a central nervous system but in many lower animals there is merely a simple network with no definite centres of activity.

NESOPITHECIDAE. Extinct monkeys from the Pleistocene of Madagascar, apparently intermediate between lemurs and true monkeys.

NEURAL. Connected with the nervous system.

NEURAL ARCH. A bony or cartilaginous arch resting on the centrum of a vertebra and forming a tunnel through which the spinal cord runs.

NEURAL FOLDS. In vertebrate embryology: two folds of ectoderm which rise up on each side of and over the neural plate, so helping to form the tubular nerve cord.

NEURAL PLATE. A part of the ectoderm in a vertebrate embryo which, during development, sinks beneath the surface and rolls up to form the beginning of the spinal cord.

NEURAL SPINE. A median dorsal process from the neural arch of a vertebra.

NEURENTERIC CANAL. In vertebrate embryology: a canal joining the primitive gut or *archenteron* with the cavity of the neural tube. It is formed when the blastopore is roofed over by the neural folds.

NEUROBIOTAXIS. An evolutionary process by which the cell-bodies of neurones tend to move towards the source of the impulses which they most habitually receive.

NEUROCRANIUM. That part of the skull consisting of the brain-case and the sense-capsules.

NEUROCYTE. A neurone.

NEUROFIBRILLAE. Minute fibrils forming a network within the cytoplasm of a nerve cell and passing into the axon and dendrites.

NEUROGLIA (GLIA). Undifferentiated supporting cells forming a packing around the nerve cells in the brain and spinal cord.

NEUROLEMMA (NEURILEMMA). A thin membrane or sheath surrounding a nerve fibre.

NEUROMAST. A system of sensory organs arranged in definite tracts on or near the surface of the body and head of a fish. They comprise the pit-organs or *Ampullae of Lorenzini* and the lateral line organs which, together with the fundamentally similar receptors of the inner ear, make up the *acustico-lateralis system.*

NEUROMERES. Embryonic segmentally arranged blocks of nervous tissue which later give rise to the segmental ganglia in arthropods and annelids.

NEURONE (NEURON). A nerve-cell with its axons and dendrites: the fundamental unit of the nervous system.

NEURONEMES. Thread-like strands of ectoplasm linking the bases of cilia and trichocysts in certain protozoa. They are capable of conducting impulses after the manner of the nerves of higher animals.

NEUROPODIUM. The ventral lobe of the parapodium of a polychaete worm; the dorsal lobe being called the *notopodium.* Parapodia are paired outgrowths of skin bearing bristles, and are present on most of the segments of free-swimming polychaete worms.

NEUROPORE. The opening by which the anterior end of the neural canal communicates with the exterior in a vertebrate embryo.

NEUROPTERA. Lacewings, Alder-flies and Snake-flies: an order of holometabolous insects having two pairs of membranous wings and a soft body. They have biting mouth-parts and are usually predators; the antennae are well developed but there are no abdominal cerci. The larvae, which are also predators, are of the *campodeiform* or long-legged type. They may be aquatic or terrestrial.

NEURO-SENSORY SYSTEM. The whole system of nerves and sensory organs, basically consisting of receptors (on the body surface), connector neurones (within the body) and effectors (in muscles etc.).

NEUTROPHIL. A cell which is neither predominantly basophil nor acidophil; that is to say it stains equally well with basic and acidic stains. The term is used with particular reference to certain types of white blood corpuscles.

NICTITATING MEMBRANE. A third eyelid, sometimes transparent, present in some reptiles, birds and a few mammals.

NIDAMENTAL GLANDS. Shell glands which secrete an egg-capsule, *e.g.* in Cephalopods such as the cuttle-fish and the octopus.

NIDICOLAE. Birds which, when hatched, are helpless and often naked and blind so that they must be fed for a while in the nest.

NIDIFUGAE. Birds which, when hatched, are able at once to leave the nest, run about or swim and generally fend for themselves. Their eyes are open; they have a downy covering and still possess a store of food-yolk.

NISSL'S GRANULES. Granules of a chromophil substance containing a nucleo-protein combined with iron in the cell-bodies of neurones.

NITIDULIDAE. Small predatory beetles often inhabiting the tunnels of bark-beetles and weevils and feeding on their eggs and larvae.

NOCTUIDAE. Owlet moths whose larvae, known as cut-worms, do much damage to young trees by feeding on the roots and young shoots. They are most active at night and hide under the soil during the day.

NODES OF RANVIER. See *Ranvier*.

NODUS. Entomology: a notch in the costal margin of the wing of a dragon-fly or similar insect; a prominent cross-vein leading to this notch.

NOMARTHRA. Pangolins or scaly ant-eaters of S.E. Asia and Aardvarks of South Africa. These were formerly grouped together as a suborder of the Edentata but their affinities are doubtful and they are now generally put into separate orders; the former *Pholidota* and the latter *Tubulidentata*.

NOSE LEAVES. Membranous outgrowths of skin near or around the nose in certain bats. They are believed to be sensitive to high frequency vibrations and changes of pressure, enabling the animal to detect the presence of obstacles without touching them.

NOTACANTHIDAE. Deep-sea eel-like fish having pelvic fins in the abdominal position and having an air-bladder with an open duct to the back of the pharynx.

NOTHOSAURIDAE. Triassic Plesiosauria with elongated limbs adapted for moving on land as well as in water.

NOTIDANI (NOTIDANOIDEA). The most primitive of living sharks characterized by the presence of six or seven pairs of gill clefts and having a single spineless dorsal fin.

NOTO-. Prefix from Greek *Notos*: back.

NOTOCHORD. An elastic skeletal rod running along the back beneath the nerve cord in the embryos or adults of all members of the phylum Chordata. In the development of a vertebrate the skeletal elements of the vertebral column surround and more or less obliterate the notochord.

NOTODONTIDAE. A family of moths including the Puss Moth and the Buff-tip, whose caterpillars do much damage to common trees such as willow, poplar, oak, elm and lime. In some species the cocoon bears a remarkable resemblance to the bark of the tree on which it lives.

NOTOGAEA. A zoological region comprising Australia and the neighbouring islands in which Monotremes and most Marsupials are to be found; bounded by Wallace's Line.

NOTONECTAL. Organisms which swim on their backs, *e.g.* Water boatmen (see below).

NOTONECTIDAE. Water Boatmen: carnivorous insects of the group Hemiptera-heteroptera which swim on their backs moving their flattened hind limbs in unison like oars.

NOTOPODIUM. The more dorsal lobe of the *parapodium* in Polychaete worms (see also under *Neuropodium*).

NOTORYCTIDAE. Marsupial moles of Australia having the general appearance and burrowing habits of moles although in no way related to them. This is a case of convergent evolution in which similarity of habit has led to similarity of structure.

NOTOSTRACA. Crustaceans of the class Branchiopoda having a broad flattened carapace, short antennae, small compound eyes, numerous trunk-limbs and two long jointed caudal filaments.

NOTOTHENIINA. Blenny-like fish of the Antarctic, having the pelvic fins in the jugular position and the dorsal fin divided into two parts.

NOTUM. A tergum or dorsal cuticular plate covering each of the three thoracic segments of an insect. According to their position they are named *pronotum*, *mesonotum* and *metanotum*.

NUCLEAR MEMBRANE. A membrane enclosing the nucleus of a cell and separating it from the surrounding cytoplasm.

NUCLEOLUS. A structure resembling a small nucleus within the normal nucleus of a cell.

NUCLEUS. A compact body which is present in most cells and contains hereditary material in the form of chromosomes surrounded and separated from the rest of the cell by a nuclear membrane. The nucleus apparently initiates and governs the division of a cell (see also *Meganucleus* and *Micronucleus*).

NUCLEUS PULPOSUS. See under *Intervertebral disc*.

NUDA. A small group of Ctenophora or 'sea-gooseberries' without tentacles.

NUDIBRANCHIATA. Sea-slugs: Gasteropod molluscs in which the shell, gills and certain other organs are completely absent. They belong to the class Opisthobranchiata in which the loss of various organs is probably brought about by the phenomenon of *torsion* followed by *detorsion* (*q.v.*).

NUMIDINAE. Guinea-fowls: turkey-like birds of Africa and Madagascar.

NUMMULITES. Large marine Foraminifera with flat spiral shells, some of the fossil forms being an inch or more in diameter. The name is given from the coin-like appearance of these.

NUPTIAL PADS. Wart-like swellings on the hands and arms of certain frogs and toads; present only on the males and used for grasping the females during copulation.

NURSES. Young worker bees whose chief occupation is the feeding of the larvae.

NUTRITION. The taking in, breaking down, absorption and assimilation of food.

NYCT-. Prefix from Greek *Nux, nuktos*: night.

NYCTERIBIDAE. A family of blind, wingless flies with long spider-like legs, living in the fur of bats and sucking their blood.

NYCTERIDAE. Large-eared carnivorous bats of the tropics and sub-tropics, having cutaneous appendages or *nose-leaves* on the margins of the nostrils.

NYCTIPITHECINAE. Nocturnal monkeys of Central and South America, having owl-like faces, large eyes and long non-prehensile tails.

NYMPH. An insect larva which hatches out in a well developed state closely resembling the adult except for the fact that the wings and reproductive organs are undeveloped. The former appear as wing-buds at an early age and gradually increase in size with each instar. There is no pupal stage.

O

OB-. Latin prefix and preposition: in the way of, against.

OBLIGATE PARASITE. A parasite which cannot live independently of its host.

OBTECT PUPA. The type of pupa or chrysalis found in most butterflies and moths in which the wings and appendages are glued down to the body and most of the abdominal segments are immovable.

OBTURATOR FISSURE. A fissure separating the hind-part of the pubis from the ventral border of the ischium in the skeleton of a bird. It corresponds to the *ischio-pubic foramen* of reptiles and the *obturator foramen* of mammals.

OBTURATOR FORAMEN. A large opening between the pubis and the is-chium on each side of the pelvic girdle of a mammal (see *Obturator fissure* above).

OCCIPITAL REGION (OCCIPUT). (1) The hind region of a vertebrate cranium (occipital bones, condyles etc.).
(2) The dorsolateral cuticular plates of the head of an insect.

OCELLUS. A simple eye consisting of a few sensory cells and a single cuticular lens; present in insects and a few other invertebrates.

OCT-, OCTO-, OCTA-. Prefixes from Latin *Octo*, Greek *Okta*: eight.

OCTODONTIDAE. A name formerly given to a group of rodents, mostly of Central and South America, including the porcupine-rat which has small spines in the fur, and the Coypu which is aquatic and has recently become a pest in certain parts of England.

OCTOPODA. Octopuses: Cephalopod molluscs with eight tentacles round the head and having a rounded body with no shell. They are closely related to cuttle-fish and squids, but the latter have a rudimentary internal shell and have ten tentacles.

OCULAR PLATES (OF ECHINODERMS). Five calcareous plates between which are five genital plates (bearing genital pores), the whole ten forming a ring round the anal region on the aboral surface of an echinoid or sea-urchin.

OCULO-. Prefix from Latin *Oculus*: eye.

OCULOMOTOR GANGLION. See *Lenticular ganglion*.

167

OCULOMOTOR NERVE. The third cranial nerve of a vertebrate, helping to move the eyeball by innervating the anterior, superior and inferior rectus muscles and the inferior oblique muscle.

ODONATA. Dragonflies: large hemimetabolous insects with long brightly coloured bodies and two pairs of membranous wings with numerous cross-veins. The jaws are large and powerful for seizing small insects in flight; the eyes are very well developed but the antennae are small. The aquatic nymph or larva is predatory and is characterized by an elongated prehensile labium forming the so-called *mask*.

The Odonata are closely related to the *Protodonata* which are found as Palaeozoic fossils and sometimes attained great size with a wing expanse exceeding 2 ft.

ODONTO-. Prefix from Greek *Odous, odontos*: tooth.

ODONTOBLASTS. (1) Cells which give rise to dentine in the vertebrate tooth.
(2) Cells which give rise to simple teeth, as for instance those on the radula or tongue of a snail.

ODONTOCETI. Toothed whales. See *Delphinoidea*.

ODONTOID PROCESS (ODONTOID PEG). A tooth-like projection of the axis vertebra, directed forwards and fitting into a hole in the atlas vertebra. By this means the atlas can rotate on the axis enabling the head to turn. Such an arrangement is found in reptiles, birds and mammals.

ODONTOLCAE. Extinct marine flightless birds of the Cretaceous Period, without a sternal keel, with reduced wings and with teeth set in grooves in the jaws.

OEDICNEMIDAE. Stone curlews: wading birds distinguished from ordinary curlews by the absence of a hind-toe (see *Limicolae*).

OEGOPSIDA. Gigantic cephalopod molluscs up to 60 ft. in length, found at great depths in the sea. They are strong swimmers and some have phosphorescent organs. The suckers on the ten tentacles are sometimes modified into hooks.

OENOCYTES. Large wine-coloured cells present in various parts of the bodies of insects, probably producing a hormone connected with the moulting process.

OENOCYTOIDS. Rounded acidophil corpuscles of unknown function in the blood of insects.

OESOPHAGEAL BULBS. A pair of glandular swellings in the wall of the oesophagus in nematode worms.

OESOPHAGEAL GLANDS. (1) Any glands which discharge secretions into the oesophagus.
(2) Of earthworms etc. Lateral glands which secrete calcium carbonate into the oesophagus, thereby getting rid of excess of calcium which has been taken in with the food.

OESOPHAGEAL POUCHES. Lateral diverticula in the oesophagus of an earthworm connected with the calciferous glands (see *Oesophageal glands* above).

OESOPHAGUS. That part of the alimentary canal leading from the pharynx to the stomach.

OESTRIDAE. Warble-flies and Bot-flies: dipterous insects whose larvae are endoparasites of mammals, doing much damage to the hides of cattle etc.

OESTROUS CYCLE. A cycle of activity of the reproductive organs of many mammals, usually consisting of the following stages:

(1) Growth of *Graafian Follicles* (*q.v.*) in the ovary and secretion of the hormone *oestrogen*.

(2) Ovulation accompanied by the urge for mating (*oestrus* or 'heat').

(3) Formation of *Corpora lutea* (yellow bodies) in the ovary and secretion of the hormone *progesterone*.

(4) Reversion to stage 1.

These stages recur at regular intervals if the female is not pregnant, but if fertilization takes place the cycle is suspended in stage 3 and the corpora go on enlarging with further hormone production. A modification of the oestrous cycle occurs in humans and some apes in which *menstruation* or destruction of the mucous membrane of the uterus occurs after the progesterone stage.

OESTRUS. The period of 'heat' coinciding with ovulation, at which time a female mammal will copulate with a male (see also *Oestrous Cycle* above).

OIL GLAND. Of Birds: see *Uropygial gland*.

OLECRANON PROCESS. A part of the mammalian ulna projecting back behind the humerus for attachment of the triceps and other muscles which straighten the arm or fore-limb.

OLFACTORY. Connected with the sense of smell.

OLFACTORY BULB. See *Olfactory lobes*.

OLFACTORY CAPSULES. See *Nasal capsules*.

OLFACTORY LOBES. The anterior parts of the fore-brain of a vertebrate, from which arise the olfactory nerves.

OLFACTORY NERVES. The first pair of cranial nerves of a vertebrate, running from the organs of smell directly into the fore-brain.

OLFACTORY PIT. See *Kölliker's Pit*.

OLFACTORY TRACTS. Tracts of nervous tissue connecting the olfactory lobes with the rest of the brain in a vertebrate.

OLIGO-. Prefix from Greek *Oligos*: small, *oligoi*: few.

OLIGOCHAETA. Annelids of the earthworm type having few chaetae and no parapodia. They are hermaphrodite and reproduce by cross-fertilization. Eggs are laid into a cocoon secreted by the saddle or *clitellum*.

OLIGOPOD LARVA. See *Campodeiform larva*.

OLIGOTRICHA. Protozoa having a gullet with an undulating membrane and an adoral wreath of cilia, but having few cilia over the rest of the body. The group includes fresh water and marine species and some which live as symbionts in the alimentary canals of mammals.

OMASUM. An alternative name for the *Psalterium* or third stomach of a ruminant.

OMENTUM (GREAT OMENTUM, OMENTUM MAGNUM). A mesentery or membrane supporting the stomach of a vertebrate. In some mammals it is very large and extends back in folds over part of the small intestine.

OMENTUM BURSA. A sac formed by folds of the large omentum in the region of the stomach, communicating with the cavity of the peritoneum by the *Foramen of Winslow*.

OMMATIDIUM. A single optical unit of the compound eye of a crustacean or insect. Each ommatidium is a cone-shaped structure consisting of a cuticular lens, a crystalline cone and retinulae or sensitive cells at the base. The whole eye is roughly hemispherical with all the ommatidia converging inwards towards the optic nerve.

OMOSTERNUM. A part of the sternum in front of the clavicles in some amphibia. In the frog, for instance, the sternum consists of four segments. From posterior to anterior these are the *xiphisternum, mesosternum, omosternum* and *episternum*.

OMPHALOIDEAN. Appertaining to the umbilicus.

OMPHALOIDEAN TROPHOBLAST. That part of the mammalian trophoblast which is on the umbilical side of the embryo and covers a part of the yolk sac. It may form a temporary placenta before the large allantoic placenta is formed.

ONCHOSPHERE (ONCOSPHERE). The embryonic form of a tapeworm consisting of a spherical chitinous shell in which is a six-hooked or hexacanth embryo.

ONYCHO-. Prefix from Greek *Onux, onuchos*: a nail.

ONYCHOPHORA. A small subphylum of primitive arthropods consisting of the single genus *Peripatus*. These creatures have a soft worm-like body with many segments each bearing two simple clawed limbs. Breathing is by tracheae and spiracles; excretion is by tubules contained in the limbs. It appears that the Onychophora are intermediate between annelids and arthropods and in some respects also resemble primitive molluscs. They are probably descended from ancestors which diverged from the main evolutionary line at a very early stage.

ONYCHOPODA. A group of small crustaceans or water fleas of the group Cladocera, related to the daphnians but differing from them in having a reduced carapace which is not fused with the hind-part of the thorax.

OO-. Prefix from Greek *Oon*: an egg.

OOBLASTEMA. A fertilized ovum.

OOCYST. A cyst formed round two conjugating gametes or round a zygote, as for instance in certain protozoa such as the malarial parasite.

OOCYTE. A cell which, on undergoing meiosis or reduction division, gives rise to an ovum. Usually a primary oocyte first produces a secondary oocyte and a small *polar body*; the secondary oocyte then divides again to form an ovum and a second polar body.

OOGAMOUS, OOGAMY. Producing sexually differentiated gametes in which the female is larger and less mobile than the male.

OOGENESIS. A series of cell-divisions and changes resulting in the production of mature ova or egg-cells. Starting from the primordial germ-cells of the ovary, the successive stages are *oogonium, primary oocyte, secondary oocyte* and *ovum*. Meiosis or reduction of chromosomes takes place when the primary oocyte divides to form the secondary oocyte.

OOGONIUM. A cell which, by division, gives rise to oocytes and hence to ova. The term is also used to denote the female sex-organs of certain lower plants.

OOKINESIS. The mitotic stages in development of an ovum.

OOKINETE. A motile zygote of certain protozoa such as the malaria parasite.

OOLEMMA. A vitelline membrane: a membrane surrounding and secreted by an ovum.

OOSPERM. A fertilized ovum.

OOSTEGITES. Plates on the thoracic limbs of certain crustaceans, forming a brood-pouch in which the young develop; they are characteristic of the group *Peracarida*.

OOSTEGOPOD. A limb bearing an oostegite.

OOTHECA. An egg-case or capsule, as of certain insects; formed by the hardening of a sticky secretion from the colleterial glands.

OOTOCOID. An animal which normally gives birth to undeveloped young, *e.g.* a kangaroo.

OOTYPE. A thickened region of the oviduct in the neighbourhood of the shell-gland in Platyhelminthes; it is well developed in the liver fluke and helps to shape the eggs before they pass into the uterus.

OOZOOID. A zooid or individual unit of a colonial organism, developing from an ovum, as distinct from a *blastozooid* which develops by budding. The terms are used with reference to colonies of Tunicates or similar organisms.

OPALINIDAE. Protozoa of the group *Prociliata*, without a mouth and having uniform ciliation over the whole body. They are usually multinucleate but without differentiation into micro- and meganuclei.

OPERCULUM. A lid or flap of skin covering an aperture, *e.g.*:
    (1) The gill-slit cover of fish and larval amphibia.
    (2) In Tubicolous Polychaete worms, an enlarged stopper-like branch of a tentacle, used for closing the mouth of the tube when the worm is retracted.
    (3) The *epiphragma*: a horny calcareous membrane secreted by a snail or similar mollusc to block the aperture of the shell when the animal is withdrawn and hibernating.

OPHIDIA. Snakes: elongated limbless reptiles covered with horny epidermal scales without underlying osteoderms. The skull has movable quadrate bones and the two halves of the lower jaw are united by an elastic ligament enabling the mouth to be opened extremely wide. Snakes have a large number of vertebrae and movable ribs, but no sternum. The forward and backward movement of the ribs and of the ventral horny plates attached to them make rapid locomotion possible.

OPHIDIIDAE (OPHIDIINA). Eel-like fish usually without pelvic fins and with a tapering tail.

OPHIOCEPHALOUS. See *Pedicellariae*.

OPHIUROIDEA. Brittle stars: Echinoderms whose long narrow arms are more flexible than those of a star-fish. Locomotion is by means of these arms and not by tube-feet as in a star-fish.

OPHTHALMIC GANGLION. See *Lenticular ganglion*.

OPHTHALMIC NERVES (SUPERFICIAL OPHTHALMIC NERVES). Branches of the Trigeminal and Facial nerves passing above the eyes to the sensory organs of the snout in a vertebrate.

OPHTHALMIC PROFUNDUS NERVE. A sensory nerve innervating the skin of the snout in many fish. It arises in front of the main root of the trigeminal nerve and embryologically is the dorsal root of the first segment of the head. The corresponding ventral root in this segment gives rise to the oculomotor nerve.

OPISTHO-. Prefix from Greek *Opisthen*: behind.

OPISTHOBRANCHIATA (OPISTHOBRANCHIA). Gasteropod molluscs which, in the course of development, first undergo torsion but afterwards

171

reverse this process. As a consequence various organs including the shell tend to be reduced or absent. The group includes the sea-hare (*Aplysia*), the sea-butterflies (*Pteropoda*), the sea-slug (*Eolis*) and the sea-lemon (*Doris*). (See *Torsion* and *Detorsion*.)

OPISTHOCOELOUS. Vertebrae which are concave behind and convex in front.

OPISTHOGLYPHA. Snakes which have one or more of the posterior maxillary teeth grooved in front and which are usually poisonous but not seriously harmful to man. They are found in most parts of the world and may be terrestrial, arboreal or aquatic.

OPISTHOGONEATA. See *Chilopoda*.

OPISTHOMI. Spiny eels of Lake Tanganyika, having median fins with spines and having no pelvic fins.

OPISTHOSOMA. The hind-part of the body of an arachnid, consisting in the most primitive forms of thirteen segments and a telson, but tending to become shortened and very variable in the majority of the more specialized forms. In spiders it forms a more or less globular and apparently unsegmented 'abdomen'. In scorpions it forms a segmented tail with a sting at the end.

OPISTHOTIC BONE. The lower posterior bone of the ear capsule of a bony fish. In higher vertebrates this is fused with other otic bones to form the *periotic* bone.

OPTIC CHIASMA. An X-shaped structure beneath the brain of a vertebrate, where some or all the fibres of the right optic nerve cross to the left side of the brain and *vice versa*.

OPTIC FORAMEN. The opening in the vertebrate cranium through which the optic nerve emerges.

OPTIC LOBES. Brain-centres from which the optic nerves arise. In vertebrates they form enlargements in the roof of the mid-brain; two in lower vertebrates and four, the *corpora quadrigemina*, in mammals.

OPTIC NERVE. The sensory nerve of the eye. In vertebrates it is the second cranial nerve, in reality an outgrowth from the wall of the brain.

OPTIC STALK. The developing optic nerve of a vertebrate: a lengthening of the optic vesicle which is an outgrowth from the fore-brain.

OPTIC THALAMI. The thickened sides of the thalamencephalon in the vertebrate brain, from which arise the primary optic vesicles.

OPTIC VESICLES. Hollow outgrowths from the thalamencephalon in the embryonic brain of a vertebrate, giving rise to the optic nerves and to the retinae.

OPTOCOELS. Cavities sometimes present in the optic lobes of the vertebrate brain. They open out from the *Iter* or passage linking the third and fourth ventricles.

ORAL. Appertaining to or adjacent to the mouth.

ORAL CONE. A cone-like structure, usually surrounded by tentacles, at whose apex is the mouth in such creatures as *Hydra* and similar polyps.

ORAL HOOD. A hood of skin extending forwards dorsally and laterally round the mouth-parts of *Amphioxus*. The sides of this hood bear sensory cirri and its inner surface bears a ciliated tract known as the *wheel-organ*.

ORAL PLATE. In a vertebrate embryo, a sheet of tissue formed from the

172

endoderm of the fore-gut where this becomes close to the overlying ectoderm and marks the position of the future mouth, though this has not yet perforated (*cf. Anal plate*).

ORAL SURFACE. The surface of an organism upon which the mouth is situated, the opposite surface being called the *aboral*. These terms are used with reference to radially symmetrical animals such as star-fish and jelly-fish in which the dorsal and ventral surfaces of the adult do not correspond with the original dorsal and ventral surfaces of the embryo or larva.

ORBIT. The eye-cavity in the skull of a vertebrate.

ORBITAL CARTILAGE. That part of a cartilaginous skull which forms the back of the orbit.

ORBITAL SINUSES. Blood-sinuses behind the eyes, leading into the anterior cardinal sinuses of a fish.

ORBITOSPHENOIDS. A pair of cartilage-bones forming part of each side of the vertebrate cranium. In mammals each orbitosphenoid is bounded by the *presphenoid* below, the *frontal* above, the *alisphenoid* behind and the *ethmoid* in front.

OREODONTIDAE. Extinct four-toed ungulates of the Eocene and Pliocene Periods.

ORGANELLES (ORGANELLAE, ORGANOIDS). Parts of a cell having a definite structure and function, *e.g.* cilia, flagella etc.

ORGANIZER. See *Spemann's Organizer*.

ORGAN OF CORTI. The auditory sense-organ within the spiral cochlea of the mammalian ear.

ORGANOGENY (ORGANOGENESIS). The formation and differentiation of the organs in an embryo.

ORIOLIDAE. Orioles: a family of passerine birds with bright yellow and black plumage, found in Europe, Asia and Australia. The American orioles belong to a different family, the *Icteridae*.

ORNITHO-. Prefix from Greek *Ornis, ornithos*: a bird.

ORNITHISCHIA (PREDENTATA OR ORTHOPODA). Large herbivorous dinosaurs, often bipedal, having no teeth on the premaxillae and having the side teeth flattened with serrated anterior and posterior cutting edges. The hind-feet had either three or four toes and the limb-bones were sometimes hollow. The name Ornithischia indicates that the pelvic girdle was bird-like in having an additional post-pubic bone stretching back on each side beneath the ischium. Included in this group are *Iguanodon, Stegosaurus* and *Triceratops*.

ORNITHODELPHIA. An alternative name for the Monotremata.

ORNITHOPODA. Large herbivorous dinosaurs mainly bipedal and having bird-like hind-feet with three toes. The best known example is *Iguanodon*.

ORNITHORHYNCHUS. The duck-billed platypus of Australia: a primitive reptile-like mammal, semi-aquatic in habits, burrowing in the banks of streams. It has thick fur, webbed feet, a duck-like bill and lays eggs (see *Monotremata*).

ORNITHOSCELIDA. An alternative name for the Dinosaurs.

ORO-NASAL GROOVE. A groove connecting the nasal apertures with the mouth in most cartilaginous fish.

ORTHO-. Prefix from Greek *Orthos*: straight.

ORTHOGENESIS. An inherent trend in certain animals or plants to evolve in a particular direction, *e.g.* the tendency to increase in size among Mesozoic reptiles; the tendency to reduce the number of bones in all vertebrate groups etc.

ORTHOGNATHISM. A condition in some mammals, such as man, in which the facial angle approaches a right angle and the jaw does not project forwards much. The opposite condition is known as *prognathism*.

ORTHOGRADE. Walking on two legs with spinal column erect as in man and a few higher apes.

ORTHOPODA. Another name for the *Ornithischia* (*q.v.*).

ORTHOPTERA. Straight-winged insects: a large order of Hemimetabola with biting mouth-parts and usually having the head bent at right angles to the body. The fore-wings are thickened to form horny tegmina which cover the folded membranous hind-wings; a few species are wingless. The order includes cockroaches, crickets, locusts, grasshoppers, stick-insects etc.

ORTHORRHAPHA. Two-winged flies whose pupae split longitudinally along the mid-dorsal line. They include the *Nematocera* (with long antennae) and the *Brachycera* (with short antennae). Gnats, mosquitoes and crane-fiies belong to the former group; horse-flies and robber-flies to the latter.

ORTHOSTICHOUS. A primitive type of fin in which the cartilaginous rays are all parallel to one another. Such a state is found in the fossil shark-like fish *Cladoselache*.

ORYCTEROPODIDAE (TUBULIDENTATA). African ant-eaters or aard-varks formerly placed in the group Edentata and the subgroup *Nomarthra* but now generally believed not to be closely related to the other Edentata.

OS. Latin. A bone.

OS CALCIS. The calcaneum. fibulare or heel-bone: a name used more particularly with reference to the human skeleton.

OS CLOACAE (HYPO-ISCHIUM). A continuation of each ischium backwards to support the cloaca in some reptiles.

OS COCCYGIS. (1) The Coccyx: the last few fused vertebrae of some mammals including man.
(2) The Urostyle of frogs etc. (see *Urostyle*).

OSCULUM. A large opening from which water passes out of a sponge.

OS INNOMINATUM. The innominate bone: the right or left half of the mammalian pelvic girdle consisting of the ilium, ischium and pubis fused together.

OSMO-REGULATION. A mechanism for keeping the water content of an organism constant and counteracting the tendency for water to pass in or out by osmosis. In *Amoeba*, for example, owing to the dilution of the salts in the surrounding medium there is a tendency for water to enter all the time. This water collects in contractile vacuoles which periodically expel it. Marine protozoa, on the other hand, have no need for such vacuoles since the greater concentration of salts in the surrounding water prevents any osmotic influx.

OSMYLIDAE. Giant lacewing-flies: insects belonging to the order Neuroptera and closely related to the green and brown lacewing-flies though considerably larger. They are to be found along banks of woodland streams where there is dense overhanging vegetation. The larvae live in wet moss or under stones.

OS PENIS. A bony support in the penis of some mammals, *e.g.* carnivora.

OSPHRADIUM. A sensory organ resembling a gill in whelks and similar molluscs.

OSPHRAMENIDAE. Brilliantly coloured spiny-rayed fish from S.E. Asia and Africa. The Paradise Fish and the Siamese Fighting Fish are examples.

OS PLANUM. The lateral portion of the mammalian ethmoid bone, covered by the maxillae.

OSSICLES. Any small bones, but particularly the Auditory Ossicles.

OSSIFICATION. The transformation of cartilage into bone. The process involves firstly a breaking down of the cartilaginous matrix by special amoeboid cells known as *chondrioclasts* and subsequently a deposition of calcareous matter by different cells known as *osteoblasts*.

OSTARIOPHYSI. Bony fish in which the air-bladder is connected to the ear by a series of small bones known as *Weberian Ossicles*. They include the majority of fresh-water fish.

OSTE-, OSTEO-. Prefix from Greek *Osteon*: bone.

OSTEICHTHYES. Bony fish: all fish whose skeleton is of true bone as distinct from cartilaginous fish or *Chondrichthyes*.

OSTEOBLASTS. Cells which secrete the calcareous matrix of bone.

OSTEOCLASTS. Multinucleate amoeboid cells which (*a*) break down cartilage prior to its transformation into bone, or (*b*) break down the calcareous substance of bone so that its shape may be remodelled as it grows. The former should more strictly be called *chondrioclasts* but the name osteoclast is often used for both.

OSTEOCYTES. Bone cells, *i.e.* osteoblasts which have become included as the bone has developed.

OSTEODERMS (OSTEOSCUTES). Bony plates covering the bodies of many reptiles and underlying the horny epidermal scales.

OSTEOGLOSSIDAE (OSTEOGLOSSOIDEA). Large fresh-water fish of the Tropics having the body covered with large mosaic-like scales.

OSTEOLEPIDAE (OSTEOLEPIDOTI). Extinct fish of the Devonian Period belonging to the subclass *Crossopterygii* and therefore closely related to the ancestors of land vertebrates. They had a well ossified skull including a ring of bones round each orbit; the frontal and parietal regions were hinged together and there was a dermal inter-parietal bone. The body was covered with cosmoid scales. The paired fins were of the *archipterygian* type and the tail was slightly heterocercal.

OSTEOLEPID SCALES. See *Cosmoid scales*.

OSTEOSCUTES. See *Osteoderms*.

OSTEOSTRACI. Extinct scaly fish-like vertebrates of Silurian and Devonian times without jaws and having a large dorsal shield covering the head (see *Ostracoderma*).

OSTIA (OSTIOLES). (1) Any small openings into a body cavity, *e.g.* small intracellular passages through which water enters the flagellated chambers of a sponge.

(2) Openings by which blood flows into the heart from the pericardial sinus of a crustacean or an insect.

OSTRACIONTIDAE. Trunk-fishes or coffer-fishes: tropical fish having the whole body covered with thick bony plates, most of which are large and hexagonal; having a small mouth, no pelvic fins and a soft dorsal and anal fin opposite to each other near the tail.

OSTRACO-. Prefix from Greek *Ostrakon*: an oyster shell.

OSTRACODA. Small crustaceans with a bivalve carapace which can be closed by an adductor muscle and usually having large antennae with fans of setae used for feeding. A typical example is the common British fresh-water genus *Cypris*.

OSTRACODERMA (OSTRACODERMI). A group of extinct fish-like vertebrates without jaws and having the body covered with bony plates. They are usually classed with the Cyclostomata as *Agnatha*.

OTARIIDAE. Eared seals or sea-lions differing from true seals in having ears with a distinct pinna and by the fact that the hind-limbs are turned forwards beneath the body making locomotion on land relatively easy. They are found in the Antarctic and up the Pacific coast of America.

OT-, OTO-. Prefix from Greek *Ous, otos*: the ear.

OTIC GANGLION. A ganglion connected with the glossopharyngeal nerve of vertebrates, from which excitor neurones innervate the parotid salivary glands.

OTIC PROCESS. A cartilaginous or bony process by which the upper jaw becomes attached to the auditory capsule during the embryonic development of the skulls of all land vertebrates and certain fish (see *Amphistylic* and *Autostylic*).

OTIDIDAE. Bustards: large marsh birds of the Old World, allied to the cranes. They have been extinct in England since 1838 but are plentiful in S.E. Europe and Persia.

OTIDIUM. A molluscan otocyst or statocyst.

OTIORRHYNCHINAE. A subfamily of weevils having short thick snouts. They include many common species most of which do considerable damage to all kinds of trees. The larvae feed on the roots and the adults on foliage, buds and young stems.

OTOCONIUM. An otolith (*q.v.*).

OTOCYST. See *Statocyst*.

OTOLITHS (STATOLITHS). Chalky particles suspended in the fluid of the utricle and sacculus in the vertebrate ear or in statocysts of certain invertebrates. As the animal puts itself into different positions, the stimulation of sensory 'hair-cells' by these particles gives a sense of balance and brings about reflex actions causing automatic adjustment of position.

OV-, OVO-, OVI-. Prefixes from Latin *Ovum*: an egg.

OVARIAN FUNNELS. Funnels formed by a widening of the anterior end of the oviduct in many animals. They serve to collect the ova which are shed into the coelomic cavity from the ovary.

OVARIAN LAMELLA. The form in which the eggs of barnacles develop as a flat mass held together by a glutinous secretion and attached to the inside of the mantle.

OVARIOLES (OVARIAN TUBULES). Separate egg-tubes or branches of the ovaries, characteristic of insects. The number of these varies considerably, being very great for instance in the queen bee. There are three types of ovariole known respectively as the *panoistic*, the *polytrophic* and the *acrotrophic* types. (See separate headings for details of these.)

OVARY. The female reproductive organ. The organ which produces eggs or egg-cells.

OVICELL. A much modified zooecium or capsular body-wall in some species of Polyzoa; shaped to serve as a brood-pouch in which the primary embryos develop.

OVIDUCAL GLAND. A gland which secretes substances into the oviduct, as for instance that which produces the shell of an egg.

OVIDUCT. The duct down which eggs or egg-cells pass from the ovary to the exterior.

OVIGEROUS FRENUM. A projecting fold inside the mantle of a barnacle on which the eggs are held by a glutinous secretion during developjment (see *Ovarian lamella*).

OVIPAROUS. Reproducing by laying eggs.

OVIPOSITOR. An external organ specially adapted for egg-laying.

OVISAC. A sac in which eggs develop or mature.

OVOLEMMA. See *Zona pellucida*.

OVOTESTIS. A hermaphrodite organ which can at times produce either eggs or sperm cells, as for instance in a snail.

OVOVIVIPAROUS. Producing eggs which are not actually laid, but undergo development in the oviduct so that the young are born as free-living individuals. Such a condition occurs in the viper.

OVULATION. The release of eggs from an ovary.

OVUM. The female gamete or reproductive cell consisting of a haploid nucleus and a varying amount of yolk surrounded by a vitelline membrane. According to the amount of yolk present, the ovum is classed as *megalecithal* (much yolk) or *microlecithal* (little yolk). If the ovum is a true egg and not just an egg-cell, it may be surrounded by albumen, a number of membranes and a shell. The main function of the latter is to prevent the egg from drying up. Eggs laid in water are therefore usually without a shell.

OXYHAEMOGLOBIN. The red pigment of blood in an oxygenated state, *i.e.* loosely combined with oxygen which it has received from the respiratory organs and which it can easily give up on reaching the tissues.

OXYNTIC CELLS. Glandular cells which secrete dilute hydrochloric acid. They are found in the stomachs of vertebrates in close association with peptic cells. The acid is necessary in order to produce an optimum pH value in which the pepsin can break down proteins to peptones.

P

PACHY-. Prefix from Greek *Pachus*: thick.

PACHYDERM. PACHYDERMATOUS. A thick-skinned animal such as the elephant or rhinoceros.

PACHYTENE STAGE. A stage immediately prior to the diplotene stage in meiosis, when the chromosomes become shorter and thicker and the two members of a homologous pair become intimately coiled round each other.

PACINIAN CORPUSCLE. A sensory nerve-ending surrounded by concentric layers of fibrous tissue forming a bulb-like structure; found in the skin and elsewhere.

PAEDO-. Prefix from Greek *Pais, paidos*: boy, child.

PAEDOGAMY (AUTOGAMY). A process in which the nucleus and cytoplasm of a cell divide and reunite, gametes being formed and syngamy occurring within a single cell.

PAEDOGENESIS. Becoming sexually mature while retaining larval characteristics, as in the Axolotl (see *Neoteny*).

PAGRINA. Sea-breams with a single series of conical teeth in front and molars at the sides; they are carnivorous and inhabit warm shores.

PALAEO-. Prefix from Greek *Palaios*: old, ancient.

PALAEODICTYOPTERA. Primitive insects of the Carboniferous Period resembling dragonflies but having a pair of short wing-like appendages on most of the segments of the body. Possibly in the first place these helped the insect to glide and at a later stage two pairs became greatly enlarged to form true wings.

PALAEO-ENCEPHALON. A primitive type of brain.

PALAEOGENETIC. Showing persistence of characteristics which are normally embryonic.

PALAEOGNATHAE. The more primitive of the two major groups of living birds. Classified on the basis of jaw structures, these have a large prevomer touching the pterygoid bones. All the *Ratites* or flightless birds are included in this group, together with *Tinamus*, an anomolous bird from Central America which can fly well, has a sternal keel, but in many ways resembles the running birds.

PALAEONEMERTINI. A primitive group of nemertine worms in which the proboscis is without stylets.

PALAEONISCOIDEA (PALAEONISCOIDEI, PALAEONISCIDAE). Extinct bony fish of Palaeozoic times having thick ganoid scales, bony plates covering the head, and paired fins consisting of short rounded lobes with divergent radials. It is supposed that these fish arose from an ancestor which was also common to the *Choanichthyes* from which land animals developed. The only remaining descendants of the palaeoniscoid stock at the present day are apparently the sturgeons and the three isolated genera *Polypterus, Lepidosteus* and *Amia* (see also *Palaeopterygii*).

PALAEOPTERYGII. Primitive bony fish with thick ganoid scales and with paired fins of a modified actinopterygian type. The group includes the extinct Palaeoniscoids and the present-day sturgeons and *Polypterus*.

PALAEOSPONDYLIA (PALAEOSPONDYLOIDEA). Extinct fish of the placoderm type (*Aphetohyoidea*), having a flattened skull, a well ossified vertebral column and a heterocercal tail.

PALAEOTHERIIDAE. Extinct ungulate mammals having three toes and including some types which were probable ancestors of the horse (see also *Hyracotheriidae*).

PALAEOZOIC. The period from the Pre-cambrian to the end of the Permian, during which aquatic forms of life were predominant.

PALAMEDEIDAE. South American 'Screamers': goose-like birds with sharp spurs on the wings and with well developed air cavities extending beneath the skin and into the fingers and toes.

PALATE. The roof of the mouth of a vertebrate (see also *False Palate*).

PALATINE BONES. Dermal bones forming part of the roof of the mouth of a vertebrate. In some fish they bear teeth but in higher vertebrates this is not so. In mammals and crocodiles the palatines form the hind-part of the palate, curving round the sides of the nasal passages and abutting on the pterygoids which form the sides of the throat. In these animals the anterior part of the palate is in reality a *false palate* formed from the maxillae and premaxillae.

PALATO-PTERYGO-QUADRATE (PALATO-QUADRATE). That part of the skull which forms the upper jaw in cartilaginous fish.

PALATO-QUADRATE. See *Palato-pterygo-quadrate* above.

PALINGENESIS. The recapitulation of ancestral characters during the development of an embryo.

PALINURA. Crawfish or spiny lobsters: a suborder of decapod crustaceans having a spine in place of the rostrum and having no pincers on any of the legs.

PALLIAL. Appertaining to the mantle of a mollusc.

PALLIAL GILLS. Respiratory organs formed from folds in the mantle of a mollusc.

PALLIAL SINUS. A prolongation of the coelom into the mantle in Brachiopoda etc.

PALLIO-VISCERAL NERVE CORDS. Nerve cords which in some molluscs go both to the mantle and to the viscera. In the group *Amphineura* the palliovisceral system forms a ring round the outer part of the body, linked in many places by transverse connectives.

PALLIUM. (1) The mantle of a mollusc: the outer soft skin next to the shell.
(2) The roof of the cerebrum of a vertebrate. In reptiles it thickens to form the *archipallium* connected with the sense of smell and in mammals thickens in the lateral regions to form the *neopallium*, the main seat of the intelligence.

PALMELLA (PALMELLAR STATE). A pseudo-colonial state sometimes occurring in the Phytomastigina (plant-like protozoa also classed as unicellular algae), in which flagella are withdrawn, cells rounded off and a cyst or jelly containing numerous cells is formed.

PALP (PALPUS). An elongated sensitive tactile organ, usually near the mouth, on many invertebrates such as polychaetes, molluscs and arthropods. In crustacea the palps arise from the basal portions of the mandibles; in insects they are on the labium and the maxillae but never on the mandibles.

PALPIFER (PALPIGER). A lobe bearing a palp, *e.g.* on the maxillae of insects.

PALUDICOLA. (1) Animals which dwell in marshes.
(2) A group comprising most of the fresh-water Turbellaria or free-swimming flatworms: ciliated leaf-like organisms, rather larger than the marine forms, having a ventral mouth and a pair of simple eyes in front (see *Turbellarians* and *Planarians*).

PANCREAS. A digestive gland of vertebrates secreting a mixture of enzymes which pass into the duodenum. It also contains small clusters of cells known as *Islets of Langerhans* which secrete the hormone *insulin*.

PANCREATIC DUCT. The duct down which enzymes flow from the pancreas into the duodenum.

PANDIONIDAE. Ospreys: large falcon-like birds also known as sea-hawks, found in most parts of the world.

PANOISTIC OVARIOLES. Ovarioles or egg-tubes which contain no special feeding cells but secrete a nutrient fluid for the developing eggs; they are found in the more primitive groups of insects.

PANSPOROBLAST. A complex spore-mother cell formed within the syncytium of certain parasitic protozoa of the group Neosporidia. Each pansporoblast contains several sporoblasts which in turn give rise to multicellular spores.

PANT-, PANTO-. Prefix from Greek *Pas, pantos*: all.

PANTODONTIDAE. West African salmon-like fish with very large pectoral fins.

PANTOPODA (PYCNOGONIDA). A small class of Arachnida with very long legs, a small body and a long suctorial proboscis; they are marine and many are ectoparasites of sea-anemones. In addition to the normal four pairs of walking legs there is an extra pair in front. These are known as ovigerous legs and are used by the males for carrying the eggs.

PAPILIONOIDEA (PAPILIONINA). Butterflies: formerly classed as *Rhopalocera*, they are now included in the *Heteroneura*. They may be distinguished from moths by their clubbed antennae and by the absence of frenulum and jugum on the wings.

PAPILLA. Any small nipple or projection on the body of an animal or plant.

PAPILLA ACUSTICA BASILARIS. A patch of sensory epithelium in the cochlea of the ear in reptiles, birds and mammals. In the well developed spiral cochlea of the latter, it is highly specialized and differentiated to form the *Basilar membrane*, the *Organ of Corti* etc.

PAPILLA ACUSTICA LAGENAE. A patch of sensory epithelium in the lagena or rudimentary cochlea of the ear in amphibia and reptiles.

PAPILLA CIRCUMVALLATA. See *Circumvallate papilla*.

PARA-. Greek prefix: by the side of.

PARABASAL BODY. A secondary basal granule: a small body, whose function is somewhat obscure, connected by threads of protoplasm to the basal granule of a flagellum in some protozoa (see *Blepharoplast*.)

PARACHORDALS. Embryonic skeletal elements consisting of a pair of cartilaginous rods parallel to the notochord and on each side of it in the region of the middle and hind-brain of a vertebrate. As the skull develops these rods become fused to the auditory capsules and form part of the base of the cranium.

PARACTINOPODA. An alternative name for the order of echinoderms usually called *Synaptida* (*q.v.*).

PARADISEIDAE. Birds of Paradise and Bower Birds of Australia and New Guinea. The former are brilliantly coloured and often have the tail in the form of two long filaments; the latter are smaller birds which construct 'bowers' in which they play as a prelude to mating. These may take the form of arbours or covered runways beautifully decorated with flowers, shells or feathers.

PARAGASTER. The chief internal cavity of a sponge.

PARAGLOSSAE. The two outer lobes of the ligular of an insect (see *Glossae*).

PARAGNATHA. The two lobes of the metastoma or lower lip of a crustacean.

PARAMERES. (1) The two halves of a bilaterally symmetrical organism.

(2) A pair of inner processes which, together with the claspers and the penis, comprise the external genitalia of a male insect.

PARAMITOME. An interfibrillar substance of protoplasm.

PARAMITOSIS. A process of nuclear division which takes place in certain protozoa and differs from normal mitosis in that the chromosomes do not arrange themselves symmetrically round the equator of the spindle.

PARAMYLUM. A starch-like substance found as granules of reserve food in some flagellate protozoa and in algae.

PARA-OESOPHAGEAL CONNECTIVES (CIRCUM-OESOPHAGEAL COMMISSURES). The two halves of the nerve-collar in annelids, arthropods and some molluscs, connecting the ventral nerve cord with the supra-oesophageal ganglia which form the main parts of the 'brain'.

PARAPHYSIS. A glandular outgrowth on the top of the cerebrum in many vertebrates.

PARAPINEAL EYE. An organ similar to the pineal eye projecting dorsally from the roof of the brain in certain vertebrates. It is visible as a degenerate structure in the lamprey, but in most higher vertebrates it disappears, only the pineal organ remaining.

PARAPODIA. (1) Paired lateral projections on the body-walls of polychaete worms; usually flattened and bearing *chaetae* or bristles. They may be used for locomotion or for respiration or may be modified for other functions, as in some of the tube-dwelling worms.

(2) Similar structures but fewer in number on the bodies of certain Mollusca, *e.g. Aplysia*, the sea-hare.

PARAPROCTS. A pair of ventral plates below the anus in some insects: the reduced sternal lobes of the eleventh abdominal segment.

PARAPSIDA. Reptiles whose skulls have a single superior temporal fossa on each side above the post-orbital and squamosal bones. To this group belong the Ichthyosauria, lizards and snakes.

PARASITE. An organism which lives and feeds in or on another. *Ectoparasites* live on the surface of the host; *endoparasites* within its body. *Facultative* parasites can live either parasitically or free; *obligate* parasites live only parasitically.

PARASITIC CASTRATION. A phenomenon occurring in crabs attacked by the parasitic crustacean *Sacculina*. Male crabs so attacked undergo a change in their secondary sex characters causing them to have a much broader abdomen and to lose their copulatory styles, thus showing a tendency to resemble the female type. In females which are attacked, the gonads disappear.

PARASPHENOID. A membrane-bone forming part of the floor of the cranium in bony fish, amphibia and reptiles.

PARASUCHIA. Large extinct crocodiles of the Triassic Period, having numerous rows of dorsal and ventral scutes on the body and having external nostrils much farther back than those of present-day crocodiles.

PARASYMPATHETIC NERVES. Nerves which bring about opposite effects to those of the sympathetic nerves, the two opposing systems being together known as the *Autonomic System*. The chief parasympathetic nerves are the visceral branch of the vagus nerve and some of the spinal efferent nerves of the sacral region.

PARASYNAPSIS (PARASYNDESIS). Conjugation lengthwise of homologous chromosomes.

PARATHYROID GLANDS. Two pairs of small endocrine organs near the thyroid gland in all vertebrates except fish. They produce a hormone which regulates the calcium and phosphate content of the blood.

PARATOIDS. Aggregations of poison glands forming swellings on the sides and back of the head in some amphibia such as the salamander.

PARAZOA (PORIFERA). Sponges: sessile aquatic animals of a very simple type having a body-cavity lined by collared flagellated cells known as *choanocytes*. Water, entering by numerous pores, is kept circulating by the movements of the flagella and passes out through one or more larger openings. The body-wall is usually strengthened with a skeleton of spicules which may be calcareous, siliceous or of a horny substance known as *spongin*.

PAREIOSAURIA. Extinct reptiles of the Permian and Triassic Periods, having a massive skeleton and stout limbs which lifted the body well off the ground. The bones of the skull were very thick having a roughly sculptured appearance sometimes with spikes and horn-like projections.

PARENCHYMA. Loose connective tissue formed of large cells packing the spaces between the organs in some simple acoelomate animals such as the Platyhelminthes. The name is also used for similar tissue in plants.

PARIDAE. Tits or titmice: small brightly coloured birds, usually omnivorous and having short sharp beaks.

PARIETAL BONES. A pair of large membrane-bones forming the upper and side regions of the vertebrate cranium and covering most of the upper surface of the brain.

PARIETAL FORAMEN (PINEAL FORAMEN). A median opening for the accommodation of the pineal organ between the two parietal bones in the skulls of certain vertebrates.

PARIETAL ORGAN. See *Pineal Body*.

PARIETAL RING. The middle part of the mammalian cranium, composed of the *parietal* bones above, the *alisphenoids* on each side and the *basisphenoid* below.

PAROCCIPITAL PROCESS (PAROTIC PROCESS). A bony projection on each side of the hind-end of a reptilian skull, formed by a fusion of the *opisthotic* and *exoccipital* bones. Abutting against each of these processes is the quadrate bone with which the lower jaw articulates.

PAROÖPHORON. A vestige of the embryonic *mesonephros* or middle kidney which may persist in an adult female mammal.

PAROTIC PROCESS. See *Paroccipital process*.

PAROTID GLANDS. Salivary glands in the cheeks of mammals.

PAROVARIUM. The degenerate vestige of the mesonephric duct in a female reptile, bird or mammal.

PARRIDAE. Long-toed Jacanas or water pheasants: wading birds similar to plovers from the warmer parts of North and South America.

PARTHENOGENESIS. Development of an unfertilized ovum into a new individual genetically identical with the parent. Such development is fairly common among insects and in many cases may be induced artificially by various chemical or physical stimuli.

PARTHENOGONIDIA. Zooids or units which perform asexual reproduction in certain colonial protozoa.

PARTURITION. The process of birth of a mammal.

PASSERIFORMES (PASSERES, PASSERINE BIRDS). The largest group of birds comprising more than half the total number of living species. They are usually small and always have the first toe directed backwards and the other three forwards; the legs are scaly and the wings are *quintocubital, i.e.* they have a complete set of quills. A diagnostic feature of the group is the *aegithognathous* skull which is characterized by having the two palatine bones completely separated from each other with a short wedge-shaped vomer between them. British birds of the group include sparrows, thrushes, larks, swallows, tits and finches.

PATAGIUM. (1) A membrane of skin stretched between the limbs to form the wing of a bat or the 'parachute' of such animals as the flying squirrel.

(2) A short wing-like outgrowth on each side of the prothorax in certain insects.

PATELLA. The knee-cap: a flat bone embedded in the tendon of the knee in birds and mammals. Such a bone formed in the tendon is known as a *sesamoid* bone.

PATHETICUS (TROCHLEAR) NERVE. The fourth cranial nerve of a vertebrate: a motor nerve which innervates the superior oblique muscle of the eye.

PAUNCH. See *Rumen.*

PECORA. A group of ruminants also known as *Cotylophora (q.v.)*, comprising deer, giraffes, cattle, sheep, goats etc.

PECTEN. Any comb-like structure or organ.

(1) A vascular comb-like process from the choroid in the eye of a bird. It projects through the retina near the optic nerve and passes obliquely through the vitreous humour to the lens. Its function is uncertain but it probably supplies oxygen to the inside of the eye.

(2) A comb-like stridulating organ on some insects.

(3) A row of spines or hairs on a palp or chelicera of various arthropods or on the respiratory tube of a mosquito larva etc.

PECTIN-, PECTINO-. Prefix from Latin *Pecten*: a comb.

PECTINATE. Comb-like, as for instance the antennae of some insects.

PECTINEAL PROCESS. A forward directed process of the pubis near the acetabulum in some primitive birds such as the Kiwi and also in the embryos of many other birds.

PECTINES. Tactile organs in the form of projecting combs of chitinous material behind the genital organs of scorpions.

PECTINIBRANCHIATA. Another name for the group of molluscs usually known as *Monotocardia, i.e.* Gasteropods whose heart has only one auricle and whose single gill is in the form of a comb. They include the whelk, the periwinkle and many others (see also *Monotocardia*).

PECTORAL FINS. Paired fins attached to the pectoral girdle of a fish.

PECTORAL GIRDLE. A skeletal support in the body wall for the attachment of the front limbs or fins of a vertebrate. In land animals it usually consists of paired cartilaginous coracoid bones on the ventral side and scapulae on the dorsal side. There may be additional membrane bones such as the clavicles. In mammals the coracoids are absent and the clavicles take their place ventrally. In birds and reptiles, however, the coracoids are prominent. Both coracoids and clavicles are often joined mid-ventrally to the breast-bone or sternum.

PEDAL. Appertaining to the foot, usually of a mollusc: (Pedal ganglia, pedal circulation etc.).

PEDAL GLAND. A gland in the foot of a snail or slug opening on the ventral surface just behind the mouth. It produces a flow of slimy mucus which acts as a lubricant as the snail moves over the ground.

PEDETIDAE. Long-legged jumping hares of South Africa (see also *Hysterico-morpha*).

PEDI-. Prefix from Latin *Pes*, *pedis*: foot.

PEDICEL. (1) A short stalk bearing an organ or a sessile organism.
(2) The second segment of the antenna of an insect.

PEDICELLARIAE. Minute pincers formed of calcareous ossicles embedded in the skins of star-fish and other echinoderms. They are found in large numbers over most of the surface and apparently have the function of protecting the animal from ectoparasites. The chief kinds are:
(*a*) The simple *forceps* having two movable jaws on a fixed basal ossicle.
(*b*) The *ophiocephalous* type with short toothed jaws on a small flexible stalk.
(*c*) The *trifoliate* type: small with three flattened leaf-like jaws.
(*d*) The *tridactyl* with three much longer jaws on a flexible stalk.
(*e*) The *gemmiform* type with a long thin stalk and a globular head. In this type each of the three jaws contains a poison sac.

PEDICULATI. Angler-fish: see *Lophiiformes*.

PEDIPALPS. Paired appendages of various types on the second segment of the prosoma or cephalothorax of arachnids. In scorpions they form large pincers; in *Limulus*, the king-crab, they are used for walking, whilst in spiders they are mainly sensory. The male spider usually has the ends of the pedipalps enlarged and modified for holding sperm during copulation.

PEDUNCLE. (1) Sometimes synonymous with *Pedicel*: a short stalk bearing an organ or a sessile organism.
(2) Peduncles of the brain. Tracts of fibres connecting various parts of the brain to the cerebellum. The *anterior peduncles* lead to the corpora quadrigemina; *middle peduncles* to the Pons Varolii; *posterior peduncles* to the medulla oblongata.

PEGASIDAE. Small flying fish of the Indian and Pacific Oceans having enlarged pectoral fins and the body covered with bony plates.

PELAGIC. Organisms living in the open sea, including both *plankton* and *nekton* (*q.v.*).

PELAGOTHURIDA. Holothuroids or sea-cucumbers of pelagic habit which swim by means of a circle of enlarged webbed tentacles.

PELECYPODA. An alternative name for Lamellibranchiata or bivalve molluscs.

PELLICLE. A hard protective outer layer of protoplasm surrounding certain protozoa. It is distinguished from a cuticle by the fact that the former is living; the latter dead.

PELMATOZOA. Sessile echinoderms including the stalked *Crinoids* or sea-lilies and some fossil types without stalks (see also *Eleutherozoa*).

PELOBATIDAE. A group of toads or frogs characterized by having a well developed tongue; teeth in the upper jaw only; separate Eustachian tubes but the tympanum is often absent or poorly developed. They include the spade-footed toad of Central Europe and several species in S.E. Asia and in Mexico. In some cases the tadpole is larger than the adult.

PELOMEDUSIDAE. Fresh-water carnivorous turtles having the pelvis fused to the carapace; having no nuchal shield and having the neck completely retractile and bending laterally. They are found in Africa, Madagascar and South America.

PELVIC FINS. The hind-pair of lateral fins of a fish, attached to the pelvic girdle and homologous with the hind limbs of a tetrapod. In some fish owing to the separation of the pelvic girdle from the vertebral column, these fins may come to be in a thoracic or even a pectoral position.

PELVIC GIRDLE. The skeletal girdle with which the hind-limbs or fins of a vertebrate are articulated. Normally it consists of a pair of *pubic* bones on the ventral side meeting in the mid-ventral line to form a symphysis, while dorsally on each side is an *ilium* and an *ischium*.

PELVIC NERVES. Nerves which leave the spinal cord in the sacral region and serve the muscles of the large intestines, bladder, urethra etc.

PELVIC PLEXUS. See *Sciatic plexus*.

PELVIC VEINS. Veins branching off from the femoral veins in amphibia and some reptiles, uniting mid-ventrally to form the anterior abdominal vein which carries blood forwards to the liver.

PELYCOSAURIA. Primitive reptiles of the Permian Period, some of which had enormous neural plates with lateral branches.

PEMPHERIDAE. Spiny-rayed fish, allied to the perches, having the air-bladder divided into an anterior and a posterior part and having a pneumatic duct.

PEN. The long pointed horny internal shell of a squid or similar cephalopod.

PENETRATION GLANDS. Glands secreting a histolytic enzyme by which a parasite breaks down the tissues of its host, *e.g.* those round the mouth of a liver fluke.

PENIS. An organ used to introduce seminal fluid from a male into a female.

PENNAE. The larger feathers of birds, *i.e.* quills and contour feathers as distinct from *plumulae* or down-feathers (see *Feather*).

PENNATULACEA. Sea-pens: colonial coelenterates of the group Alcyonaria, having a single long polyp with a horny axis and secondary polyps which are budded off on either side giving the whole colony a feather-like appearance. The basal stalk has a limited power of movement enabling the creature to burrow into the sand.

PENTA-. Prefix from Greek *Pente*: five.

PENTADACTYL LIMB. A limb having five digits.

PENTASTOMIDA. Segmented worm-like parasites sometimes found in the nasal passages of carnivorous animals. Although bearing some resemblance to annelids they are usually classed with the Arachnida on account of the fact that the larvae resemble certain parasitic mites. These larvae frequently infest rabbits or other herbivores where they encyst and only undergo further development on being eaten by a carnivorous animal.

PENTATOMIDAE. Shield-bugs: large and sometimes brightly coloured herbivorous insects of the group Hemiptera-heteroptera. They are characterized by five-segmented antennae and by a very large triangular dorsal shield or scutellum formed by the tergum of the mesothorax extending back over the abdomen.

PEPSIN. An enzyme produced by glands in the wall of the stomach of a vertebrate. In the presence of dilute hydrochloric acid, produced by the oxyntic cells, it breaks down proteins to peptones.

PEPTIC CELLS. PEPTIC GLANDS. See *Pepsin* above.

N   185

PEPTONEPHRIDIA. Nephridia or excretory tubules which, instead of opening to the exterior in the normal way, open into the alimentary canals of certain types of earthworms. They may possibly have a digestive function.

PER-. Prefix from Latin preposition: through, by means of.

PERACARIDA. Crustaceans of the Malacostracan class characterized by the possession of a brood-pouch formed from plates known as *oostegites* on the thoracic limbs. The group includes wood-lice, slaters and sand-hoppers as well as many shrimp-like forms.

PERAMELIDAE. Bandicoots: burrowing insectivorous marsupials of Australasia.

PERCESOCES (MUGILIFORMES). A miscellaneous group of fish including grey mullets, barracudas, sand-eels and many others which are usually pike-like and often carnivorous, but differ from pikes in having spinous fins.

PERCIDAE. Perches: fresh-water fish of temperate regions having the general characters of the Perciformes (see below), having two nostrils on each side and having four pairs of gills.

PERCIFORMES (PERCOMORPHI). Perch-like fish having well developed spinous dorsal fins and having the pelvic fins in the thoracic position. The group includes many marine and fresh-water species.

PERCOPSIDAE. Pike-like fish from fresh water of North America, having the body covered with ctenoid (comb-like) scales and having anal and dorsal fins with very few spines.

PEREIOPODS. Locomotory thoracic limbs of a crustacean.

PERENNIBRANCHIATA. Amphibia in which the gills persist throughout life, as in the cave-dwelling newt *Proteus* and certain others.

PERFORATE FORAMINIFERA. Those forms of Foraminifera in which the shell is perforated by numerous small pores through which the pseudopodia project.

PERI-. Greek prefix: around, about.

PERIBLAST. The tissue surrounding the embryonic disc or blastoderm and merging with the yolk in a meroblastic egg such as that of a bird (see *Meroblastic*).

PERIBRANCHIAL CAVITY. A cavity surrounding the gills, *e.g.* the atrial cavity of *Amphioxus* and of tunicates.

PERICARDIAL CAVITY. (1) A part of the coelom surrounding the heart of a vertebrate and enclosed by the pericardial membrane.

(2) A part of the haemocoel surrounding the heart in Arthropoda and Mollusca. Blood flows from this cavity directly into the heart by way of a number of *ostia* or valvular openings in the wall of the heart.

PERICARDIAL CELLS (INSECTS). Nephrocytes in the heart-region of an insect; believed to be connected with excretion as they can absorb nitrogenous waste substances from the blood, but the whole process is not understood.

PERICARDIO-PERITONEAL CANALS. A pair of small openings connecting the pericardial with the perivisceral cavity in Elasmobranch fish. In all other vertebrates the two cavities are completely separated from one another.

PERICARDIUM. A membrane forming the wall of the pericardial cavity in vertebrates.

PERICHAETINE. Having chaetae or bristles arranged in complete rings round the body, as in some kinds of earthworms.

PERICHONDRAL OSSIFICATION. The initial process in the transformation of cartilage into bone, in which a thin calcareous sheath is formed beneath the perichondrium or outer membrane before the main ossification takes place in the centre of the bone. The formation of this bony shell is necessary to strengthen the structure during the subsequent breaking down of the cartilage and its replacement by bone.

PERICHONDRIUM. A fibrous membrane surrounding a cartilage. If the latter becomes ossified. this membrane is known as the periosteum.

PERICHORDAL VERTEBRAE. Vertebrae which are formed and remain completely outside the notochord and do not invade its sheath. The centrum of each vertebra is therefore a ring of bone. Such a state is found in some fish.

PERI-ENTERIC PLEXUS. A network of capillaries round the intestine, as for instance in an earthworm.

PERIHAEMAL COELOM (OF ECHINODERMS). Circular and radial vessels in which lie the principal 'blood strands' of star-fish and other echinoderms (see *Lacunar system*).

PERILYMPH. Lymphatic fluid surrounding the membranous labyrinth of the inner ear and separating it from the bony auditory capsule which encloses it.

PERIMYSIUM. A sheath of connective tissue which surrounds each fasciculus or bundle of fibres in a voluntary muscle.

PERINEAL GLANDS. Glands present in most mammals close to the anus, their secretion having a characteristic odour by which an animal can recognize its own kind and the opposite sex.

PERINEUM. The part separating the anal aperture from the renal and reproductive apertures in Eutheria or higher mammals.

PERINEURIUM. A sheath of connective tissue surrounding each bundle of nerve fibres within a nerve trunk.

PERI-OESOPHAGEAL BAND. See *Peripharyngeal bands*.

PERIOSTEUM. The fibrous sheath of a bone.

PERIOSTRACUM. (1) The outermost of the three layers which make up the shell of a mollusc, composed of a horny substance, *conchiolin*, beneath which are the calcareous middle layer (prismatic layer) and the innermost (nacreous) layer.

(2) A similar layer on the shell of a Brachiopod.

PERIOTIC BONE. A bone forming the main part of the auditory capsule in higher vertebrates. It is formed by the fusion of five bones which in bony fish are distinct. These are the *pro-otic, opisthotic, sphenotic, pterotic* and *epiotic* bones.

PERIPATUS. See *Onychophora*.

PERIPHARYNGEAL BANDS. Ciliated tracts running round the pharynx of *Amphioxus* and of Tunicates, conveying microscopic food particles from the ventral endostyle towards the dorsal epipharyngeal groove and hence to the opening of the oesophagus (see also *Endostyle*).

PERIPHERAL NERVOUS SYSTEM. The nerves which extend outside the central nervous system, receiving impulses from the sense organs or transmitting them to the various muscles or other effectors. In a vertebrate the peripheral system comprises the cranial and spinal nerves and the autonomic system, while the central nervous system consists of the brain and spinal cord.

PERIPNEUSTIC. The respiratory system of an insect in which lateral spiracles occur on all the segments of the abdomen (*cf. Amphipneustic* and *Metapneustic*).

PERIPROCT. The part of an organism lying around the anus.

PERIPYLAEA (SPUMELLARIA). Radiolaria whose pseudochitinous central capsule is perforated by pores which are uniformly distributed.

PERISARC. A tubular horny covering secreted by certain colonial coelenterates of the orders Calyptoblastea and Gymnoblastea. In the former it forms cups (hydrothecae and gonothecae) round each polyp of the colony; in the latter the polyps are naked but the perisarc surrounds the connecting parts of the colony. In certain corals (Hydrocorallinae) the perisarc is massive and calcified forming the matrix of the coral.

PERISSODACTYLA. Odd-toed ungulates: hoofed mammals whose feet usually have one or three digits, the axis of symmetry of the foot passing through digit number three. In the course of evolution the first and fifth digits have disappeared, and in horses the second and fourth are incomplete (splint bones), leaving only the third which is enormously enlarged. In the rhinoceros all feet have three digits; in the tapir the hind-feet have three and the front feet four. Some of the fossil species show various stages in the reduction of the digits. Perissodactyla were formerly classed as a suborder of the Ungulata but are now regarded as a separate order.

PERISTALSIS. PERISTALTIC MOVEMENT. Waves of muscular contraction which push food along the alimentary canal.

PERISTEDIIDAE. Deep-sea gurnards: spiny-rayed fish with elongated scaly body.

PERISTOME. PERISTOMIUM. The parts of an animal surrounding the mouth, *e.g.*:

    (1) The spiral ciliated groove round the mouth of *Vorticella, Stentor* and similar protozoa.

    (2) The leathery region round the mouth of an echinoderm.

    (3) The oral cone of a hydra or similar organism.

    (4) The first segment of an earthworm containing the opening of the mouth.

PERITONEAL CAVITY (PERIVISCERAL COELOM). The main part of the body-cavity of a vertebrate, containing the principal viscera other than the heart.

PERITONEUM. A membrane lining the body-cavity and hanging in folds or mesenteries containing and supporting the principal visceral organs.

PERITRICHA (PERITRICHIDA). Protozoa of the class Ciliophora, usually sessile and stalked, having a wide spiral gullet round the rim of which is a circle of cirri known as an *adoral wreath*. The rest of the body is unciliated. *Vorticella*, the 'Bell animalcule' is one of the best known examples.

PERITROPHIC MEMBRANE. A thin chitinous tube projecting back from the mid-gut into the hind-gut of most insects and some crustacea. It protects the epithelial cells from damage by the passage of food, but is permeable to enzymes and to digested foods.

PERIVISCERAL COELOM. That part of the body-cavity in which the principal viscera lie.

PERLARIA. Stone-flies: an alternative name for the Plecoptera.

PERRADIAL. A term applied to tentacles and other organs of a medusa or jelly-fish to denote positions in line with the four radial canals on which the gonads lie (*cf. Interradial* and *Adradial*).

PES. The hind-foot of a tetrapod vertebrate.

PESSULUS. A vertical bony septum in the trachea of a bird between the anterior apertures of the two bronchi. It forms part of the syrinx or vocal organ (see also *Syrinx* and *Tympanum*).

PETROMYZONTIDAE (PETROMYZONTIA, PETROMYZONTES). Lampreys. Cyclostomata or fish-like creatures without jaws but with a circular suctorial mouth and a ring of horny teeth at the anterior end. They have seven pairs of gills and two dorsal fins. The latter fact distinguishes them from the *Myxinoidei* or Hag-fish.

PETROSAL NERVES. Parts of the autonomic nervous system which excite the lachrymal, salivary and nasal glands.

(1) *Greater superficial petrosal nerve*. A part of the palatine branch of the facial nerve going to the spheno-palatine ganglion and hence to the lachrymal glands and the glands of the nose.

(2) *Lesser superficial petrosal nerve*. An autonomic nerve running from the Glossopharyngeal nerve to the Otic ganglion and hence to the parotid salivary glands.

(3) *Deep Petrosal nerve*. A branch of the Internal Carotid Nerve (anterior part of the sympathetic chain) joining the Palatine nerve to form the *Vidian nerve* and running to the spheno-palatine ganglion.

PEYER'S PATCHES. Masses of lymphatic tissue in the walls of the small intestines of vertebrates. From them, lymphocytes pass into the cavity of the intestine.

PHAEODARIA (TRIPYLAEA). Deep-sea radiolaria whose pseudochitinous central capsule has three openings or 'oscula', one of which is surrounded by a mass of coloured granules called the *phaeodium*.

PHAEODIUM. See above.

PHAGO-. Prefix from Greek *Phagein*: to eat.

PHAGOCYTE. A type of white blood corpuscle able to engulf bacteria or other foreign bodies by flowing round them after the manner of feeding of *Amoeba*.

PHALANGEAL FORMULA. A series of figures giving the number of phalanges in each digit of a hand or foot. In humans, for instance, it is 2, 3, 3, 3, 3. In some vertebrates such as whales there is a state of *Hyperphalangy* in which the normal number of phalanges is increased.

PHALANGERIDAE. Arboreal marsupials of Australasia having a prehensile tail and often a membrane stretched from the front to the hind limbs, acting as a parachute when the animal is jumping or gliding.

PHALANGERINAE. The chief subfamily of the Phalangeridae comprising all the species which have a membrane or patagium for flying.

PHALANGES. The small bones forming the skeleton of a finger or toe (sing. Phalanx or Phalange).

PHALANGIDA. Harvestmen: long-legged arachnids resembling spiders in many respects but having no silk glands. There is no waist between the prosoma and the opisthosoma; the chelicerae are three-jointed and bear pincers; the pedipalps are leg-like. Breathing is entirely by means of tracheae and there are no lung-books like those of spiders. Most phalangida are predatory, feeding on small insects and other arthropoda.

PHALLUS. The penis or clitoris.

PHANERO-. Prefix from Greek *Phaneros*: visible.

PHANEROCODONIC. A term applied to a nydroid colony which reproduces by means of free-swimming medusae or jelly-fish bearing sexual organs.

PHANEROGLOSSA. Frogs and toads having a tongue and separate Eustachian tubes; the tadpoles have only one spiracle and, except in the Discoglossidae, this is on the left side.

PHARYNGEAL APERTURES. Gill slits (see under *Gills*).

PHARYNGEAL GLANDS. Any glands which secrete mucus, enzymes etc. into the pharynx.

PHARYNGO-. Prefix from Greek *Pharunx, pharungos*: the throat.

PHARYNGOBRANCHIAL. The uppermost of the cartilaginous or bony elements supporting the gills and surrounding the pharynx of a fish. Each gill-arch is formed of a *pharyngobranchial*, an *epibranchial*, a *ceratobranchial* and a *hypobranchial* element.

PHARYNX. The part of the alimentary canal between the mouth and the oesophagus.

PHASCOLARCTINAE. Koalas: tailless marsupial tree-bears of Australia.

PHASCOLOMYIDAE. Wombats: short-tailed burrowing marsupials of Australia and Tasmania having teeth like those of a rodent and feeding mostly on roots. Present-day species are small but some fossil species of Pleistocene times were nearly as large as a horse.

PHASIANIDAE (PHASIANINAE). Galliform birds including fowls, turkeys, pheasants, peacocks and many others. They are terrestrial birds which often roost in trees but build nests on the ground. They are polygamous, the males often being larger and more brightly coloured than the females.

PHASMIDAE (PHASMOIDEA). Stick-insects and leaf-insects: members of the order Orthoptera common in tropical forests and occurring to some extent in temperate regions. Stick-insects are elongated thin-legged creatures whose body, stiffly stretched out on leaves or twigs and coloured to match them, becomes almost invisible. Leaf-insects are remarkable in having their wings shaped and coloured to resemble the foliage amongst which they live.

PHENOTYPES. A term used in Mendelism to denote individuals having the same visible characteristics but not necessarily the same genetical constitution. They may, for instance, have inherited different recessive factors but these will not always show themselves and the actual visible qualities will be those of the dominant factors (*cf. Genotypes*).

PHOBOTAXIS. An 'avoiding movement' of protozoa or other simple organisms in which an unfavourable stimulus causes withdrawal. In some cases the movement of withdrawal makes a definite angle with the original direction of approach. If a second approach again brings the individual into an unfavourable position the process is repeated again and again until eventually the obstacle is avoided.

PHOCIDAE. Seals: aquatic carnivorous mammals with fin-like limbs, the hind pair being directed backwards and united to the tail. This fact and the absence of external ears distinguishes them from the *Otariidae* or sea-lions which are less well adapted to an aquatic life.

PHOCINAE. A subfamily of seals with well developed claws. They include among others the common grey seal of the Atlantic.

PHOLIDOPHORIDAE. Extinct fish of the Lower Lias having the general aspect of a herring but having ganoid scales and other primitive features.

PHOLIDOTA. Scaly Ant-eaters (see *Manidae*).

PHORONIDEA. A small group of marine invertebrates very similar to Polyzoa but differing from them in having a vascular system and blood containing haemoglobin.

PHOROZOOIDS. A special type of zooid found in the Doliolida, an order of pelagic tunicates. These zooids attach themselves to the back of the parent, at the same time carrying and nourishing another type called the gonozooid which subsequently breaks free to reproduce sexually.

PHOSPHORESCENT PROTOZOA. There are numerous species of these, chiefly among the Dinoflagellates and the Radiolarians. They emit a cold light which apparently depends upon the oxidation of fatty substances for its source of energy.

PHOTO-. Prefix from Greek *Phos, photos*: light.

PHOTOGENIC ORGANS. Light-producing organs of glow-worms and similar creatures. Little is known of the chemistry of these organs but it is believed that a substance called *Luciferin* is oxidized with the help of an enzyme called *Luciferase*; a case of chemiluminescence with very little production of heat and therefore no unnecessary expenditure of energy.

PHOTO-RECEPTOR. A sensory cell or collection of cells which are stimulated by the presence of light. Usually such cells contain a pigment which absorbs light of certain frequencies and at the same time undergoes some chemical change which stimulates the appropriate nerve.

PHRACTAMPHIBIA. An alternative name for the *Stegocephalia* (*q.v.*).

PHRAGMOCONE. The chambered part of the internal shell of a cuttle-fish or other cephalopod; probably a reduced form of the spiral chambered shell found in *Nautilus* and the Ammonites.

PHRAGMA. One of a series of parts of the endoskeleton of an insect or crustacean in the dorsal region, consisting of *apodemes* or ingrowths of the integument between the segments. They are hardened and serve for the attachment of muscles and other organs (see *Apodeme* and *Endophragmal Skeleton*).

PHRENIC NERVES. Nerves which control the movements of the diaphragm in mammals; branches of the cervical plexus formed from the fourth and fifth cervical nerves running back to the diaphragm on each side of the heart.

PHYLACTOLAEMATA. Fresh-water Polyzoa of the group *Ectoprocta* having a horse-shoe arrangement of oral tentacles (see *Ectoprocta*).

PHYLLO-. Prefix from Greek *Phullon*: a leaf.

PHYLLOBRANCHIAE. The type of gill found in crabs consisting of a central axis on each side of which is a row of plates set closely together like the leaves of a book.

PHYLLOPODIUM. A broad flattened type of crustacean limb having lateral lobes known as exites and endites. A phyllopodium may be used for swimming or for sending a respiratory current of water under the carapace; sometimes the first endite is modified into a *gnathobase* or jaw-like structure for feeding.

PHYLLOSTOMIDAE (PHYLLOSTOMINAE). Bats of Central America and the West Indies having *nose-leaves* or cutaneous processes close to the nostrils. The group includes many genera, some fruit-eating, some insectivorous and some blood-sucking.

PHYLLOXERIDAE. A family of plant-lice which includes the *Phylloxerinae* and the *Adelginae* (*q.v.*).

PHYLLOXERINAE. A subfamily of plant-lice which undergo a complex life-history similar to that of the Aphis. They differ from the latter, however, in that their reproduction is never viviparous. The genus *Phylloxera* is an important enemy of the vine.

PHYLUM. One of the larger groups used in classifying plants and animals. There are about twenty phyla in the animal kingdom, the largest being *Protozoa, Coelenterata, Platyhelminthes, Annelida, Arthropoda, Mollusca, Echinodermata* and *Chordata*. Each of these is divided into a number of classes, some of which may be grouped into subphyla (see Appendix).

PHYSETERIDAE. Whales with functional teeth in the lower jaw only, but sometimes having small teeth embedded in the gums of the upper jaw. The skull is slightly asymmetrical and its upper anterior end or *cranial basin* is filled with fat known as *spermaceti*. The family includes sperm-whales and bottle-nosed whales.

PHYS-, PHYSO-. Prefix from Greek *Phusa*: a bladder.

PHYSOCLEISTOUS. A term used of fish which have the air-bladder disconnected from the gut (*cf. Physostomi*).

PHYSOSTOMI (PHYSOSTOMOUS). Fish which have a pneumatic duct leading from the air-bladder to some part of the alimentary canal (usually the oesophagus). They include members of a number of suborders and the name is no longer used in systematic classification.

PHYTO-. Prefix from Greek *Phuton*: a plant.

PHYTOMASTIGINA. Flagellate protozoa, some of which are regarded as intermediate between plants and animals by reason of the fact that they possess chloroplasts and can carry out photosynthesis. Some of them such as *Chlamydomonas* and *Volvox* have been classed by botanists among the algae.

PHYTOMONADINA. See *Volvocina*.

PHYTOPHAGA. A large group or superfamily of beetles, mostly very destructive to crops and trees. They include the Longhorns (Cerambycidae) and the leaf-beetles (Chrysomelidae). The well known Colorado Beetle, the pest of the potato plant, is included in the latter group.

PHYTOPHAGOUS. Herbivorous: feeding on plants.

PIA MATER. The inner of the two meninges, a thin vascular membrane surrounding the brain and spinal cord. Outside it is the tougher membrane called the *Dura Mater*.

PICI. A group of birds of the tribe Coraciiformes, characterized by having the first and fourth toes pointing backwards. They include Puff-birds and Toucans of South America, Honeyguides of India and Malaya as well as various kinds of woodpecker from most parts of the world.

PICIDAE (PICINAE). Woodpeckers: powerfully built birds having a strong beak with which they bore holes for their nests in hollow trees. The tongue, which is long, horny and barbed at the tip, can be extended for a considerable distance. The sharp claws and the large tail are used for climbing.

PIERIDAE. A family of butterflies which are nearly all white or yellow and are characterized by having two forked claws on each foot. They include the well known Cabbage White whose caterpillars are major pests of cruciferous plants.

PIGEON'S MILK. A nutrient secretion from the lining of the crops of pigeons, used for feeding the young.

PIGMENT LAYER OF THE EYE. A layer of black pigmented cells between the retina and the choroid of the vertebrate eye or in a corresponding position in the eyes of some invertebrates. Its function is to absorb surplus light and to prevent internal reflections.

PIGMENTS. (1) Blood pigments. *Haemoglobin, Haemocyanin* etc. which can form loosely combined compounds with oxygen and so carry it round the body.

(2) Skin pigments. *Melanin* and *Xanthophyll*, found in the Malpighian layer of the vertebrate skin, *e.g.* in many human races. In some animals the skin pigments are contained in specialized cells or *chromatophores* which can expand and contract under the influence of the nervous system and of the pituitary hormones. In this way many amphibia and reptiles are able to vary the colour of the skin (see *Chromatophores*).

(3) Visual pigments, *e.g. Visual purple* formed by the retinal rods of the vertebrate eye. The chemical change (bleaching) undergone by this pigment under the influence of light is the basis of the light-sensitive properties of the rods. Other pigments acting in a similar manner are found in the eyes or 'eye-spots' of many animals.

PILAE ANTOTICAE. A pair of cartilaginous pillars rising from the para-chordals in front of the auditory capsules and joining on to the orbital carti-lages in the embryonic cranium of a vertebrate (see also *Fenestra pro-otica*).

PINACOCYTES. Flattened cells forming an outer covering on sponges. In appearance they resemble pavement epithelium but can change their shape.

PINEAL BODY. An outgrowth from the roof of the fore-brain primitively forming two lobes one behind the other, though in most vertebrates only the posterior one remains. In some reptiles and in the lamprey it has an eye-like structure with lens and retina, but in most vertebrates its function is obscure. It may secrete a hormone.

PINEAL EYE. See *Pineal Body* above.

PINNA. (1) A bird's feather or the vane of a feather.

(2) The funnel-shaped external part of the mammalian ear.

PINNIPEDIA. Seals, sea-lions, walruses etc. Aquatic carnivorous mammals with pentadactyl fin-like limbs whose digits are united by a membrane. They were formerly classed with the terrestrial Carnivora but are now regarded as a separate order.

PINNULARY OSSICLES. Small dermal ossicles covering the pinnules of a Crinoid or sea-lily (see *Pinnule* below).

PINNULE. Any feather-like structure, *e.g.* the slender arms of the Feather-star and other Crinoids each bearing a double row of alternate branchlets.

PIROPLASMIDEA. Parasitic protozoa which infest the red blood corpuscles of vertebrates and cause such diseases as red-water fever of cattle. In some respects their life-history resembles that of the malaria parasite. They are transmitted, however, not by mosquitoes but by ticks.

PISCES. See *Fish*.

PISIFORM BONE. A small round sesamoid bone (*i.e.* a bone formed in a tendon) at the junction of the ulna and the ulnare in some mammals.

PITHEC-. Prefix and root-word from Greek *Pithekos*: an ape.

PITHECANTHROPUS. An early type of man found as a fossil in the Pliocene of Java. It had a cranial capacity considerably greater than that of the apes, had thick bony brow-ridges above the orbits and probably walked erect.

PITHECIINAE. New World monkeys whose tails are not prehensile and whose incisor teeth are inclined forwards. They are chiefly found in the forests of the Amazon.

PIT ORGANS. Sensory organs sunk in small pits on various parts of the head and body of a fish. They are supplied by the facial, vagus and trigeminal nerves (see *Ampullae of Lorenzini*).

PITUITARY BODY. An endocrine organ beneath the brain of a vertebrate. It produces a large number of hormones many of which control the activities of other endocrine organs. For this reason it is sometimes called the 'master-gland' of the body. Embryologically it is developed from an upgrowth of the stomodaeum (the *hypophysis*) fused with a downgrowth from the thalamen-cephalon (the *infundibulum*).

PLACENTA. An organ consisting of embryonic tissue in close contact with the uterus of the parent, by means of which the mammalian embryo is nourished. Usually the placenta is a composite structure formed from parts of the chorion and the allantois, having numerous villi embedded in the wall of the uterus. By a breaking down of tissues the maternal blood vessels are in close contact with those of the foetus. In this way plasma containing dissolved foods and oxygen can pass from the parent to the offspring.

PLACENTALIA (EUTHERIA). Mammals which have a true placenta, *i.e.* all except Monotremes and Marsupials.

PLACO-. Prefix from Greek *Plax, placos*: a flat plate.

PLACODES. Sensory organs formed by proliferation of the lower epidermis which subsequently sinks beneath the surface. Such sensory organs are the lateral line organs of fish, the olfactory sacs, the auditory sacs and the lens of the eye.

PLACODONTIA. Fossil reptiles of the Triassic Period having large pavement-like crushing teeth on the jaws and palate, indicating that they probably fed on hard-shelled molluscs.

PLACOID RECEPTORS. Sensory organs, possibly olfactory, on the antennae of some insects. They consist of a number of sensory cells beneath a very thin plate of cuticle.

PLACOID SCALES. Dermal denticles of cartilaginous fish: small tooth-like structures formed of dentine covered by a layer of enamel and having a flattened base embedded in the skin.

PLAGIAULACIDAE. Fossil marsupials found in strata from the Trias to the Eocene in both Europe and America. They had one pair of large rodent-like incisors in the lower jaw and the other teeth were sharp and tuberculated. They are included in the group Allotheria, considered to be intermediate between Monotremes and Marsupials.

PLAGIOSTOMI. An alternative name for *Selachii* or sharks.

PLANARIANS. Free-swimming Turbellarians of the order Tricladida having flat ciliated bodies, a ventral mouth, well developed sense-organs and an alimentary canal divided into three main branches. These flatworms show remarkable powers of regeneration, almost any part being able to grow into a complete new organ if injured. If the head is cut in half, for example, two new heads will develop. The planarians are classified according to their habitats as *Maricola* (marine), *Paludicola* (fresh water) and *Terricola* (inhabiting dry land).

PLANIPENNIA. A large suborder of Neuroptera including the brown and green lacewing-flies and many other insects having two pairs of membranous wings with complex veins. The predatory larvae are terrestrial or aquatic and have piercing mouth-parts with the maxillae and mandibles interlocking to form a tube. They weave cocoons through anal spinnerets.

PLANKTON. Animal and plant life, mostly microscopic, which floats and drifts near the surface of a lake, river or sea. The chief constituents are unicellular algae, protozoa, small crustacea and various larvae of mollusca, crustacea and other invertebrates. The green algae manufacture carbohydrates by photosynthesis and are the first stage in a food-chain which supports all the other forms of aquatic life.

PLANOBLAST. A free-swimming medusa or jelly-fish.

PLANT-. Prefix from Latin *Planta*: the sole of the foot.

PLANTIGRADE. Walking with the whole of the lower surface of the foot on the ground.

PLANTULAE. Adhesive pads on the tarsal sclerites of some insects.

PLANULA. The flat ciliated larva of a coelenterate having a solid core of endoderm and an outer layer of ectoderm but no body-cavity.

PLASMA. The liquid medium in which blood corpuscles are suspended, consisting of an aqueous solution of mineral salts, absorbed food substances, waste products, carbon dioxide, hormones, antibodies etc.

PLASMO-. Prefix from Greek *Plasma*: form.

PLASMODI-TROPHOBLAST. In mammalian embryology, the part of the trophoblast which is in contact with the wall of the uterus and in which the cell-walls break down to form a syncytium.

PLASMODIUM. A multinucleate mass of protoplasm formed by the union of many amoeboid cells, as for instance in the Mycetozoa.

PLASMOGAMY (PLASTOGAMY). Fusion of cytoplasm but not of nuclei when two or more cells join to form a plasmodium.

PLASMOLEMMA. The bounding 'membrane' of a mass of naked protoplasm such as *Amoeba*.

PLASMOPHAGOUS. Feeding on protoplasm.

PLASMOTOMY. The division of a multinucleate cell or plasmodium to form multinucleate offspring.

PLASTIDS. Small protoplasmic bodies, often coloured, found in the cells of plants and some animals.

PLASTIN. An acidophil or acid-staining substance of which nucleoli are made.

PLASTO-. Prefix from Greek *Plastos*: moulded, formed.

PLASTOGAMY. See *Plasmogamy*.

PLASTRON. The ventral bony plate of the Chelonians (tortoises and turtles), consisting of about nine flat dermal bones attached to the marginal plates of the dorsal carapace. The plastron and carapace thus together form the 'shell' of a tortoise. Both are usually covered by horny epidermal plates.

PLATANISTIDAE. Small toothed whales, often blind, found in the larger estuaries of Asia and South America.

PLATY-. Prefix from Greek *Platus*: flat.

PLATYBASIC (PLATYTRABIC). A term used to describe a skull whose trabeculae (*q.v.*) are widely separated from one another as in the frog (*cf. Tropibasic*).

PLATYCTENEA. A group of tentaculate Ctenophora (sea-gooseberries) which differ from the normal globular type by being flattened dorsoventrally. The flattening is produced by an expansion outwards of the stomodaeum. Some are free-swimming and others live on rocks and seaweed.

PLATYGASTERIDAE. A family of small parasitic hymenopterous insects exhibiting the phenomenon of polyembryony to a marked degree. Most of them are parasitic on small gall-flies.

PLATYHELMINTHES (PLATYHELMIA). Flat worms: a phylum of bilaterally symmetrical acoelomate invertebrates without an anus and usually with complex hermaphrodite reproductive organs. The group is divided into three classes: *Turbellaria* or free-swimming flat worms, *Trematoda* or parasitic flukes and *Cestoda* or tapeworms.

PLATYPODIDAE. Pinhole-borers: beetles with a snout like that of a weevil which bore minute holes in the bark of trees. The Oak Pinhole Borer (*Platypus cylindrus*) has in recent years become a serious timber pest in Europe; many others occur in the tropics.

PLATYRRHINA (PLATYRRHINAE, PLATYRRHINI). Monkeys of Central and South America distinguished from the Old World monkeys and apes in having a broad cartilaginous internasal septum and in having long tails which are usually prehensile.

PLATYSOMIDAE. Extinct fish of Carboniferous times having a deep body covered with rhomboid ganoid scales.

PLATYSTERNIDAE. Aquatic tortoises of S.E. Asia having nine plastral bones and having the temporal region of the skull completely roofed over.

PLATYTRABIC. See *Platybasic*.

PLECOPTERA. Stone-flies: a somewhat primitive order of hemimetabolous insects having much in common with mayflies. There are four membranous wings, the hind pair having exceptionally large anal lobes. The body is soft and both the antennae and cerci are long and slender. The adult, as in the case of the mayfly, does not feed and lives only for a short time. The aquatic larvae, however, are predators.

PLECTOGNATHI. Tropical fish with rough scales or spines; with a narrow mouth and usually without pelvic fins. Some kinds such as the Globe-fish and the Porcupine-fish can inflate the body and erect the spines.

PLEOPODS. Swimmerets: small biramous appendages on the abdomens of most crustacea.

PLERERGATE. A worker ant with stomach enormously distended with food.

PLEROCERCUS (PLEROCERCOID). A worm-like but unsegmented larval stage of a tapeworm (*cf. Cysticercus*).

PLESIOSAURIA (SAUROPTERYGIA). Amphibious or marine reptiles of the Triassic Period and later, having a long neck and two pairs of five-toed paddles with many phalanges. Most were purely aquatic but a few with more elongated limbs could probably also move on land.

PLESIOSAURIDAE. Those plesiosauria whose limbs were short and paddle-shaped for an almost entirely aquatic existence.

PLETHODONTIDAE. Salamanders mostly found in America, having a very long protrusible tongue with a soft knob on the end.

PLEURA. A membrane in which the lungs of a vertebrate are suspended from the wall of the thorax.

PLEURO-. Prefix from Greek *Pleuron*: side.

PLEURACANTHIDAE (PLEURACANTHODII). See *Ichthyotomi*.

PLEUROBRANCHIAE. Gills situated on the sides of the body under the carapace in certain crustacea. They are above the normal gills, which are on the legs, and are believed to have come to be in this position through incorporation of the precoxae with the body.

PLEUROCENTRUM. That part of the centrum of a vertebra which is formed from the *interventrals* and *interdorsals* (*q.v.*). During the course of evolution two main types of vertebra have developed. In primitive amphibia such as the Labyrinthodonts each vertebra has two centra: an anterior *hypocentrum* and a posterior *pleurocentrum*. In later amphibia the hypocentrum enlarges at the expense of the pleurocentrum but in reptiles, birds and mammals the opposite has occurred. In the latter, therefore, the centrum is in fact a pleurocentrum.

PLEURODIRA. Tortoises and turtles having the pelvis fused to the shell and having the neck retractile by bending laterally. Most of them are fresh-water forms and are almost entirely carnivorous. They are found in the Southern Hemisphere.

PLEURODONT. Having the teeth attached by their sides to the rim of the jaw or set in a groove, as in certain lizards.

PLEURON. The name given to the lateral part of the cuticle in each segment of an insect or crustacean. A typical segment has four main cuticular plates, *viz.* a dorsal *tergum*, two lateral *pleura* and a ventral *sternum*. These may be further differentiated into tergites, sternites and pleurites.

PLEURONECTIDAE. See *Flat fish*.

PLEUROPODITE (PRECOXA). A segment, additional to the normal number, preceding the coxopodite in the limbs of some crustacea.

PLEUROPTERYGII. See *Cladoselachii*.

PLEURORACHIC. An arrangement of the cartilages or bones in the paired fins of certain fish whereby there is an axis near one side, having peripheral elements branching from it mostly on the other side. Such fins are found in the extinct shark-like *Pleuracanthus*.

PLEXUS. (1) A network of nerves which branch and rejoin: *e.g.* the pectoral and pelvic plexus of a vertebrate.
    (2) A number of ganglia in close proximity or fused together from which numerous nerves proceed: *e.g.* the solar plexus.
    (3) A network of blood vessels branching and rejoining.

PLEXUS MIRABILE. A network of nerves and blood vessels in the limbs of a vertebrate embryo.

PLICA SEMILUNARIS. A vestigial nictitating membrane or third eyelid present in some mammals.

PLICATE CANALS. Ribbon-like organs conveying blood along the roof of the mantle to the kidney in a lamellibranch mollusc.

PLOCEIDAE. Weaver-birds: passerine birds from India, Africa and Australia, so called because of the elaborate woven nests which many of them make.

PLUMULAE. See *Down feathers*.

PLUTEUS. A free-swimming Echinoderm larva having ciliated 'arms' sometimes supported by calcareous rods.

PNEUM-, PNEUMO-, PNEUMATO-. Prefixes from Greek *Pneuma, Pneumatos*: wind.

PNEUMATICITY OF BONES. The presence of air spaces in the bones of birds, connecting with the lungs and with air-sacs in the body. These spaces are most developed in the larger flying birds such as the albatross and are almost absent in small birds.

PNEUMATOPHORE. A large bladder-like polyp filled with gas acting as a float in colonial coelenterates such as the 'Portuguese Man-of-war' (*Physalia*). It is the first-formed polyp of the colony, from which hang down numerous other polyps specialized for stinging, catching and digesting their prey.

PNEUMOGASTRIC NERVE. An alternative name for the Vagus or tenth cranial nerve (*q.v.*).

PNEUMOSTOME. A small opening allowing the passage of air into a respiratory cavity such as that of a snail.

PODARGIDAE. Nocturnal, wide-mouthed, owl-like birds of Australia, Malaya, Borneo etc.

PODEON (PODEUM). A narrow stalk or petiole connecting the thorax with the abdomen in some insects.

PODEX (PYGIDIUM). The posterior terminal or anal region of an invertebrate such as the earthworm.

PODICAL PLATES. The reduced parts of the eleventh abdominal segment in certain insects. They may bear the anal cerci.

PODICIPEDIDAE. Grebes: aquatic birds with webbed feet, short upright body, short tail and fairly long neck. They construct floating nests of reeds and when feeding they dive and swim for some distance under water.

POD-, PODO-. Prefix from Greek *Pous, podos*: a foot.

PODITE. A segment of the limb of an arthropod, *e.g. protopodite, endopodite, exopodite* etc.

PODIUM. An organ acting as a foot, as for instance the tube-feet of a star-fish.

PODOBRANCHIAE. Gills consisting of thin-walled tufts of skin on the limb-bases of most crustaceans. Sometimes additional series of gills are also present, *viz. Pleurobranchiae* on the body-wall and *arthrobranchiae* on the joints.

POIKILOCYTE. A misshapen, diseased red blood corpuscle.

POIKILOTHERMOUS (POIKILOTHERMAL). Cold-blooded: having a body temperature which varies with that of its surroundings, as is the case in all living creatures except birds and mammals. A natural consequence of this variable state is that the rate of metabolism slows down in cold conditions and the animal may hibernate or at least become very inactive.

POISON FANGS. The long sharp maxillary teeth of snakes, having a groove or channel down which poison flows from the labial poison glands in the upper jaw.

POLAR BODIES. Minute cells produced during oogenesis when the primary oocyte undergoes reduction division to form a secondary oocyte and when the latter divides again to produce an ovum.

POLE CAPSULES. Structures resembling nematocysts in the multicellular spores of certain parasitic protozoa of the order Cnidosporidia. They consist of thread-like organs which are coiled up in vesicles and can be shot out. Their function is apparently to anchor the spores to the lining of the host's gut.

POLIAN VESICLES. Small stalked sacs or reservoirs leading off from the water-vascular ring in some Echinoderms.

POLLEN BASKET. A part of the hind leg of a bee specially modified for carrying pollen. The tibia is flattened and bordered by long hairs making a trough into which the pollen can be pushed. In leaf-cutting bees and some other solitary species the pollen basket is situated underneath the abdomen.

POLLEX. The thumb or first digit on the fore-limb of a tetrapod vertebrate.

POLY-. Prefix from Greek *Polus*: many.

POLYBORINAE. American carrion hawks.

POLYCHAETA. Marine annelid worms having sensory organs and tentacles on the head and having numerous chaetae or bristles borne on flat outgrowths of the body called *parapodia*. Males and females are separate and fertilization is external. The reproductive organs are arranged segmentally but in some cases are concentrated in the posterior part of the worm which may be cast off during the breeding season and regenerated the following year. The basic structure of the body may be modified according to the way of life and this leads to a division of the class into three groups, *viz. Errantia* or free-swimming worms; *Tubicola* which live in tubes made of sand grains etc., and *burrowing* worms such as the lugworm.

POLYCLADIDA. Marine Turbellaria (free-swimming flatworms) in which the mouth is at the hind end and the gut has a main stem with numerous branches. The larvae which are planktonic are known as *Müller's larvae*.

POLYDISC STROBILATION. The method of reproduction by which a number of saucer-like jelly-fish or *ephyrae* are budded off from a hydra-like polyp known as a *scyphistoma*. The parent polyp first develops numerous constrictions and the segments so formed then become flattened and detach themselves one by one from the top of the polyp (see *Ephyra* and *Scyphistoma*).

POLYEMBRYONY. The production of two or more individuals from a single egg by division of the embryo at a very early stage. This happens in many animals but is most notable among parasitic hymenopterous insects, some of which produce many hundreds of embryos from one egg.

POLYENERGID NUCLEI. Nuclei possessing several sets of chromosomes formed by repeated mitosis within a nuclear membrane which remains intact throughout the process.

POLYGONEUTIC. Having many broods in a season.

POLYMASTIGINA. The most highly organized of the flagellate protozoa, usually having more than three flagella, one or more undulating membranes, a gullet and an axostyle. There is also an extra-nuclear division centre formed from the united basal granules of the flagella. They are completely holozoic, and some are parasites of humans.

POLYMASTIGOTE. Having many flagella in a bunch.

POLYMASTODONTIDAE. Large extinct marsupial mammals from the Eocene of North America. They belong to the group Allotheria and had rodent-like incisors in the lower jaw.

POLYMORPHISM. Having several forms, *e.g.* the various types of polyp in a coelenterate colony.

POLYMORPHS (POLYMORPHONUCLEAR LEUCOCYTES). The commonest type of white corpuscle in vertebrate blood, having granular cytoplasm and a nucleus which is drawn out and divided into three or four expanded lobes. Corpuscles of this type are frequently phagocytic, *i.e.* able to engulf bacteria or other foreign bodies in the blood.

POLYNEMIDAE. Mullet-like fish inhabiting sandy shores and rivers of the tropics, used much as food and yielding isinglass from the air-bladder.

POLYP. A cylindrical organism with a thin body-wall consisting of two single layers of cells, the *ectoderm* and the *endoderm* separated by a gelatinous noncellular layer or *mesogloea*. At one end there may be a mouth which is usually surrounded by tentacles. Polyps may be single (*e.g. Hydra*) or colonial (*e.g.* the coral-forming organisms). The latter are formed by repeated budding from a parent polyp. In such colonies a tube or *coenosarc* links the body-cavities of all the individuals with one another. There may be many different kinds of polyp in a colony, each specialized for some different function such as feeding, reproduction, protecting etc. (see *Coelenterata* and *Diploblastica*).

POLYPARIUM. The common base and investment of a polyp colony.

POLYPHAGA. The largest suborder of beetles including a great diversity of families all of which can be distinguished from the other suborder, the *Adephaga*, by the fact that the larvae are either legless or have legs with only one claw. The antennae are of various forms and there is generally an absence of cross-veins in the hind-wings. Included in the Polyphaga are the Rove beetles (*Staphylinae*), the Carrion beetles (*Silphidae*), the Ladybirds (*Coccinellidae*), the weevils (*Curculionidae*) and many others.

POLYPHYODONT. Having many successive sets of teeth, as in most lower vertebrates (*cf. Diphyodont*).

POLYPLACOPHORA. Molluscs of the class Amphineura having the shell formed of a number of dorsal transverse plates, as opposed to the *Aplacophora* which have no shell. The best known example is *Chiton* (see *Amphineura*).

POLYPIDE. The polyp-like body of a member of a Polyzoan colony. It has a U-shaped alimentary canal oral tentacles and a tentacle-sheath. The whole structure is retractile and held by a strand of connective tissue (the *funiculum*) to the cup-like body-wall which is called the *zooecium*.

POLYPLOID. An individual whose chromosome number is a multiple of the normal number.

POLYPOD LARVA. See *Eruciform larva*.

POLYPROTODONTIA. Those marsupials which have numerous incisors, large canines and cuspidate molars. They are usually carnivorous or insectivorous. The group includes the bandicoots of Australia and the opossums of America.

POLYPTERINI. Primitive fish from the fresh waters of Central Africa, having rhombic ganoid scales, a series of small dorsal fins, a nearly symmetrical tail and paired fins with fleshy lobes. These fish and the sturgeons are believed to be survivors from the old Palaeoniscid stock of the Devonian Period.

POLYTHALAMIA. Marine Foraminifera, usually with a calcareous shell of several chambers, and with reticulate pseudopodia (*cf. Monothalamia*).

POLYTHELY. Having supernumerary nipples on the mammary glands.

POLYTOCOUS. Producing many at a birth.

POLYTRICHA. Protozoa having uniform ciliation over the surface of the body and having a permanent gullet with an undulating membrane and an adoral wreath. The group includes the well known fresh-water *Stentor* as well as several marine forms and some intestinal parasites.

POLYTROPHIC. (1) Deriving food from more than one type of organic substance.

(2) Of parasites: nourished by several different hosts.

POLYTROPHIC OVARIOLE. A type of egg-tube present in most insects, consisting of a successive series of chambers each of which contains trophocytes for nourishing the developing egg-cells (see also under headings *Panoistic* and *Acrotrophic*).

POLYZOA (BRYOZOA). Minute sedentary animals usually living in colonies and having a superficial resemblance to coelenterate polyps. Like the latter, they form asexual buds and each individual has a ring of tentacles round the mouth. Here, however, the resemblance ends; polyzoa are considerably higher in the scale of evolution. They are triploblastic with a coelom and an alimentary canal.

PONS VAROLII. A band of nervous tissue connecting the two sides of the cerebellum in mammals.

PORE PLATE (OF RADIOLARIANS). See *Monopylaea*.

PORE-RHOMBS. Diamond shaped patterns of pores perforating the dermal plates of some Crinoids or sea-lilies (see also *Diplopores* and *Cystoidea*).

PORIFERA (PARAZOA). Sponges: the name *Porifera* is used for the phylum and the name *Parazoa* when they are regarded as a separate sub-kingdom as distinct from Protozoa and Metazoa (see under *Parazoa*).

POROCYTES. Cells whose intracellular passages form the pores through which water enters a sponge.

PORTAL VEIN. A vein through which blood flows from one system of capillaries to another, as for instance from the intestines to the liver or from the tail to the kidneys (see *Hepatic portal system* and *Renal portal system*).

POST-. Latin prefix and preposition: after.

POST-AXIAL SIDE. The side of a pentadactyl limb on which the fifth digit is situated; in the fore-limb this is on the same side as the ulna and on the hind-limb it is on the same side as the fibula. The term post-axial refers to the primitive arrangement in which the axis of the limb was at right-angles to that of the body. The post-axial margin was then behind the axis of the limb when considered in relation to the body as a whole. In many animals, however, the primitive position has been changed by varying degrees of twisting so that post-axial and pre-axial sides are not always obvious.

POST-CLEITHRUM. A membrane-bone between the cleithrum and the supra-cleithrum in the pectoral girdle of a bony fish. These three bones are of dermal origin and are superimposed upon the original cartilaginous pectoral girdle which consists of the scapulae and coracoids.

POSTERIOR CARDINAL SINUS. See *Cardinal Veins*.

POSTERIOR RECTUS. See *Eye muscles*.

POSTERIOR TEMPORAL ARCADE. See *Post-temporal bar*.

POSTERIOR VENA CAVA (INFERIOR VENA CAVA. POST-CAVAL VEIN). The chief vein which returns blood from the middle and posterior region of the body to the right auricle of the heart (or to the sinus venosus if one be present) in tetrapod vertebrates.

POST-GANGLIONIC FIBRE. An excitor neurone whose cell-body and nucleus are in a sympathetic ganglion and whose axon goes from there to some part of the viscera.

POST-MENTUM. The proximal part of the labium of an insect, usually divided into two parts known respectively as the *mentum* and *submentum* (see (*Mentum*).

POSTNOTUM (POST-SCUTELLUM). The hindmost dorsal sclerite on either of the wing-bearing segments of an insect. A typical notum or tergum is divided into three parts, *viz.* the *prescutum*, *scutum* and *scutellum*. The post-scutellum when present is immediately behind the scutellum and is formed by hardening of the intersegmental membrane.

POST-OESTRUS. The sexual phase of mammals which follows oestrus (*see Oestrus cycle*).

POST-ORBITAL BAR. A bar of bone separating the orbit from the temporal fossae in certain reptiles and mammals.

POST-ORBITAL BONE. A membrane-bone behind the orbit in certain fish and reptiles; it may form part of the post-orbital bar (*q.v.* above).

POST-PUBIS. An additional bone present in the pelvic girdle of some dinosaurs extending back beneath the ischium.

POST-SPIRACULAR LIGAMENT. One of the ligaments which, together with the hyomandibular cartilage or bone, suspends the upper jaw from the neurocranium in most fish (*hyostylic* suspension).

POST-SPIRACULAR NERVE RAMUS. A branch of the seventh cranial nerve in cartilaginous fish, passing down behind the spiracle to innervate the sensory organs of the skin of the lower jaw and the muscles of the hyoid arch.

POST-TEMPORAL BAR (POSTERIOR TEMPORAL ARCADE). A bar of bone separating the superior and inferior temporal fossae from the post-temporal fossa in crocodiles and certain other reptiles. It is formed from the *supratemporal* and the *squamosal* bones.

POST-TEMPORAL BONE. A membrane-bone which, in many fish, connects the pectoral girdle with the back of the skull behind the auditory capsule.

POST-TEMPORAL FOSSA (POSTERIOR TEMPORAL FOSSA). An opening on each side of the hind-part of the skull in certain reptiles, by which the temporal cavity (the space between the cranium and the overlying membrane-bones) opens to the exterior.

POTAMOGALIDAE. Large insectivorous mammals of Madagascar without clavicles or zygomatic arches and having the urinogenital organs and the anus opening within the same fold of integument.

POTOROINAE. Small kangaroo-rats of Australia and Tasmania with long hairy tails.

PRAECOCES. Birds which hatch out of the egg in a relatively mature state, covered with down and able to walk or swim.

PRE-. Prefix from Latin *Prae*: before.

PREANTENNAE. Mobile sensitive appendages situated on the first body-segment of the primitive arthropod *Peripatus*. In all higher arthropods the first segment disappears in the adult and the antennae are borne on the second segment.

PRE-AXIAL SIDE. The side of a pentadactyl limb on which the first digit is situated. This digit is usually recognized by its only having two phalanges and being on the same side as the *radius* of the fore-limb and the *tibia* of the hind-limb (see also *Post-axial*).

PRECIPITIN TEST. A means of estimating the relative similarity between the blood of different animals by comparing the degree of precipitation caused by a specific antibody. *e.g.* If human serum is injected into a rabbit the latter produces in its serum an antibody which will precipitate human blood 100%, gorilla 64%, orang-utan 42%, baboon 29%, sheep 10%, horse 2%, marsupial nil.

PRECORACOID (PROCORACOID). A bone or cartilage parallel to the coracoid and anterior to it, forming part of the pectoral girdle in some amphibia and reptiles.

PRECOXA (PLEUROPODITE). The basal sclerite of some of the limbs of certain crustacea. It is only present in a few genera, the coxopodite being in most cases the first segment.

PRE-DENTARY BONE. A bone in front of the dentary in the lower jaws of some dinosaurs.

PREDENTATA. See *Ornithischia*.

PRE-GANGLIONIC FIBRES. Nerve fibres or axons running outwards from the central nervous system in the *Rami communicantes* to the sympathetic ganglia. Their cell-bodies are in the spinal cord but their axons make connection by synapses with the sympathetic neurones whose cell-bodies are in the ganglia.

PREFRONTAL BONES. Paired bones forming part of the roof of the cranium anterior to the frontal bones in most vertebrates other than mammals.

PRE-ISCHIOPODITE. See *Metabasipodite*.

PRELACHRYMAL FOSSA. An opening in the skull between the orbit and the nostril in birds and certain reptiles.

PREMANDIBULAR SEGMENT (PREMANDIBULAR SOMITE). The first somite of the head in the vertebrate embryo, innervated by the *oculomotor* and the *ophthalmicus profundus* nerves.

PREMAXILLAE. The most anterior of the bones forming the upper jaw of a vertebrate; in mammals they bear the incisor teeth.

PREMENTUM. The free distal part of the labium of an insect, bearing the labial palps and the ligula.

PREMOLARS. The teeth in front of the molars, *i.e.* in the sides of the jaws of a mammal. They may be similar in appearance to the molars but differ from them by reason of the fact that molars only appear in the second set of teeth, whereas premolars arise as milk teeth also.

PRE-OPERCULAR BONE. The foremost of four membrane bones which support the operculum or gill-cover of a bony fish. The others are the *opercular, subopercular* and *interopercular* bones.

PRE-ORAL. In front of the mouth, *e.g.* with reference to body-segments, ganglia etc.

PRE-ORAL GUT. A small pocket formed of the embryonic gut of a vertebrate by reason of the fact that perforation of the mouth does not occur at the extreme front end.

PREPUCE (PRAEPUTIUM, FORESKIN). A fold of skin, richly glandular and vascular, surrounding the tip of the penis in a mammal.

PREPUPA. The stage before the pupa in the development of a holometabolous insect. During this stage feeding usually ceases and sometimes a cocoon is produced. The prepupa may resemble the larva but is often shrunken and less pigmented.

PREPYLORIC OSSICLE. A calcareous thickening of the wall of the proventriculus or 'stomach' of a crustacean in the median dorsal position between the large anterior *cardiac* chamber and the smaller posterior *pyloric* chamber. The prepyloric ossicle bears a large median tooth which helps to grind food by moving to and fro between two rows of lateral teeth.

PRESCUTUM. The anterior of three sclerites which typically form the notum or dorsal cuticular covering over each of the three thoracic segments of an insect (see also *Scutum* and *Scutellum*).

PRESPHENOID. A narrow triangular bone in front of the basisphenoid, forming part of the base of the cranium in some vertebrates.

PRESPIRACULAR. In front of the spiracle, *e.g.* relating to the nerves of a fish.

PRESPIRACULAR CARTILAGES. Small cartilages sometimes present in the anterior wall of the spiracle in certain cartilaginous fish.

PRESTOMUM. A mouth-like aperture in front of the true mouth in insects such as the house-fly. It leads into a tube formed by the labrum and the labium, whilst on each side of it are the two large suctorial pads characteristic of the mouth-parts of flies.

PRESUMPTIVE AREA. A part of an embryo consisting of cells which will in the normal course of development, become a particular organ or tissue, *e.g.* presumptive epidermis.

PRETARSUS. A name sometimes given to the clawed distal segment of the tarsus of an insect.

PRETREMATIC NERVE. A branch of the facial or seventh cranial nerve passing in front of the spiracle and innervating the sense-organs of the side of the head in elasmobranch fish.

PRIMARY FEATHERS. Quill feathers attached to the manus or 'hand' of a bird. Those which are attached to the arm are known as secondary feathers (see *Feather*).

PRIMARY GILL SLITS. The first formed gill-slits of an aquatic chordate, not necessarily corresponding to the gill-slits of the adult since secondary bars of cartilage may subdivide the original slits into secondary slits. Such a process takes place, for example, in *Amphioxus* and in Tunicates.

PRIMATES. An order comprising monkeys, apes and humans: mammals characterized by having greatly enlarged cerebral hemispheres usually going with a high degree of intelligence. The eyes are directed towards the front and are separated from the temporal fossa by a bony partition. The limbs are pentadactyl, the thumbs and sometimes the big toes being opposable. This arrangement is an adaptation for the arboreal life which is characteristic of most primates. There are two main groups, *viz.* the *Platyrrhini* or New World monkeys usually having prehensile tails, and the *Catarrhini* or Old World apes and monkeys in which the tail is never prehensile and is often absent. The much more primitive groups *Lemuroidea* and *Tarsioidea* are sometimes also included with the primates.

PRIMITIVE CHARACTERS. Fundamental features showing little specialization and indicating an early stage in evolution.

PRIMITIVE GROOVE. See under *Primitive Streak*.

PRIMITIVE KNOT. In vertebrate embryology: a swelling at the anterior end of the primitive streak (*q.v.*), representing the dorsal lip of the blastopore. It is composed of cells which proliferate and later migrate inwards beneath the surface to form the notochord.

PRIMITIVE STREAK. In vertebrate embryology: a line of cells lying in the centre of the embryonic disc of a reptile, bird or mammal and marking the axis of the future embryo. Running along the centre of the streak is the *primitive groove* into which cells migrate to give rise to the future notochord and mesoderm.

PRIMORDIAL GERM CELLS. The germinal epithelial cells of the testis or ovary which give rise by a series of changes and cell-divisions to spermatozoa and ova respectively.

PRISMATIC LAYER. The middle layer of the shell of a mollusc consisting largely of crystalline calcium carbonate. A normal shell is composed of three layers, an outer which is horny and a middle and inner which are calcareous (see also *Periostracum* and *Nacreous Layer*).

PRISTIDAE. Saw-fish: cartilaginous fish with somewhat flattened shark-like body and with an elongated snout or rostrum bearing lateral teeth.

PRISTIOPHORIDAE. Shark-sawfish having a saw-like snout but a more elongated body than that of the Pristidae (see above).

PRO-. Greek prefix: before.

PRO-AMNION. A term used in vertebrate embryology for the part of the blastoderm lying immediately in front of the head and later rising up in a fold to form the beginning of the amnion.

PRO-ATLAS. A small vertebral bone between the atlas and the skull in some reptiles.

PRO-BASIPODITE. The more proximal of the two segments of the basipodite in those crustacea which have the full number of joints in their limbs.

PROBOSCIDEA. Elephants, mammoths etc. An order of mammals characterized by the presence of a trunk or proboscis.

PROBOSCIS. (1) The trunk of an elephant.
(2) The elongated mouth-parts of certain insects, worms etc.

PROBOSCIS SHEATH (NEMERTEA). A muscular sheath surrounding the long eversible proboscis of a nemertean worm.

PROCELLARIIFORMES (PROCELLARIIDAE). Petrels, albatrosses and similar ocean birds with great powers of flight. They breed on rocky precipitous coasts, one egg being laid at a time and both parents taking turns at incubation.

PROCEREBRUM. The fore-brain of an arthropod consisting of the fused ganglia of the first segment (*protocerebrum*) together with a pair of primitive presegmental ganglia connected with the eyes and other sense organs.

PROCESSUS FALCIFORMIS. An older name for the *Retractor lentis* muscle by means of which the lens of the eye is moved backwards for varying the focal distance; present in many vertebrates.

PROCILIATA. Protozoa with cilia distributed uniformly over the body; having many nuclei all of the same size and having no mouth. Asexual reproduction occurs by simple fission of the protoplasm, some nuclei going into each half.
    A well known example is *Opalina* which occurs in the rectum of frogs and toads.

PROCOELOUS. A term used to describe vertebrae whose centra are concave in front and convex behind, as in many reptiles and amphibia.

PROCORACOID. See *Precoracoid*.

PROCRYPTIC. Having protective coloration for concealment.

PROCTODAEUM. The hind-gut: that part of the alimentary canal which is lined by ectodermal tissue (usually stratified epithelium), formed in the embryo by an invagination or folding in of the ectoderm at the anus (*cf. Stomodaeum*).

PROCTOTRYPOIDEA. Minute parasitic hymenopterous insects which attack the eggs or small larvae of other insects. They differ from ichneumons, chalcids and other members of the group Parasitica in having the ovipositor in a terminal position instead of issuing further forward beneath the abdomen.

PROCYONIDAE. Mammals of the order Carnivora, but often omnivorous in habit, having poorly developed carnassial teeth. Most are arboreal and many have long tails. They include the panda of the Himalaya and the kinkajous and racoons of America.

PRO-EPIPODITES. Epipodites or lateral appendages borne on the precoxa (first segment) of a crustacean limb; those borne on the second segment are known as *metepipodites*.

PROFUNDUS GANGLION. A ganglion near the root of the ophthalmic profundus nerve (see below).

PROFUNDUS NERVE (OPHTHALMICUS PROFUNDUS). A sensory nerve present in some fish, running from the snout to the dorsal root of the first (premandibular) segment of the cranium. The corresponding motor nerve (*i.e.* from the same cranial segment) is the oculomotor.

PROGLOTTIS. One of the numerous segments of a tapeworm which bud off from the scolex or head. Each segment is a complete unit containing hermaphrodite reproductive organs, but the two parallel nerve cords and the excretory ducts are common to all the segments. Each proglottis becomes filled with eggs and when mature breaks off from the parent worm, to be cast out with the faeces of the host (see *Strobilation*).

PROGNATHOUS. PROGNATHISM. Having a forward projecting face and jaws as in anthropoid apes and some human races (see *Cranio-facial angle*).

PROGONEATA. See *Diplopoda*.

PROJAPYGIDAE. A primitive family of Diplura or bristle-tails closely approximating to the supposed ancestral type of insect and forming a link between insects and Myriopoda.

PROLEGS. Short unjointed limbs on some of the abdominal segments of caterpillars. Except for the 'loopers', which have only two pairs, most caterpillars of butterflies and moths have five pairs of prolegs. Those of sawflies usually have seven pairs. Prolegs are used for grasping and this is often helped by numerous small hooks.

PROLEUCOCYTE. A type of blood corpuscle common in insects, having deeply staining cytoplasm and a large nucleus. They are thought to be young developing corpuscles.

PRONATION. The typical position of the mammalian fore-arm in which the radius is crossed over the ulna and the palm of the hand is facing down. Many mammals have the two bones fused together in this position so that the palm cannot be turned upwards. In man and most primates, however, the bones may be uncrossed and the palm turned upwards (supination).

PRONEPHRIC DUCT (MÜLLERIAN DUCT). See *Pronephros* below.

PRONEPHROS. The 'head-kidney': a set of tubules forming the first excretory organ in a vertebrate embryo. Later it is replaced by a *mesonephros* (middle kidney) or a *metanephros* (hind-kidney). Its duct, the Müllerian duct, degenerates in adult males but enlarges to form the oviduct in females.

PRONOTUM. The cuticular covering of the first thoracic segment of an insect.

PRONUCLEUS. The nucleus of a male or female gamete containing the haploid number of chromosomes. When two gametes combine to form a zygote, the pronuclei combine to form a single diploid nucleus.

PRO-OESTRUS. The sexual phase of mammals which immediately precedes oestrus or 'heat' (see *Oestrous cycle*).

PRO-OSTRACUM. The anterior plate of the internal shell in a cuttle-fish or similar cephalopod: an extension forwards of the chambered *phragmocone*.

PRO-OTIC BONE. A bone forming part of the front of the auditory capsule in bony fish; in higher vertebrates becoming fused with others to form the *periotic* bone.

PRO-OTIC GANGLION. A ganglion formed by fusion of the *Gasserian* and the *Geniculate* ganglia (on the fifth and seventh cranial nerves respectively) in the head of a frog or similar amphibian.

PROPHASE. The first stage of mitosis or meiosis during which chromosomes appear within the nucleus and the nuclear membrane breaks down.

PROPODITE. The sixth segment of the leg in such crustaceans as the crayfish, crab or lobster. In order from the base outwards the segments are known as the *coxopodite, basipodite, ischiopodite, meropodite, carpopodite, propodite* and *dactylopodite*.

PROPOLIS. Bee-glue: a resinous material of vegetable origin used by bees in addition to wax in the building of their nests.

PROPRIOCEPTORS. Nervous receptors within muscles, tendons, joints etc., by which an animal receives unconscious information with regard to the state of these organs and so automatically co-ordinates its movements or adjusts its posture.

PROPTERYGIUM. The first of the three basal cartilages in the pectoral fins of sharks, dogfish etc., the other two being the *mesopterygium* and the *metapterygium*.

PROSCOLEX. The scolex or head of a bladder-worm (larval tapeworm), in all essentials similar to the scolex of the adult worm but having hooks and suckers invaginated into its hollow bladder-like structure (see *Cysticercus*).

PROSENCEPHALON. The fore-brain of a vertebrate consisting of the olfactory lobes, the cerebral hemispheres and the thalamencephalon.

PROSICULA. A cone-shaped structure at the apex of a Graptolite (fossil colonial coelenterate); apparently the horny covering of the first zooid of the colony (see *Graptolite, Sicula, Metasicula*).

PROSIMIAE. An alternative name for *Lemuroidea* (lemurs).

PROSOBRANCHIATA (PROSOBRANCHIA). An alternative name for the group of molluscs usually called *Streptoneura (q.v.)*.

PROSOMA. The cephalothorax or anterior part of an arachnid such as a spider, usually bearing a pair of prehensile *chelicerae*, a pair of sensory or prehensile *pedipalps* and four pairs of walking legs.

PROSOPYLES. Openings through which water flows from the inhalent canals to the flagellated chambers of the more highly developed types of sponge (see also *Apopyle* and *Leucon*).

PROSTATE GLAND. A gland which helps to produce seminal fluid in male mammals.

PROSTOMIUM. The pre-oral lobe of an earthworm or other annelid. It may take various forms: in the earthworm it overhangs the mouth and acts as a stopper when the worm is burrowing; in free-swimming polychaetes it bears tentacles, palps and simple eyes; in leeches it forms part of the oral sucker and in Echiuroids it forms a long mobile proboscis.

PROTASPIS. The larva of a Trilobite, similar to the nauplius type of larva of other crustaceans.

PROTEASE. A protein-splitting enzyme.

PROTECTIVE COLORATION. Patterns or markings on an animal's body enabling it to blend with its background so that when motionless it is not easily seen. Most animals show such coloration which has an obvious survival value.

PROTEIDAE. Newts whose external gills persist throughout life. Many inhabit caves and subterranean waters. The skin is often white but turns black on exposure to light.

PROTEINS. Very complex nitrogenous compounds whose molecules consist of numerous amino-acid molecules linked together. They are the basic constituents of all living things and are a necessary part of the food of all animals. Some of them contain many thousands of amino-acid molecules and have a total molecular weight running into millions. The number of such combinations is practically infinite and it is not surprising therefore that few have been completely analysed.

PROTELIDAE. Burrowing nocturnal carnivora of South Africa, which feed on insects and on carrion; somewhat wolf-like and often placed in the same family as the hyaenas.

PROTEPHEMEROPTERA. Fossil insects of the Permian Period having affinities with present-day mayflies (*Ephemeroptera*) and also with the primitive fossil dragonflies (*Protodonata*). Both are characterized by the very complete pattern of venation of the wings with numerous cross-veins forming a network or *archidictyon*.

PROTEROGLYPHA. A group of extremely poisonous snakes having the anterior maxillary teeth grooved or perforated forming channels down which poison comes from the labial glands. Behind the fangs is a series of smaller solid teeth.

PROTEROSAURIDAE. Extinct lizard-like reptiles of the Permian and Triassic Periods, belonging to the primitive class Rhynchocephalia. They had numerous small conical teeth either implanted in shallow pits or fused to the jaws. Some species were up to 7 ft. long.

PROTEROTHERIIDAE. Extinct horse-like ungulates of South America, some with three toes and some with one.

PROTHORAX. The first of the three thoracic segments of an insect.

PROTHROMBIN. A protein in mammalian blood which, in the presence of calcium ions, gives rise to *thrombin*; the latter in turn reacting with fibrinogen to bring about the clotting of the blood.

PROTISTA. A term used to denote all unicellular creatures whether animal or plant. Some unicellular algae, for instance, are obviously related to similar species of protozoa and only differ from them by the presence of chlorophyll enabling photosynthesis to take place. The distinction between plant and animal at this level is often difficult and artificial so that the use of a single term seems desirable.

PROTO-. Prefix from Greek *Protos*: first.

PROTOBLAST. A naked cell, *i.e.* a mass of protoplasm without a cell wall.

PROTOBRANCHIATA (PROTOBRANCHIA). The most primitive type of lamellibranch mollusc in which the gills consist of short filaments hanging down from the dorsal side of the mantle on each side of the body, within the two valves of the shell.

PROTOCERCAL. Primitively symmetrical pointed tail-fins of certain fish (see *Diphycercal*).

PROTOCEREBRUM. That part of the brain of an arthropod formed from the fused ganglia of the first somite (see also *Deutocerebrum* and *Tritocerebrum*).

PROTOCHORDATA (PROTOCHORDA, ACRANIA). Simple chordates (animals with a notochord), distinguished from higher members of the phylum by having no true brain and in some cases no heart. They include tunicates or sea-squirts (*Urochordata*), *Amphioxus* (*Cephalochordata*) and acorn-worms (*Hemichordata*). The latter are sometimes put into a separate subphylum.

PROTOCILIATA. See *Prociliata*.

PROTOCOCCACEAE. Unicellular organisms containing chlorophyll and frequently living as symbionts (zoochlorellae) in sponges, coelenterates, worms etc. They are classed by botanists among the Algae and by zoologists among the Phytomastigina.

PROTOCOLEOPTERA. Fossil insects with thickened tegmina resembling those of beetles; possibly they were the ancestors of both the Coleoptera and the Strepsiptera.

PROTOCONCH. The first formed part of the shell of a mollusc.

PROTODONATA. Primitive dragonflies of Palaeozoic times with wing-venation more simplified than that of present-day species. Possibly the Protodonata and the true Odonata are both derived from a common ancestor approximating to the *Palaeodictyoptera* (*q.v.*).

PROTOHYMENOPTERA. Fossil insects resembling bees and wasps, found in the Permian and Jurassic rocks. They had two pairs of wings of equal size, without coupling apparatus and with venation of a generalized type.

PROTOMERITE. The middle one of three segments in certain Gregarine protozoa. The other parts are the *epimerite* or fixing organ and the *deutomerite* which contains the nucleus.

PROTOMONADINA (PROTOMONADIDA). Protozoa with one or two flagella and sometimes also with amoeboid properties. The flagella may be partly united to the body of the organism by an undulating membrane. The group includes many of the most deadly parasites of man and of animals, the best known being *Trypanosoma* which is carried by the tse-tse fly (*Glossina*) and is the cause of African Sleeping Sickness.

PROTOMYXA. An alternative name for the *Mycetozoa* or *Myxomycetes*.

PROTONEPHRIDIUM. A primitive excretory organ consisting of a bunch of flame cells which discharge waste products into a tubule leading to the exterior. Such organs are found chiefly in the Platyhelminthes or flatworms and flukes (see *Flame cell*).

PROTOPERLARIA. Fossil insects which were probably ancestral to present-day stone-flies (Plecoptera) and also showed some resemblance to Orthoptera.

PROTOPLASM. The living material of a cell consisting of a complex colloidal mixture of water, proteins, lipoids, carbohydrates, mineral salts etc.; having the power of changing its state from a thick liquid to a jelly-like solid and *vice versa*; having also the fundamental powers of living matter to absorb and assimilate food, to respire and produce energy and to respond to external stimuli. Protoplasm is extremely variable and is probably not exactly alike in any two species. In any cell a distinction is made between *nucleoplasm* in the nucleus and *cytoplasm* surrounding it. In some organisms such as *Amoeba* the cytoplasm is further differentiated into outer *ectoplasm* and inner *endoplasm*.

PROTOPLAST. The protoplasmic content of a single cell.

PROTOPODITE. The basic portion of the biramous limb of a crustacean. Normally it consists of two parts, the proximal *coxopodite* and the distal *basipodite*. In certain crustaceans, however, it may consist of four parts known respectively as the *precoxa, coxa, probasipodite* and *metabasipodite*.

PROTOPOD LARVA. The most primitive type of insect larva with unsegmented abdomen and rudimentary appendages. Such larvae are usually endoparasites and can only survive where food is abundant making locomotion unnecessary.

PROTOPTERUS. A lung-fish from the swamps and rivers of tropical Africa; one of the only three living genera of lung-fish and differing from the Australian variety in having two lungs (see *Dipnoi*).

PROTORTHOPTERA. Fossil insects of the Palaeozoic Period, probably ancestral to both Orthoptera and Coleoptera.

PROTOSPONDYLI. An order of fish represented at the present day by a sole living species *Amia calva*, the Bow-fin, from the fresh waters of North America. It has a long continuous dorsal fin, a practically homocercal tail, a bilobed lung and many other distinctive features some primitive and some specialized. The order dates from the Triassic Period and some of the members may be considered as ancestors of the Teleosts or modern bony fish.

PROTOTHERIA. An alternative name for the *Monotremata*.

PROTOTROCH. A ring of cilia in a preoral position encircling a trochosphere larva (see *Trochosphere*).

PROTOZOA. The subkingdom which comprises all unicellular animals (by some regarded as non-cellular). Included in the group are some which may also be classed as plants, the distinction being that these possess chlorophyll.

PROTURA. Minute insects without eyes or antennae and with piercing mouthparts contained in pockets from which they can be extruded when feeding (see also under *Anamorphosis*).

PROVENTRICULUS. The gizzard or 'stomach' of crustacea and insects, usually thick and muscular with chitinous teeth or ossicles.

PSALTERIUM. The third compartment of the stomach of a ruminant

PSEUD-, PSEUDO-. Prefix from Greek *Pseudos*: false.

PSEUDAPOSEMATIC. Imitating the warning coloration or other features of a noxious animal and thereby obtaining a greater degree of protection from predators.

PSEUDOBRANCH. A vestigial gill with no respiratory function in the spiracle of certain fish; the spiracle itself being a shrunken gill-cleft.

PSEUDOCAUDAL FIN. A tail-fin which is merged with portions of the dorsal and ventral median fins in certain fish such as the cod.

PSEUDOCHROMIDAE. Small marine perches of the Atlantic, Indian and Pacific Oceans.

PSEUDOCOEL. A body-cavity which is not part of the true coelom, *e.g.* the so-called *fifth ventricle* of the mammalian brain which is in the membrane separating the two cerebral hemispheres.

PSEUDOCOLONY. A number of individual protozoa united by stalks or tubes of dead material; distinct from a true colony in which the individuals are united by living protoplasm. Pseudocolonies occur among the Mastigophora and Vorticellidae.

PSEUDO-HEARTS. Enlarged commissural blood-vessels of certain annelids, containing valves and acting as hearts. In the earthworm there are five pairs of these organs which pump blood from the main dorsal vessel to the ventral vessel.

PSEUDONAVICELLA. The sporocyst of certain gregarine protozoa such as *Monocystis*; so named on account of its resemblance to the small boat-shaped diatom *Navicella*.

PSEUDOPHYLLIDEA. A suborder of tapeworms whose scolex or head has no clearly marked neck and no hooks, but usually has two suckers. Members of this group are usually parasites of fresh-water fish but some species are found in birds and in man.

PSEUDOPODIA. 'False feet': temporary protrusions of protoplasm acting as locomotory and feeding organs of certain protozoa. They may be of various types, *viz*. *Lobopodia* (blunt), *Filopodia* (fine threads), *Rhizopodia* (branching and rejoining to form a network) and *Axopodia* (with an internal supporting rod).

PSEUDOPODIOSPORES. Amoeboid spores (see *Amoebulae*).

PSEUDOSUCHIA. Extinct crocodiles of the Triassic Period having large orbital cavities, anterior laterally placed nostrils, and teeth only on the anterior part of the jaw.

PSEUDOTRACHEAE. Food channels resembling tracheae in the membrane covering the labellar lobes of flies and similar insects. They converge into collecting ducts which lead to the prestomum and hence to the mouth.

PSEUDOVELUM. A membrane forming an internal flange round the rim of a jelly-fish (scyphomedusa). It differs from the true velum, found on the medusae of the Hydrozoa, in that it contains no muscles or nerve-ring (see *Velum*).

PSEUDOVILLI (UTERINE VILLI). Structures resembling villi by which a mammalian placenta is embedded in the wall of the uterus.

PSITTACI. Parrots and similar birds noted for their remarkable power of speech; usually brightly coloured and having a bent beak with the upper half overlapping the lower at the tip. They inhabit most of the warmer parts of the world and are monogamous, but usually roost and feed in company.

PSITTACIDAE. Smooth-tongued parrots including cockatoos, macaws, grey parrots and budgerigars.

PSITTACINAE. All parrots except the cockatoo.

PSOCOPTERA. Book-lice: small soft-bodied insects with long thread-like antennae. They may be winged or wingless and have biting mouth-parts. They feed on the dried paste of old books. Many related species are found among vegetation or on the bark of trees.

PSOPHIIDAE. Trumpeters: crane-like birds of tropical South America.

PSYCHIDAE. A family of moths in which the female is wingless and lives in a caddis-like case made by the caterpillar. The male is fully winged.

PSYLLIDAE. Small reddish-yellow plant-lice (Homoptera) which infest a variety of trees and shrubs, curling the leaves or attacking the twigs.

PTER-, PTERO-. Prefix from Greek *Pteron*: a wing.

PTERERGATE. An exceptional form of worker-ant having the rudiments of wings.

PTERICHTHYOMORPHI. An alternative name for the *Antiarcha* (*q.v.*).

PTERINE PIGMENTS (PTERINS). White, yellow or red pigments in the bodies and wings of insects; chemically related to uric acid and possibly waste products of metabolism.

PTEROBRANCHIA. Small worm-like Hemichordata living at great depths in the sea in tubular dwellings which they secrete for themselves by means of glands on the proboscis.

PTEROCARDIAC OSSICLES. Two lateral wing-like ossicles in the anterior chamber of the stomach or proventriculus of a crustacean (see *Gastric Mill*).

PTEROCLIDAE. Sand-grouse: desert birds of Africa and Asia having the feet and toes covered as far as the claws with hairy plumage.

PTERODACTYLS. See *Pterosauria*.

PTEROPHORIDAE (ALUCITIDAE). Plume-moths: insects whose wings are deeply cleft and fringed like a bunch of feathers.

PTEROPIDAE. Large fruit-eating bats, some with a wing-span of up to 5 ft. and usually without tails. They are found in tropical and subtropical regions of the Old World.

PTEROPODA. Sea-butterflies: pelagic molluscs of the order Opisthobranchiata, usually having a transparent vase-shaped shell from which projects the foot in the form of two 'wings' or fins. Locomotion is by a slow flapping of these.

PTEROSAURIA (PTERODACTYLS). Flying reptiles of the Mesozoic Period having membranous wings supported by the enormously elongated fourth finger. Although very different from birds, they had some features in common with them, notably the large keeled sternum and hollow limb-bones. These are adaptations to flying which have probably evolved independently in the two groups.

PTEROSTIGMA. A wing spot: a small darkly pigmented area on the wing of a dragon-fly and on some bees and wasps.

PTEROTIC BONE. One of the bones forming the upper part of the ear capsule in bony fish; in higher vertebrates it becomes fused with other bones to form part of the *periotic* bone.

PTERYGO-. Prefix from Greek *Pterux, pterugos*: a wing.

PTERYGOIDS. Paired membrane-bones of the vertebrate skull, forming the lateral walls of the pharynx.

PTERYGOPODIUM. See *Mixipterygium*.

PTERYGO-QUADRATE (PALATO-PTERYGO-QUADRATE). Paired skeletal elements of cartilage or bone forming the main part of the upper jaw in fish.

PTERYGOTA. All insects which have wings or which may have lost them in the course of evolution, but not including the primitively wingless insects such as the spring-tails and bristle-tails (*Apterygota*).

PTERYLAE. The lines along which are arranged the contour feathers of a bird. Between these are the *apteria* or spaces which are either naked or only covered with down.

PTILINUM. A sac or vesicle on the head of a fly by which the insect pushes open the puparium as it emerges. Afterwards the sac is withdrawn into the head leaving a U-shaped suture.

PTYALIN. An enzyme produced by the salivary glands of some vertebrates and having the power of breaking down starch into maltose.

PUBES (PUBIC BONES). A pair of bones forming the ventral part of the pelvic girdle in the vertebrate skeleton.

PUBIC SYMPHYSIS. The junction or suture of the two pubic bones in the mid-ventral line of the pelvic girdle of a vertebrate.

PULMO-, PULMON-. Prefix from Latin *Pulmo, pulmonis*: a lung.

PULMO-CUTANEOUS ARTERIES. A pair of arteries in frogs and other amphibia, taking blood from the heart to be oxygenated through the moist skin and through the lungs. Each pulmo-cutaneous artery divides into a pulmonary branch and a cutaneous branch, the latter dividing into numerous smaller branches going all over the skin.

PULMONARY ARTERY. An artery carrying blood from the heart to the lungs. In mammals and birds it leads out of the right ventricle but in amphibia and reptiles, which have only one ventricle, it is a branch of the truncus arteriosus.

PULMONARY VEIN. A vein which carries freshly oxygenated blood from the lungs to the left auricle of the heart in air-breathing vertebrates.

PULMONATA. Gasteropod molluscs such as the snail in which the mantle cavity has become a 'lung', *i.e.* a chamber with a small aperture (the pneumostome) and a highly vascular roof through which oxygen can be absorbed into the blood.

PULP CAVITY. The space in the centre of a tooth containing blood-vessels and nerve-endings.

PULSATILE ORGANS. Contractile regions of certain arteries in insects. In the cockroach and the bee, for example, pulsatile organs are found at the bases of the wings and antennae and are able to pump blood into these organs.

PULVILLUS. A pad or cushion between the claws of an insect.

PUNCTA LACHRYMALIA (PUNCTA LACRIMALIA). Points where the nasal ducts receive the secretions from the lachrymal glands at the inner angle of each eye in a mammal.

PUPA. The third stage in the life of an insect: the stage between the larva and the adult during which the insect appears to be resting but in reality is undergoing many internal changes. There are three types of pupa, the commonest kind being the *exarate* or free pupa whose wings and other appendages are clearly visible and movable. In butterflies and moths the pupa is of the *obtect* type in which the wings and appendages are stuck to the body and cannot be moved; in the house-fly there is a hard barrel-shaped case or *puparium* which protects the exarate pupa within. This type is said to be *co-arctate*.

PUPARIUM. A hard barrel-shaped case enclosing the pupa of certain dipterous insects such as the house-fly (see also *Co-arctate pupa*).

PYCNO-. Prefix from Greek *Puknos*: thick, dense.

PYCNODONTIDAE. Extinct fish of the Jurassic and Cretaceous Periods having a high oval, laterally compressed body covered with rhomboidal ganoid scales articulated together.

PYCNOGONIDA. Sea-spiders. See *Pantopoda*.

PYCNONOTIDAE. Bulbuls: Persian singing birds closely related to thrushes and larks.

PYGIDIUM. (1) The last or anal segment of an annelid, formed from the hind-part of the larva and not arising by segmentation.

(2) The posterior part of the abdomen in certain arthropods, formed by the fusion of several body segments.

PYGOSTYLE. The last four fused tail-vertebrates of a bird.

PYLORIC. Appertaining to the pylorus or posterior end of the stomach.

PYLORIC CAECA. Diverticula or caeca arising from the pyloric region in many invertebrates, *e.g.* insects, echinoderms etc.

PYLORIC OSSICLE. A median ossicle forming part of the *gastric mill* in the hind-part of the stomach of a crustacean.

PYLORIC SAC. A five-sided sac immediately above the stomach of an echinoderm.

PYLORIC SPHINCTER. A sphincter at the pyloric or posterior end of the vertebrate stomach, separating it from the duodenum.

PYRALIDAE (PYRALIDOIDEA). A family of moths, many of which are important on account of the damage they do to crops and food products. They include the flour-moth *Ephestia*, the sugar-cane moth *Chilo*, and the wax-moth *Galeria* which inhabits beehives.

PYRAMIDS (OF KIDNEY). Conical structures formed from a number of collecting tubules bound together near the point where they lead out of the kidney into the ureter.

PYRENOIDS. Small protoplasmic or protein granules which act as centres of starch-formation in plants and in the plant-like protozoa (Phytomastigina).

PYRIFORM. Pear-shaped: a term used to describe a number of organs, *e.g.* a sensory organ on a Polyzoan larva; a type of silk gland in spiders etc.

PYRIFORM CORTEX (PYRIFORM LOBES). Lateral parts of the cerebral hemispheres concerned with nervous impulses from the nose. They are well developed in reptiles and form part of the *archipallium*. In mammals, however, they are overshadowed by the *neopallium* which is developed out of all proportion to the rest of the cortex and has to do with general correlation of functions other than olfactory.

PYROSOMATIDA. See *Luciae*.

PYRRHOCORIDAE. Red bugs: a family of plant-bugs of the group Gymnocerata including the so-called 'stainers' which discolour cotton fibres.

Q

QUADRATE BONE. A cartilage-bone at the hind-end of the upper jaw in most vertebrates. In bony fish, amphibia, reptiles and birds it forms an articulation with the lower jaw. In mammals, however, it is very much reduced and becomes the *incus* or 'anvil' of the ear.

QUADRATO-JUGAL BONE. A bone between the quadrate and the jugal in the skulls of some vertebrates.

QUADRUMANA. Apes in which both the thumb and the large toe are opposable. This fact distinguishes them from the *Bimana* (*e.g.* humans) in which only the thumb is opposable.

QUEEN. A fertile female, the only individual capable of laying eggs, in a colony of bees, wasps, ants or termites. In the latter the queen may attain a length of four inches or more. In a colony of hive-bees the queen develops in a special 'queen-cell' and is produced by feeding a female larva on partly digested food known as *royal jelly*.

QUILL. (1) The *calamus* or hollow stalk of a feather; the term is also sometimes used for a whole wing- or tail-feather.

(2) A hollow spine or stiffened hair of a porcupine.

QUINTOCUBITAL. A word used to denote birds which have a fifth cubital feather on the wing, as opposed to the *aquintocubital* birds which have a gap between the fourth and sixth.

R

RACHI-. Prefix from Greek *Rachis*: a spine.

RACHIGLOSSA. Predatory gasteropod molluscs having a long narrow tongue or radula with teeth arranged in rows of three. An example is the whelk *Buccinum*.

RACHIODONTINAE. Harmless snakes of South Africa which feed on birds' eggs, breaking the shell in its passage down the oesophagus by pressure from the much enlarged hypapophyses of the thoracic vertebrae.

RACHIOSTICHOUS. A term used to describe the paired fins of certain primitive fish in which the skeleton of the fin has a central axis or rachis with branches placed like the barbs of a feather.

RACHIS (RHACHIS). (1) The spinal column.

(2) The axis of a feather.

(3) Any central axis or midrib.

RADI-, RADIO-. Prefix from Latin *Radius*: a ray.

RADIAL CANALS OF COELENTERATES (PERRADIAL CANALS). Four canals which run outwards from the central gastric cavity to the circular canal at the rim of a medusa or jelly-fish. In some types the gonads are situated on these radii.

RADIALE. The proximal carpal bone adjacent to the radius in the limb of a normal tetrapod vertebrate. In humans and some other animals it is shaped like a boat and known as the *scaphoid*.

RADIALIA (SOMACTIDS). Cartilaginous or bony elements forming the basal fin-rays of most fish.

RADIAL NERVES (ECHINODERMATA). Nerve fibres running along the ambulacral grooves in the 'arms' of a star-fish or other echinoderm (see also *Nerve ring*).

RADIOLARIA. Amoeboid protozoa, usually marine and planktonic, having an internal siliceous skeleton of complex and fantastic shape. Around this is an external layer of highly vacuolated buoyant protoplasm from which radiate numerous fine pseudopodia.

RADIO-ULNA. A fusion of the radius and the ulna, as for instance in the skeleton of a frog.

RADIUS. The anterior or pre-axial of the two bones forming the fore-arm of a tetrapod vertebrate.

RADULA. The 'tongue' of a mollusc: a ribbon-like organ whose surface is studded with rows of horny teeth.

RAII (RAJI, RAIIDAE, BATOIDEI). Skates and rays: cartilaginous fish with the body flattened dorso-ventrally; with the gill-clefts on the ventral surface and with enormously expanded pectoral fins.

RALLIDAE. Rails, coots and water-hens: marsh birds inhabiting reedy lakes and ponds in most parts of the world. They belong to the Gruiformes or crane-like birds, but the legs are shorter than those of a crane.

RAMICORN. A branched antenna or horn.

RAMUS. Latin: a branch.

RAMUS COMMUNICANS. A short nerve connecting each sympathetic ganglion with the corresponding spinal nerve in a vertebrate. (Pl. *Rami communicantes*.)

RANIDAE. Frogs having a protrusible tongue, a distinct tympanum, separate Eustachian tubes, epicoracoids fused together and precoracoids always present. The vertebrae are procoelous and the sacral vertebra has cylindrical transverse processes. The tadpoles have a single spiracle which is on the left side.

RANINAE. Common frogs having all the above characters of the Ranidae and having teeth on the upper jaw only. Most are aquatic or semi-aquatic and have webbed feet. Some arboreal forms have much enlarged webbed hands and feet which act as parachutes; others have adhesive discs on the digits.

RANVIER, NODES OF. Constrictions in the sheath of a medullated nerve fibre at regular intervals along its length.

RAPHIDIIDAE. Snake-flies: insects of the order Neuroptera having a long stalk-like thorax resembling a neck. The larvae live under the bark of trees and feed on wood-boring insects.

RATHKE'S POCKET. The cavity of the pituitary hypophysis: embryonically a diverticulum of the buccal cavity. In some animals it remains as a cleft separating the anterior pituitary lobe from the posterior; it may be obliterated in the adult.

RATITES (RATITAE). Flightless birds with no sternal keel: ostriches, rheas, kiwis, emus etc.

RAUBER CELLS. In mammalian embryology: the cells of the trophoblast which overlie the embryonic disc and later break down so that the embryo comes to lie on the surface (see *Trophoblast*).

RE-. Latin prefix: back, again.

RECAPITULATION. The theory of Von Baer and Haeckel that any animal during its embryonic development recapitulates the main stages of its ancestral evolution.

RECEPTACULUM SEMINIS. See *Spermatheca*.

RECEPTOR. A sensory organ or cell which receives stimuli from outside or inside an animal and passes impulses on to the nervous system. Receptors may be sensitive to light, temperature, pressure, chemicals etc. and are classified according to their position as follows:

    (1) Exteroceptors: those on the outside of the body.

    (2) Interoceptors: those inside the body which are often stimulated unconsciously and cause appropriate reactions, *e.g.* those in the stomach, bladder etc.

    (3) Proprioceptors: those in muscles, tendons, joints etc. which give the animal unconscious information relating to position, tension etc.

RECESSIVE QUALITY. A quality which, in Mendelian inheritance, does not show itself in hybrids. The opposite of *dominant*. If for instance a pure-bred tall pea plant is crossed with a pure-bred short plant, the resulting offspring will all be tall. Tallness is therefore said to be dominant and shortness recessive.

RECIPROCAL HYBRIDS. Two hybrids, one of which is descended from the male of species A and the female of species B; the other from a male of species B and a female of species A.

RECIPROCAL ORGANS. Organs exhibiting secondary sexual characters which are opposite in the male and female, *e.g.* large horns in a male and short horns in a female; large breasts in a female and small breasts in a male etc.

RECT-. Root word from Latin *Rectus*: straight.

RECTAL CAECA. Diverticula or blind ducts leading off the rectum, as for instance in birds.

RECTAL GLANDS. Glands opening into the rectum in many vertebrates. Their function is uncertain.

RECTAL PAPILLAE (OF INSECTS). Thickened bands in the lining of the rectum, able to absorb water from the faeces and so prevent undue loss of water from the insect as a whole.

RECTRIX (RECTRICES). The tail feathers of a bird.

RECTUM. The last part of an alimentary canal leading to the anus.

RECTUS MUSCLES. Four muscles at right angles to one another which help to move the eye of a vertebrate. They are attached to the back of the eyeball and are known as the inferior, superior, anterior and posterior recti. In addition to these there are two oblique muscles making six in all.

RECURRENT LARYNGEAL NERVE. A branch of the vagus nerve of a mammal, looping round the ductus arteriosus and going forwards beside the trachea to the larynx. The peculiar course of this nerve is brought about by the lengthening of the neck.

RED BONE MARROW. Highly vascular fatty tissue where red blood corpuscles are made in the central cavities of the larger bones.

RED GLAND. See *Rete mirabile*.

REDIA. The third larval stage of a liver-fluke, developing from the sporocyst larva. It is an elongated sac-like organism with a small mouth, a suctorial pharynx and a simple gut. Within the body cavity are germinal cells which may give rise to secondary rediae or to the last type of larva which is known as a *cercaria*.

REDUCTION DIVISION. See *Meiosis*.

REDUVIIDAE. A family of heteropterous bugs, some of which prey on other insects and some of which are blood-suckers of man and other mammals. Many tropical forms can transmit trypanosomes and other parasites. The nymphal stage resembles an ant.

REFLEX ACTION. A form of animal behaviour in which a certain stimulus produces a definite automatic and often unconscious response. In its simplest form it involves the passage of a nervous impulse from a sensory receptor along an afferent nerve to the central nervous system (spinal cord or brain); thence by way of one or more connector neurones to an efferent nerve which brings about a response such as movement, secretion etc.

REGENERATION. The re-growth of an organ which has been lost or injured: a power which is greatest in some of the lower or less specialized animals and least in mammals. Experiments with planarians have shown that in some cases a whole animal may be regenerated from a small part.

P

REGULATION EGGS. Eggs which at first show little or no differentiation and have no definite organ-forming areas until cleavage is well advanced (*cf. Mosaic eggs*).

REISSNER'S FIBRE. A thread-like bundle of nerve fibres running from the roof of the mid-brain through the length of the cerebrospinal canal. It is present in most chordates but not in man.

REISSNER'S MEMBRANE. A membrane running the length of the spiral cochlea in the mammalian ear and separating the upper chamber or *scala vestibuli* from the middle chamber or *scala media*.

RELAY NEURONE. A neurone linking two others by means of synapses; particularly one linking a sensory with a motor nerve fibre in a simple reflex arc.

REMEX (Pl. REMIGES). The wing-quill of a bird. These are classified according to their position as primary (on the manus) and secondary or *cubital* (on the ulna). (See *Feathers*.)

RENAL. Appertaining to a kidney.

RENAL ARTERY AND VEIN. The artery and vein leading to and from a kidney.

RENAL PORTAL VEINS (RENAL PORTAL SYSTEM). Veins taking blood from the tail or from the hind limbs directly to the kidneys. Such a system is only present in fish and amphibia, *i.e.* in those vertebrates whose kidney is a *mesonephros* (*q.v.*).

RENNIN. A substance contained in the gastric juice of some vertebrates and having the power to curdle milk.

RENO-PERICARDIAL APERTURE. A coelomic canal or passage connecting the kidney with the pericardium in some molluscs.

REPRODUCTION. The production of young; in animals it may be by any of the following means:
(1) Simple or multiple fission as in Protozoa.
(2) Gemmation or budding as in such organisms as *Hydra*.
(3) Sexual reproduction: the union of a male and a female gamete to form a zygote.
(4) Parthenogenesis: development of an unfertilized ovum into a new individual which is genetically identical with the parent.
In the case of sexual reproduction and of parthenogenesis the actual birth may be *oviparous*, *ovo-viviparous* or *viviparous*. Oviparous animals lay normal eggs with membranes and a shell; ovo-viviparous animals produce eggs which hatch before birth; viviparous animals give birth to live young which have developed inside the body of the parent (*e.g.* attached by a placenta).

REPTILIA. Reptiles: cold-blooded, scaly lung-breathing vertebrates which lay large eggs usually having a shell. The young develop within an amnion and an allantois. The blood circulatory system is similar to that of an amphibian having a right and a left aortic arch and, except in crocodiles, a three-chambered heart. Reptiles were the largest group of vertebrates in the Jurassic and Cretaceous Periods, but are now restricted to the few surviving groups, namely: *Chelonia* (tortoises), *Squamata* (lizards and snakes), *Crocodilia* (crocodiles) and *Rhynchocephalia* (the tuatara of New Zealand).

RESPIRATION. (1) The taking in of oxygen and the giving out of carbon dioxide.
(2) Internal or tissue respiration: the oxidation of a substance such as a

carbohydrate or fat within the tissues to form carbon dioxide and water with liberation of energy. This process is usually brought about with the help of specific enzymes such as *oxidase* or *dehydrogenase*.

The mechanism by which gases are exchanged is usually called breathing or *external respiration* to distinguish it from the internal process which is true respiration. Normally free oxygen is necessary for respiration and the process is called *aerobic*. Sometimes, however, the necessary oxygen can be obtained by the decomposition of other substances in the absence of air. This is known as *anaerobic respiration* and takes place, for example, in muscle contraction when glycogen is oxidized to lactic acid.

RESPIRATORY ORGANS. Organs usually having a large surface through which interchange of oxygen and carbon dioxide can take place between the body-fluids and the surrounding medium. The chief types of respiratory organs are:

(1) *External gills* (amphibian larvae, crustacea etc.).

(2) *Internal gills* with gill-slits leading from the pharynx to the exterior (fish and later amphibian larvae).

(3) *Lungs* (reptiles, birds and mammals).

(4) *Spiracles* and *tracheae* (insects).

(5) *Lung-books* (spiders).

(6) *Gill-books* (the king-crab *Limulus*).

(7) *Respiratory 'trees'* (Holothuroidea).

(8) *Gill-membranes* (Lamellibranch molluscs).

RESPIRATORY PIGMENTS. Coloured substances contained in blood and capable of forming temporary compounds with oxygen, thus carrying it from the respiratory organs to the tissues. The best known examples are *haemoglobin* in the blood of all vertebrates and a few invertebrates, and *haemocyanin* in certain mollusca and crustacea.

RESPIRATORY TREES. The main respiratory organs of Holothurians or 'sea-cucumbers': complex branched tubes leading from the cloaca to various parts of the body enabling dissolved oxygen to pass from the water into the coelomic fluid.

RESTIFORM BODY (CORPUS RESTIFORME). A part of the brain of a vertebrate forming two lobes behind the cerebellum at the sides of the fourth ventricle.

RETE MIRABILE. (1) Of fish: the 'red gland', a network of small blood-vessels inside the air-bladder, able to secrete or absorb oxygen so that the amount of gas in the bladder varies and the fish can maintain its buoyancy at different depths.

(2) Of Echinoderms: a complex network of haemal strands or 'blood-vessels' by which the dorsal vessel in Holothurians hangs in a fold of peritoneum from the intestine.

RETICULO-ENDOTHELIAL SYSTEM. Tissues which contain or produce phagocytes capable of destroying bacteria and other foreign bodies and converting damaged red corpuscles into bile pigments. The connective tissue of the lungs, spleen, liver etc. forms part of this system.

RETICULUM. (1) A network, *e.g.* a protoplasmic network or *mitome*.

(2) The second compartment in the stomach of a ruminant.

RETINA. The layer of sensitive cells and nerve fibres lining the interior of an eye. In many vertebrates there are two types of cell in the retina, *viz. Rods* and *Cones*. Only the latter can distinguish colour.

RETINACULUM. (1) A hook-like structure by which the fore-wing of a moth engages with a bristle or *frenulum* on the hind-wing, thus coupling the wings together during flight.

(2) A hook beneath the abdomen of insects of the order Collembola (springtails) by which the furcula of the tail is held down beneath the insect. When released the tail propels the insect forwards into the air.

RETINULAE. Sensory cells in the compound eyes of arthropods. The eye is made up of many visual units or *ommatidia* each having pigmented retinular cells round the base of a crystalline cone.

RETRAL PROCESSES. Pockets projecting back from the hind edge of each chamber in the spiral shells of certain foraminifera.

RHABD-. Prefix from Greek *Rhabdos*: a rod.

RHABDITES. Crystalline rods in some of the epidermal cells of Turbellarians; their function is unknown.

RHABDITOID LARVA. The first larval form of a Nematode worm, frequently living in the soil and only becoming a parasite at a later stage of development.

RHABDOCOELIDA. A group of Turbellaria or free-swimming flatworms in which the mouth is near the anterior end and the gut is a straight tube with or without lateral pouches; they are found in both fresh and salt water.

RHABDOLITHS. Calcareous rods forming an outer protective case for certain marine flagellate protozoa (see *Coccolithophoridae*).

RHABDOM (RHABDOME). A transparent rod passing down the centre of each ommatidium or optical unit in the compound eye of an insect or crustacean.

RHABDOMERE. One of a number of segments which make up the rhabdom (see above).

RHACHIS. See *Rachis*.

RHAMPHASTIDAE. Toucans: extraordinary birds with a huge beak and a horny brush-like tongue. They inhabit the tropical forests of Central and South America.

RHAMPHOTHECA. The horny covering ensheathing the jaws of a bird and forming the beak.

RHEIDAE. Rheas: large South American flightless birds with partially feathered head and neck and with three-toed feet.

RHENANIDA (RHENANIDI). Extinct fish of the Devonian Period having a poorly developed cranium. Some species living on the bottom have, by a convergent adaptation, acquired very much the shape of the present-day skate although no relation to it.

RHEOTACTIC RECEPTORS. Sensory cells which detect changes in the flow of water over the surface of a body; found for example in the epidermis of certain flatworms.

RHIN-, RHINO-. Prefix from Greek *Rhis, rhinos*: nose.

RHINAL FISSURES. Fissures separating the olfactory tracts from the cerebral hemispheres in the brain of a mammal.

RHINENCEPHALON. An alternative name for the olfactory lobe of a vertebrate brain.

RHINOBATIDAE. Flattened cartilaginous fish (Raiidae) having a long tail with two dorsal fins and without serrated caudal spines.

RHINOCEROTIDAE. Rhinoceroses: large unwieldy pachyderms belonging to the Perissodactyla or odd-toed ungulates, having three toes on each hind foot. They may have one or two horns on the nose but these, unlike the horns of other animals, consist of a hard compact mass of agglutinated hairs.

RHINOCHETIDAE. Crane-like birds of New Caledonia about the size of a fowl.

RHINOCOEL. A cavity sometimes present in each olfactory lobe of the brain of a vertebrate. It communicates with the other cavities and like them contains cerebro-spinal fluid.

RHINODONTIDAE. Whale-sharks: large sharks of the Indian and Pacific Oceans exceeding 50 ft. in length, having wide gill-clefts and having the mouth with small teeth at the tip of the snout.

RHINOLOPHIDAE. Small insectivorous bats with large ears and well developed 'nose-leaves'; found in most parts of the world.

RHIPIDO-. Prefix from Greek *Rhipis, rhipidos*: a fan.

RHIPIDOGLOSSA. A group of molluscs whose radula has rows of teeth diverging like the ribs of a fan (see *Radula*).

RHIPIDOSTICHOUS. Fins having the skeletal elements arranged in a fan-like manner as in most cartilaginous fish.

RHIPTOGLOSSA. Chamaeleons and similar lizards with no clavicles or inter-clavicles; with well developed limbs, independently movable eyes and a very long worm-like tongue.

RHIZO-. Prefix from Greek *Rhiza*: a root.

RHIZOCEPHALA. A group of crustaceans including the well known *Sacculina* which is parasitic on crabs. They are extremely degenerate and are not easily recognizable as crustacea. There is no alimentary canal, no segmentation and no limbs or other appendages. Root-like processes from the parasite penetrate deep into the tissues of the host causing the phenomenon known as *parasitic castration (q.v.)*.

RHIZOMASTIGINA. A small order of protozoa, mostly fresh-water species, having an amoeboid body with one or two flagella.

RHIZOPLASTS. Protoplasmic threads connecting the basal granule of a flagellum with the nucleus or with the parabasal body in certain protozoa.

RHIZOPODA (SARCODINA). Amoeboid protozoa: unicellular organisms which usually feed and move by means of pseudopodia and have no cilia or flagella. They may be naked or may secrete shells of silica or of calcium carbonate. Reproduction may be by binary or multiple fission or by syngamy. The class includes Amoebina, Foraminifera, Radiolaria and Heliozoa.

RHIZOPODIA. Thin pseudopodia which branch and anastomose, as for instance in Foraminifera.

RHIZOSTOMEAE. Jelly-fish (Scyphomedusae) in which the four lips are enormously enlarged and the central mouth is reduced or absent. There are thousands of tiny sucking mouths situated along grooves in the lips. Micro-organisms are digested externally and absorbed by these mouths.

RHODENIA. Bitterlings and similar fresh-water fish related to the roach; characterized by having a very long tubular ovipositor with which eggs are laid into the mantle cavities of mussels or other molluscs.

RHODOPSIN. See *Visual purple*.

RHOMBENCEPHALON. An alternative name for the hind-brain of a vertebrate.

RHOMBIFERA. Extinct Crinoids or sea-lilies which had diamond-shaped patterns of pores perforating their dermal plates (see *Pore-rhombs*).

RHOPALIA. Marginal sense-organs of jelly-fish.

RHOPALOCERA. The name given, according to the old system of classification, to butterflies as distinct from *Heterocera* or moths. A modern system of nomenclature depending on the wing-venation has now superseded these names.

RHYNCHO-. Prefix from Greek *Rhunchos*: a snout.

RHYNCHOBDELLIDAE. Marine and fresh-water leeches having a protrusible proboscis without jaws; usually ectoparasites on fish or molluscs.

RHYNCHOCEPHALIA. A primitive group of reptiles including *Sphenodon* of New Zealand which is noted for having a well developed pineal eye in the roof of the skull. Other primitive features include abdominal ribs and persistent remains of the notochord. The skull is diapsid with immovable quadrate bones.

RHYNCHOCOEL. A cavity within the proboscis sheath of a nemertean worm.

RHYNCHODAEUM. The part of the proboscis in front of the brain in a nemertean worm.

RHYNCHOPHORA. Beetles whose head is extended to form a rostrum or snout in front of the eyes; they include weevils, pinhole-borers and bark-beetles.

RHYNCHOPINAE. Skimmers: gull-like birds of the Indian Ocean and the Western Atlantic.

RHYNCHOSAURIDAE. Extinct reptiles of the Triassic Period, up to six feet in length, belonging to the primitive group Rhynchocephalia. The skull was massive with a toothless snout but having several rows of pyramidal teeth farther back.

RHYNCHOTA (HEMIPTERA). Bugs: a large order of hemimetabolous insects characterized by mouth-parts formed of long stylets for piercing and sucking. The wings are variously developed or absent (see also under *Homoptera* and *Heteroptera*).

RING BONE. See *Sphenethmoid*.

RING CANAL. (1) A circular water vessel such as that present in Echinoderms and some Coelenterates.
(2) A circular body-cavity with extensions leading into the tentacles in Polyzoa.

RING VERTEBRAE. Vertebrae whose centrum forms a ring round the constricted notochord, as for example in most fish.

RODENTIA. Rodents: gnawing mammals having a pair of large sharp chisel-shaped incisors in each jaw and without canines. They include rats, mice, guinea-pigs, porcupines, squirrels and many others. Rabbits and hares were formerly grouped with these but are now placed in a separate order, the *Lagomorpha*.

RODS (OF EYE). Rod-shaped retinal cells in the eyes of vertebrates. They are sensitive to light but, unlike the cones which may also be present, they cannot perceive colour. The action of light on the rods is to decompose and bleach the pigment *rhodopsin* or *visual purple* which is continually being regenerated. The break-down products of this substance produce nervous impulses which go by way of the optic nerve to the brain.

ROSENMÜLLER'S ORGAN (EPOÖPHORON). The atrophied mesonephros or middle kidney in female reptiles, birds and mammals. In these vertebrates the functional kidney is the *metanephros* or hind-kidney; the mesonephros, the original excretory organ, forms the tubules of the testis in males but usually disappears in females.

ROSETTE OSSICLE. A dermal plate formed by the fusion of five others in the centre of a *crinoid* or sea-lily. These five surround a larger *centrodorsal* plate to which the stalk of the crinoid is attached.

ROSTELLUM. A projection crowned with a ring of hooks on the end of the scolex or head of a tapeworm; characteristic of *Cyclophyllidea*, the largest group of tapeworms occurring in mammals.

ROSTRUM. (1) A pointed projection of the carapace between the eyes of a crab or similar crustacean.

(2) A forward-directed process on the anterior end of the skull in certain cartilaginous fish, *e.g.* the many-toothed snout of the sawfish (*Pristis*).

(3) Of Mollusca: the projecting hind-part of the internal shell of a cuttlefish, homologous with the larger pointed 'guard' on the shells of the now extinct Belemnites.

ROTIFERA. Wheel-animalcules: minute unsegmented animals, the smallest of all metazoa, characterized by an anterior ring of cilia with which they move and feed. They usually live in fresh water and some are sedentary. The body has no true coelom and in many cases the cell-walls break down to form a syncytium.

ROTULAE. Five radial ossicles uniting the five jaws of a sea-urchin (see *Aristotle's Lantern*).

ROUGET CELLS. Cells in the walls of capillaries, able to contract like muscle fibres and to cause the constriction of the tubes.

ROYAL JELLY. A predigested food substance on which the larvae of bees must be fed in order to develop into queen bees. It is produced from the pharyngeal glands of the workers and contains a high proportion of proteins.

RUMEN. The paunch or first stomach of a ruminant in which the food is partly digested before being regurgitated into the mouth for a second mastication as the animal 'chews the cud'.

RUMINANTIA. Artiodactyla or even-toed ungulates which ruminate or 'chew the cud'. Typically they have four compartments in the stomach, *viz.* the *rumen*, the *reticulum*, the *psalterium* and the *abomasum*. Food passes into the first two, then goes up again into the mouth to be re-chewed before going to the third and fourth. The group includes cattle, sheep, goats, antelopes, camels and giraffes.

RUPICAPRINAE. Ruminants intermediate between goats and antelopes, found in Europe, Asia and North America.

RUT. The period of testicular activity and sexual excitement in male mammals, corresponding to *oestrus* in females.

S

SABELLIDAE. Fan-worms: tube-dwelling polychaetes whose anterior bears long brightly coloured feathery arms which can be rapidly withdrawn into the tube. These convey small organisms to the mouth by ciliary action and also act as respiratory organs.

SACCOPHARYNGIDAE. Eel-like fish from the depths of the Atlantic, having a narrow filamentous tail and with a series of sensory appendages in place of the lateral line.

SACCULINA. See *Rhizocephala* and *Parasitic castration*.

SACCULUS. (1) Any small sac or vesicle.

(2) The lower chamber of the membranous labyrinth in the vertebrate ear, from which the endolymphatic duct and the cochlea proceed.

SACCULUS ROTUNDUS. A small round swelling of the ileum where it joins the colon and the caecum in the alimentary canal of a mammal. It contains the ileo-colonic valve.

SACCUS ENDOLYMPHATICUS. See *Endolymphatic sac.*

SACCUS VASCULOSUS. A glandular dilatation of the floor of the brain on each side of the pituitary organ of a fish.

SACRUM (SACRAL VERTEBRAE). Thick strong vertebrae which are generally fused together and firmly attached to the ilia of the pelvic girdle, thus giving great strength to the hip region in a tetrapod vertebrate.

SALAMANDRIDAE (SALAMANDRINAE). Salamanders and newts: urodele amphibians without gills or gill-slits in the adult; having teeth in both jaws, the palatine teeth being in two longitudinal rows. Many of the land species are viviparous and some exude a poisonous white fluid from cutaneous glands.

SALIVARY GLANDS. Glands which secrete fluid (saliva) into the mouth. In terrestrial vertebrates this fluid usually contains mucus and ptyalin, an enzyme which breaks down starch. In some snakes the glands are modified to form poison sacs. In blood-sucking insects, leeches etc. the saliva may contain an anti-coagulant.

SALMONI-CLUPEIFORMES (MALACOPTERYGII). Soft-rayed bony fish such as the salmon and the herring with scales covering the body but not the head; having four pairs of gills; having no barbels; having the pectoral girdle suspended from the skull and having the pelvic girdle, if present, in the abdominal position.

SALMONOIDEA (SALMONIDAE). Salmon, trout etc. Sea and fresh-water fish with soft fin-rays and with scales covering the body but not the head; having a large simple air-bladder with a pneumatic duct; having four pairs of gill-slits and jaws without barbels.

SALPIDA. See *Hemimyaria.*

SALTATORIA. A suborder of Orthopterous insects whose hind-legs are long and modified for jumping. They include locusts, grasshoppers and crickets.

SAPROZOIC (SAPROPHAGOUS). Living on decomposing organic matter.

SARCO-. Prefix from Greek *Sarx, sarkos*: flesh.

SARCODICTYON. Reticulated protoplasm.

SARCODINA. See *Rhizopoda*.

SARCOLEMMA. The sheath of a striated muscle fibre.

SARCOMERE. A small segment of striated muscle fibril between two successive Z-discs (*q.v.*).

SARCOPHAGINAE. Flesh-flies: a subfamily of cyclorrhaphous Diptera closely resembling blow-flies but having larvae which feed almost exclusively on meat or other decaying flesh. Many species are viviparous.

SARCOPLASM. The contractile protoplasm of muscle-fibres.

SARCOSPORIDIA. Parasitic sporozoa living in the muscle fibres of mammals including man. In the adult stage the organism is a syncytium with radial stripes in the ectoplasm. It reproduces by means of sickle-shaped spores.

SARCOSTYLE. A fibril of striated muscle.

SAROTHRUM. The pollen-brush on the hind-leg of a bee.

SATURNIIDAE. A family of large moths characterized by the silky cocoon which encloses the pupa. They include the Silk Moth of China, the Emperor Moth of Britain and the giant Atlas Moth of the tropics.

SAUR-, SAURO-. Prefix from Greek *Saura, sauros*: a lizard.

SAURIANS. A somewhat indefinite name used to denote lizard-like reptiles, especially such ancient types as the *Dinosaurs, Ichthyosaurs, Plesiosaurs* etc.

SAURISCHIA. Dinosaurs having a normal pelvic girdle without post-pubic bones; they include the large four-footed herbivorous types such as the 90 ft. long *Diplodocus* and also the smaller carnivorous bipeds (*cf. Ornithischia*).

SAUROPODA. Herbivorous, quadrupedal, plantigrade dinosaurs having five toes on each foot. Many were of enormous size having long neck and tail but were remarkable for the minute size of the skull. The group includes *Atlantosaurus* (115 ft. long), *Brontosaurus* (60 ft.) and *Diplodocus* (85 ft.).

SAUROPSIDA. A group of vertebrates comprising tortoises, lizards, snakes, crocodiles, ichthyosaurs, dinosaurs and birds, all of which, as shown by their types of skull, apparently belong to the same line of evolution while the rest of the reptiles and the mammals belong to another, the *Theropsida*. The name Sauropsida was originally suggested by Huxley for a less restricted group including all reptiles and birds.

SAUROPTERYGIA. Extinct reptiles of the Plesiosaurian type, secondarily adapted for life in water. They belong to the group *Theropsida* which had some mammalian features as distinct from the *Sauropsida* which had some bird-like features (see *Plesiosauria*).

SAURURAE. An alternative name for the *Archaeornithes* or fossil toothed birds with long tails (see *Archaeornithes* and *Archaeopteryx*).

SAVI'S VESICLES. Sensory organs sunk in small pits round the electric organs of *Torpedo* and similar skate-like fish.

SCALA MEDIA. The middle chamber of the spiral cochlea in the mammalian ear. It is an extension of the sacculus and is filled with endolymph.

SCALA TYMPANI. The lower chamber of the spiral cochlea in the mammalian ear. It is filled with perilymph and is bounded at its lower end by the *Fenestra rotunda*.

SCALA VESTIBULI. The upper chamber of the spiral cochlea in the mammalian ear. It lies between *Reissner's membrane* and the wall of the bony canal and is filled with perilymph. At the tip of the cochlea it connects with the lower chamber or *scala tympani* by means of a narrow passage called the *helicotrema*.

SCALE INSECTS. Members of the family Coccidae, the females of which often lose their limbs and become covered by a scale formed of cast-off skins glued together. In this state it is difficult to recognize them as insects. Many are harmful parasites of trees and shrubs; some yield useful products such as shellac and cochineal (see *Coccidae*).

SCALES. (1) *Fish*. These may have any of the following types of scales:

(*a*) *Placoid scales* or *Dermal denticles* made of dentine covered by enamel; characteristic of cartilaginous fish.

(*b*) *Cosmoid scales*. Found in *Osteolepids* of the Devonian Period and in early forms of lung-fish. In these scales the dermal denticles are fused together to form a layer of *cosmin* resting on an underlying bony plate or *isopedin* layer.

(*c*) *Ganoid scales*. Typical of many primitive 'Ganoid fish'; like cosmoid scales but covered with a thick layer of enamel-like substance called *ganoin*.

(*d*) Scales of modern bony fish. These are very thin and are derived from the ganoid type by loss of the ganoin layer.

(2) *Reptiles*. These have two types of scale, *viz.* epidermal *corneoscutes* sometimes resting on or fused with underlying bony scales or *osteoscutes* formed from the dermis.

(3) *Insects*. The minute scales found on the wings of Lepidoptera and some other insects are modified cuticular hairs.

SCAPE. The enlarged first segment of an insect's antenna.

SCAPHO-. Prefix from Greek *Skaphe*: a boat.

SCAPHOGNATHITE. The enlarged flattened exopodite on the maxilla of a crab, lobster or similar crustacean. Its function is to send a stream of water through the gill-chamber formed by the overhanging carapace.

SCAPHOID. Boat-shaped: a term used to denote certain carpal and tarsal bones with particular reference to their shape, but not necessarily used for bones which are homologous with one another in different animals.

SCAPHOPODA. A small class of mollusca having a tubular shell open at both ends, a reduced foot for burrowing and a head with many prehensile tentacles. In many respects this class is regarded as being near to the primitive ancestral molluscs.

SCAPULA. The shoulder-blade: the dorsal part of the pectoral girdle of a vertebrate.

SCAPULAR GIRDLE. That part of the pectoral girdle which is composed of cartilages or cartilage-bones, *viz.* the *scapulae* and *coracoids* (*cf. Clavicular girdle*).

SCAPUS. The stem of a feather including the hollow *calamus* or stalk and the solid *rachis* or vane-shaft.

SCARABAEIDAE. Scarab-beetles: large insects belonging to the group Lamellicornia whose antennae have clubs made of separate movable plates. The larvae, which live for several years and are very destructive to plants, may be recognized by the fact that the head is large and horny but the abdomen is soft and inflated with conspicuous spiracles. The group includes chafers and dung beetles.

SCARIDAE. Parrot-wrasses: spiny-rayed fish having jaws in the form of a sharp beak, the teeth being fused together. They are herbivorous and are abundant in coral reefs.

SCENT GLANDS. Cutaneous glands which produce strong smelling secretions. They may be on a number of different parts of the body, *e.g. occipital glands* of the camel, *temporal glands* of the elephant, *facial glands* of the bat, *pedal glands* of ruminants, *preputial glands* of the musk deer and the beaver and *perineal glands* of most mammals.

SCENT SCALES (LEPIDOPTERA). Tufts of modified scales or hairs with scent glands at their bases, occurring near the genital apertures of many butterflies and moths. By means of these the male may be attracted to the female over a considerable distance.

SCHINDYLESIS. A joint in which a flat bony process fits into a cleft in an adjacent bone.

SCHIZO-. Prefix from Greek *Schizein*: to split, cleave.

SCHIZOCOEL. A coelom or body-cavity which arises by the splitting of the mesoderm.

SCHIZOCHROAL EYES. Closely packed groups of simple eyes each with a separate lens and cornea, *e.g.* in some Trilobites.

SCHIZOGAMY. Fission with production of sexual and asexual zooids.

SCHIZOGENESIS. Reproduction by fission.

SCHIZOGNATHOUS. A term used to denote birds in which the vomer is small or absent and the maxillo-palatine bones do not unite with it or with each other. Plovers, gulls, fowls and pigeons are included in this group.

SCHIZOGONY. Asexual reproduction of protozoa by multiple fission into *schizozoites*, as for instance in many of the Sporozoa.

SCHIZOGREGARINARIA. Gregarine protozoa which undergo schizogony; many are parasites of insects (see *Gregarinidea*).

SCHIZOKINETE. The motile phase in Haemosporidia (*e.g.* in the Malaria Parasite).

SCHIZONT. A cell (*e.g.* in some protozoa) which by repeated asexual division or *schizogony* gives rise to numerous small cells known as *schizozoites*.

SCHIZOPODA. A name formerly used to denote many small prawn-like crustacea now distributed in the two orders *Mysidacea* and *Euphausiacea*.

SCHIZOPOD LARVA. A prawn-like stage ('Mysis stage') in the development of a crab or similar crustacean.

SCHIZORHINAL. A term used to describe a bird's skull in which the external nostrils are elongated, the posterior border being behind the posterior ends of the premaxillae (see also *Holorhinal*).

SCHIZOZOITES (MEROZOITES). Spores which are usually mobile and are formed by *schizogony* or multiple asexual fission, as in many protozoa.

SCIAENIDAE. Spiny-rayed fish of warm sandy shores, often large and usually edible. Some species make a peculiar grunting noise apparently from the air bladder.

SCIATIC ARTERIES. Arteries taking blood into the hind-limbs of vertebrates.

SCIATIC NERVE. The principal nerve going to the muscles of the hind-leg of a vertebrate.

SCIATIC PLEXUS (PELVIC PLEXUS). A network of nerves proceeding to the hind-limbs, formed by the branching and rejoining of spinal nerves in the region of the pelvic girdle.

SCINCIDAE. Skinks: viviparous lizards with strongly developed osteoscutes on the head and body and with a scaly tongue. The limbs may be reduced or absent. They are found in most warm regions of the world.

SCIURIDAE. Squirrels: simplicident rodents, usually arboreal, having long hairy tails.

SCIUROMORPHA. Squirrel-like rodents including beavers and several other forms in addition to true squirrels.

SCLERITES. Thick chitinous plates forming units separated by thinner membranes in the exoskeleton of an insect or crustacean.

SCLERO-. Prefix from Greek *Skleros*: dry, hard.

SCLEROBLAST. Any cell which secretes a hard skeletal substance.

SCLEROCOELS. Cavities partly surrounding the nerve cord and notochord in a vertebrate embryo, the walls of the cavities being the *sclerotomes* (*q.v.*).

SCLERODERMI. Tropical fish with rough bony scales or spines and with a narrow mouth sometimes having the teeth fused to form a beak.

SCLEROPAREI (TRIGLIIFORMES). Mail-cheeked fish: spiny-rayed fish having the cheek strengthened by an extension of the third suborbital bone.

SCLEROTIC COAT (SCLERA). The white fibrous outer coat of the eye of a vertebrate.

SCLEROTIN. A hard nitrogenous substance impermeable to water, found together with chitin in the cuticles of insects, crustaceans etc. Little is known of its composition.

SCLEROTOMES. Parts of the mesodermal somites from which the axial skeleton is derived in the vertebrate embryo.

SCOLEX. The head-like structure of a tapeworm, provided with hooks or suckers for fixing itself in the tissues of the host's intestine. Since it bears neither mouth nor sense organs it cannot be regarded as a true head.

SCOLO-. Prefix from Greek *Skolos*: bent.

SCOLOPACINAE. Snipe and woodcock: plover-like birds in which the beak is long and its maxillary part soft and covered by sensitive skin.

SCOLOPALE. A sensory rod forming the apex of a chordotonal receptor on an insect. Movement or vibration of the scolopale sets up impulses in the sensory cell beneath it (see *Chordotonal receptor*).

SCOLYTIDAE. Bark beetles: pests which can attack and kill healthy trees by tunnelling out complex galleries beneath the bark and by spreading the spores of injurious fungi.

SCOMBRESOCIDAE. Perch-like fish, nearly all marine, having two rows of keeled scales on the ventral side. Some species have the jaws elongated into a beak (gar-pipe); others have enlarged pectoral fins (flying-fish).

SCOMBRIDAE. Mackerel and tunny: marine fish with two small dorsal fins and with small scales or a smooth skin. They spawn in the open sea (see also *Scombriformes* below).

SCOMBRIFORMES. Fish of the mackerel type, mostly marine, having scales poorly developed or absent; having the pelvic fins forward in the thoracic position and having the tail-fin formed of numerous branching rays. The group includes mackerel, tunny, sword-fish, the 'pilot-fish' of sharks and many others.

SCOPELIDAE. Marine pike-like fish sometimes phosphorescent and normally living at great depths but occasionally coming to the surface at nights only.

SCORPAENIDAE. Spiny-rayed fish with large mouth and numerous sharp teeth, having the pelvic fins forward in the thoracic position and sometimes having the skin growing out into fronds resembling seaweed, a form of camouflage among the rocks which they inhabit. Some species have poison glands in the dorsal spines.

SCORPIONOIDEA. Scorpions: arachnids with a dorsal carapace over the front part (the *prosoma*) and with the *opisthosoma* or hind-part segmented and divided into two distinct regions. The hindmost of these, the *metasoma*, forms a flexible tail with a terminal sting. There are two large pincers in front corresponding to the pedipalps of a spider. Breathing is by means of lung-books.

SCROLL BONES. See *Turbinal bones*.

SCROTUM (SCROTAL SAC). A small sac behind the penis in most male mammals. The testes, which are at first in the abdomen, descend into this sac through the inguinal canal and usually remain there throughout life. In a few cases they remain there only during the breeding season and pass back into the abdominal cavity afterwards.

SCUTELLUM. The hindmost part of the notum of an insect, usually forming a shield-like plate which projects backwards over the dorsal side of the thorax.

SCUTUM. Latin: a shield.
(1) The middle and largest of three sclerites which typically form the dorsal cuticular covering (notum) of each thoracic segment of an insect.
(2) One of the two anterior lateral plates forming the carapace of a barnacle.

SCYLLIDAE. Dogfishes: small shark-like cartilaginous fish with a ventral crescent-shaped mouth and numerous small teeth. Usually there are two spineless dorsal fins near the hind-end.

SCYMNIDAE. Sharks having two dorsal fins without spines and having sharp oblique teeth in the lower jaw with much smaller teeth in the upper jaw. They include the Greenland Shark, 25 ft. long, which attacks and can bite pieces out of whales.

SCYPHISTOMA. The polyp-like larva of a jelly-fish which, by discoid strobilation, buds off a number of saucer-like *ephyra* larvae (see *Ephyra* and *Hydratuba*).

SCYPHO-. Prefix from Greek *Skuphos*: a bowl or cup.

SCYPHOMEDUSAE. See *Scyphozoa*.

SCYPHOZOA. A class of coelenterates which includes the common jelly-fish of temperate seas, some of which attain a great size. They are saucer-shaped with a large number of tentacles round the rim and with a mouth on the end of a stalk or *manubrium* hanging from the middle of the under side.

SEBACEOUS GLANDS. Cutaneous glands which secrete an oily liquid or *sebum* into the hair-follicles of mammals. The secretion keeps the hair greasy and probably protects it from any bacteria which may enter the follicle and infect the root of the hair.

SECONDARY CHARACTERS. Specialized features which are off the main evolutionary line and which in many cases are adaptations to particular conditions, *e.g.* the fish-like features of whales.

SECONDARY CHOANA. The opening of the nasal passage into the back o the throat in mammals.

SECONDARY FEATHERS (CUBITAL FEATHERS). Wing-quills which are attached to the ulna of a bird, as distinct from the primaries which are attached to the manus.

SECONDARY GILL BAR. See *Tongue-bar*.

SECONDARY SEX CHARACTERS. Characters shown by a male or female individual as a result of the effect of hormones secreted by special cells in the gonads. Examples are the antlers of stags; hair on the face and a deep voice in male humans; differences in colour, shape and size in the two sexes of many animals, birds, fish, etc.

SECOND VENTRICLE. See under *First and Second Ventricles*.

SECRETIN. A hormone, the production of which from the lining of the small intestine is stimulated by the presence of chyme entering it from the stomach. This hormone passes in the blood-stream to the pancreas and there brings about an increase of the flow of pancreatic juice.

SECTORIAL TEETH. An alternative name for the carnassial teeth in carnivorous mammals.

SEGMENTAL DUCT. An alternative name for the *Archinephric duct*, the embryonic precursor of the *pronephric* and *mesonephric* excretory ducts (*q.v.*).

SEGMENTATION (METAMERIC SEGMENTATION, METAMERISM). The formation or possession of a series of segments arranged along the length of an animal with repetition of the principal organs in each segment, as for instance in Annelids and Arthropods. In vertebrates the embryo shows segmentation particularly of the muscular, skeletal and nervous systems, the segments being known as *somites*. As the animal develops, however, specialization and enlargement of particular organs tends to obliterate the original segmental arrangement.

SELACHII (EUSELACHII, PLAGIOSTOMI). Sharks and Rays: cartilaginous fish with heterocercal tail; having pectoral fins with three basal cartilages; and the males having a pair of claspers or copulatory organs attached to the pelvic fins.

SELENICHTHYES. Bony fish having the mouth terminal and toothless, and having the pelvic fins attached to the scapular arch. They include the Opah or King-fish, 4 ft. long, of the North Atlantic and Mediterranean.

SELENODONT TEETH. Molar or premolar teeth having crescent shaped ridges of enamel traversing the crown. Such teeth are found in many Ruminants and other ungulates.

SELLA TURCICA. The 'Turkish saddle': a depression in the floor of the mammalian skull in which the pituitary body is lodged.

SEMEN (SEMINAL FLUID). A nutrient fluid containing spermatozoa: the product of the testis and its accessory glands.

SEMICIRCULAR CANALS. Three arch-shaped tubes containing fluid (endolymph) and forming part of the membranous labyrinth of the vertebrate ear. The plane of each canal is at right angles to the planes of the other two and all three lead into the *utricle*. The function of the semicircular canals is to detect acceleration and change of position and to give a sense of balance. Any rapid movement causes a flow of the fluid and this stimulates sensory 'hair-cells' in the ampullae at the base of each canal. The body responds automatically to such stimuli by adjusting itself in relation to the gravitational pull.

SEMILUNAR VALVES. Valves present in many vertebrates at the entrances to the main arteries as these leave the heart. Each valve consists of a number of half-moon shaped pouches of loose skin round the wall of the artery on its inner side. If blood tries to flow backwards these pouches fill out, blocking the artery and so preventing the backward flow.

SEMINAL FLUID. See *Semen.*

SEMINAL GROOVES. A pair of grooves on the ventral surface of an earthworm along which, during mutual copulation, sperm passes from the male apertures to the region of the clitellum. From here it enters the spermathecae of the other worm.

SEMINAL RECEPTACLE. See *Spermatheca.*

SEMINAL VESICLE (VESICULA SEMINALIS). A pouch or sac for storing seminal fluid: usually a widening or a diverticulum of the *vas deferens.*

SEMINIFEROUS TUBULES. A mass of coiled tubules bound together by connective tissue and forming the essential part of the testis in a vertebrate. These tubules are lined with germinal epithelium from which the spermatozoa arise. The latter are shed into the tubules whence they pass to the *vasa efferentia* and out by way of the *vas deferens* or main seminal duct.

SENSE CAPSULES. Capsules enclosing the olfactory, optic and auditory organs of a vertebrate. The olfactory and auditory capsules may be of cartilage or bone and become fused on to the cranium; the optic capsules give rise to the sclerotic coats of the eyeballs.

SENSILLA. Any small sense organ or receptor.

SENSORY CELLS. See *Receptor.*

SEPIA. The common cuttle-fish and also the name given to the dark brown paint or dye obtained from its ink-sac (see *Sepioidea* below).

SEPIOIDEA. Cuttle-fish: Cephalopod molluscs with ten tentacles surrounding the head, two of which are specially long and can be retracted into pits. The body is roughly cylindrical and contains a flattened internal shell. The mouth has a parrot-like beak and the eyes and brain are well developed. When disturbed the animal moves violently backwards by the ejection of water from its ventral siphon, at the same time surrounding itself with a cloud of brown 'ink' from the ink-sac.

SEPTI-. Prefix from Latin *Septum*: a partition.

SEPTIBRANCHIATA. A small group of lamellibranch (bivalve) molluscs in which the gills, instead of being in the form of lamellae or of filaments, are reduced to small horizontal partitions lying transversely across the gill chambers.

SEPTOMAXILLARY BONES. Small bones on each side of the anterior part of the nasal region of certain reptiles.

SEPTUM. A partition separating two tissues or separating two cavities, as for instance those separating the segments of an earthworm.

SEPTUM PELLUCIDUM. A partition separating the two cerebral hemispheres in the region between the *corpus callosum* and the *fornix* of a mammalian brain.

SERO-AMNIOTIC CONNECTION (EMBRYOLOGY). The point of fusion of the two folds of ectoderm, known as the amniotic folds, which rise up, spread over and enclose the embryo in reptiles, birds and mammals.

SEROSA (INSECTS). A layer of epithelium formed beneath the vitelline membrane during the development of an insect's egg, later laying down a second cuticle known as the serosal cuticle within the original one.

SEROSAL CUTICLE. See *Serosa* above.

SERPENTARIIDAE. African Secretary Birds: long-legged falcon-like birds about 4 ft. high which feed on insects and reptiles.

SERPULIDS. Tube-dwelling polychaete worms whose tube is made of calcium carbonate and mucin and is continually being added to by a secretion from a collar-like structure round the worm's mouth. At the anterior end are a number of tentacles, one of which bears a kind of 'stopper' which closes the tube when the animal is withdrawn into it.

SERRANIDAE. Sea-Bass and Sea Perches: a large family of carnivorous spiny-rayed fish found in all warm seas and sometimes in fresh water.

SERRASALMONINA. Caribes: voracious fresh-water fish from tropical America, having an elongated dorsal fin and a serrated belly.

SERTOLI CELLS. Large cells found among the germinal cells in the testis of a vertebrate; believed to supply nourishment to the developing spermatozoa.

SERUM. Liquid which separates out from clotting blood, *i.e.* plasma without corpuscles or fibrinogen.

SESAMOID BONES. Bones formed within a tendon and strengthening it, as for instance the patella.

SESSIIDAE. Clear-wing moths: a family of moths with semi-transparent wings. The caterpillars do considerable damage to trees by boring into the stem and branches.

SETAE. Bristles or 'hairs' of invertebrates.

e.g. (1) The hairs and scales of insects, formed from hollow projections of cuticle each enclosing an enlarged epidermal cell or part of one.

(2) The *Chaetae* or bristles of annelids composed of chitin embedded in and secreted by pits of ectoderm.

SEX CHROMOSOME. A chromosome which is present in a gamete and which carries the factor for producing a male or female offspring. Such chromosomes are usually designated by the letters X and Y. In humans the spermatozoa contain either an X or a Y chromosome; the ovum always an X chromosome. When X combines with Y at fertilization a male is produced and when X combines with X a female is produced.

SEX DETERMINATION. See *Sex chromosome.*

SEX LINKAGE. The phenomenon whereby a particular character is linked with the sex of the individual by reason of the fact that the gene for that character is carried on one of the sex chromosomes (*q.v.*).

SEX RATIO. The relative proportion of males and females in a population. The number of either sex may be expressed as a percentage of the whole or sometimes the number of males is expressed as a percentage of the number of females.

SEXUALES. A term used with reference to plant lice and similar insects to denote the generation which contains male and female individuals capable of normal sexual reproduction.

SEXUAL REPRODUCTION. Reproduction by the union of a male and a female gamete to form a zygote from which the new embryo develops. During the formation of gametes the number of chromosomes in the cells is reduced to half (*haploid*); when two gametes combine the original or *diploid* number is restored. The main advantage of sexual reproduction is that different combinations of hereditary factors may take place, so making possible the adaptation of the race to new environments and helping to bring about evolution by natural selection.

SEXUPARAE. A term used with reference to plant lice and similar insects for the generation of parthenogenetically produced females whose offspring can develop into normal males and females.

SHELL. (1) Of *Chelonia*: the dorsal carapace and the ventral plastron of a tortoise or turtle.

(2) Of *Crustacea*: the exoskeleton or hardened cuticle.

(3) Of *Echinoidea*: the corona or globe of dermal plates and spines enclosing the body of a sea-urchin.

(4) Of Eggs: the thick membranous or calcareous covering of an egg. In the case of reptiles and birds the shell is one of the tertiary membranes secreted by glands in the wall of the oviduct. In some invertebrates the egg may have a shell produced by a specialized *shell-gland*.

(5) Of *Foraminifera*: a chitinous, calcareous or siliceous secretion forming a protective coat with numerous perforations through which pseudopodia project.

(6) Of *Mollusca*: a hard secretion made by the mantle of most molluscs. It usually consists of three layers, *viz.* an inner pearly or nacreous layer mostly of calcium carbonate; a middle or *prismatic* layer also of calcium carbonate; an outer *periostracum* of horny material known as *conchiolin*. The shell may be a simple spiral or cone (univalve) or may consist of two lateral plates (bivalve). In a few mollusca such as the cuttle-fish, the shell is internal.

SHELL GLANDS. Glands in the wall of, or opening into an oviduct and secreting the substance which on hardening forms the shell of an egg.

SHELL MEMBRANE. One or more membranes immediately inside the shell of an egg. In birds' eggs the shell, the membranes and the albumen are all secreted by the walls of the oviduct and are known as *tertiary membranes*.

SHIELD-SHAPED TENTACLES. Retractile tentacles such as those round the mouth of a Holothuroid (sea-cucumber). Each tentacle consists of a stalk with a mass of branches in the shape of a shield at its end.

SIALIDAE. Alder-flies: a family of insects belonging to the order Neuroptera and having aquatic larvae with seven or eight pairs of hair-fringed abdominal appendages.

SICULA. A cone-shaped structure from which a colony of polyps develops in the Graptolites (fossil colonial coelenterates). It consists of a *prosicula* and a *metasicula* which together form the sheath or perisarc of the first formed individual of the colony (see *Graptolite, Prosicula, Metasicula*).

SILICO-FLAGELLATA. Plant-like protozoa (Phytomastigina) with one flagellum, numerous yellow chromatophores and a lattice-work case of hollow siliceous bars. Most species are marine and planktonic.

SILK. See *Bombycidae* (silkworm-moths).

SILLAGINIDAE. Small plum-coloured spiny-rayed shore fishes common in S.E. Asia and Australia.

SILPHIDAE. Burying and carrion beetles (see *Necrophoridae*).

SILURIDAE. Fresh-water cat-fishes having a skin without external scales but sometimes with bony scutes; sometimes having accessory respiratory sacs attached to the gill-arches, and having long sensory 'whiskers' or barbels round the mouth.

SIMIAN FISSURE. A groove in the occipital lobe of the brain, well developed in lower apes and slightly in higher apes and humans.

SIMIIDAE. Tailless apes and humans (see *Anthropomorphidae*).

SIMPLICIDENTATA. Rodents with only one pair of upper incisors: all the rodents except rabbits and hares which have an extra pair of incisors behind the first (see *Duplicidentata*).

SIMULIIDAE. Buffalo-gnats: blood-sucking flies with a hump-backed appearance and with broad wings. The spindle-shaped larvae live in running water and have an anal pad with setae for clinging to rocks etc.

SINU-AURICULAR NODE (SINO-AURICULAR NODE). The vestige of the sinus venosus which acts as the pacemaker of the heart in birds and mammals: an area of nervous tissue behind the right auricle, from which each heart-beat is initiated.

SINU-AURICULAR VALVE (SINO-AURICULAR VALVE). A valve between the sinus venosus and the single auricle in a fish or between the sinus venosus and the right auricle in amphibia.

SINUS. (1) A cavity within a bone, *e.g.* in the cranial and facial bones of many mammals.

(2) A large blood-filled cavity, as for instance the enlarged veins of fishes and the haemocoels of arthropods.

SINUSOIDS. Small venous cavities which, although not true veins, nevertheless allow blood to flow through them, *e.g.* the radial passages through the lobules of the liver.

SINUS TERMINALIS. A peripheral blood vessel marking the limit of the *area vasculosa* or external vascular system surrounding the developing embryo of a bird, mammal etc.

SINUS VENOSUS. The first chamber of the heart in fish, amphibia and reptiles. In fish it receives blood from the Cuvierian ducts and the hepatic veins, and leads into the single auricle. In amphibians and reptiles it receives blood from the three *venae cavae* and leads to the right auricle. Birds and mammals have no sinus venosus.

SIPHON. A tube through which water passes: a name given to a variety of organs of this nature in different animals.

e.g. (1) *Elasmobranch fish:* a sac beneath the ventral surface in the pelvic region of the male, from which water is squirted to flush seminal fluid from the grooves of the claspers into the cloaca of the female.

(2) *Lamellibranch molluscs:* openings by which water enters and leaves the branchial chamber.

(3) *Gasteropods:* a respiratory opening for the passage of water into the mantle cavity in certain aquatic gasteropods.

(4) *Cephalopods:* a ventral tube from which a squid or cuttle-fish forcibly squirts water and so propels itself backwards when disturbed.

(5) *Tunicates:* the oral and atrial openings by which water respectively enters and leaves the body of a tunicate and in doing so passes through the perforated basket-like wall of the pharynx.

SIPHONAPTERA. See *Aphaniptera*.

SIPHONIUM (SIPHONEUM). A membranous air-tube leading from the cavity in the quadrate bone to another in the articular bone of the lower jaw in certain reptiles such as the crocodile. These and other cavities in the neighbouring bones form part of a complex system of Eustachian tubes connecting the pharynx with the tympanic cavity.

SIPHONOGLYPH. A ciliated groove in the stomodaeum of a sea-anemone by means of which water is circulated throughout the body cavity.

SIPHONOPHORA. Pelagic colonial coelenterates characterized by a great diversity of zooids or polyps. These include *gastrozooids* for feeding, *gonozooids* for reproduction, *dactylozooids* with long tentacles for trapping food, and leaf-shaped *hydrophyllia* for covering and protecting the other zooids. The whole assemblage is known as a *cormidium* and is kept afloat by a large zooid in the form of an air bladder known as a *pneumatophore*. The best known example of the Siphonophora is the 'Portuguese Man-of-war' (*Physalia*).

SIPHONOZOOIDS. Specialized polyps in the form of bell-shaped organisms without tentacles and with an enlarged siphonoglyph or ciliated tract. They are found in certain colonial coelenterates, their function being to maintain the circulation of water throughout the whole colony.

SIPHUNCLE. A slender tubular prolongation of the visceral hump passing through all the compartments of the spiral shell in cephalopod molluscs such as *Nautilus* and ammonites. It contains blood vessels and can secrete gases which enable the mollusc to float (see *Nautiloidea*).

SIPHUNCULATA. Blood-sucking lice: flattened wingless insects having a tubular proboscis which can be anchored into the skin of the host by means of a number of denticles. They are a suborder of the Anoplura and are usually ectoparasites of mammals. *Pediculus*, the body-louse of humans, is a well known example; it is important as being the carrier of typhus and other diseases.

SIPUNCULOIDEA. Marine burrowing annelids without chaetae and with no obvious segmentation. The anterior part, with a frilled membrane round the mouth, can be invaginated into the posterior part of the body. The anus is near the front and is dorsal.

SIRENIA. Sea-cows: short-necked aquatic herbivorous mammals with little hair; with fin-like anterior limbs, a flattened caudal fin and no posterior limbs. They include manatees and dugongs which, through their habit of basking on sand banks and suckling their young, are said to have given rise to the idea of mermaids.

SIRENOIDEI. A group comprising the three living types of lung-fish, *Ceratodus*, *Protopterus* and *Lepidosiren* from Australia, Africa and South America respectively. They have long single dorsal and anal fins continuous with the diphycercal tail; have thick scales covering the body and a few large scutes covering the head. They are now classified as *Monopneumona* and *Dipneumona* according to whether they have one or two lungs.

SIRICIDAE. Wood-wasps: large insects up to nearly 2 in. in length, belonging to the symphytous Hymenoptera. The larvae tunnel into the wood of coniferous trees and do much damage. The commonest British species, *Sirex gigas*, is recognized by its black and yellow striped body and its large pin-like ovipositor.

SISYRIDAE. A family of Neuropterous insects whose aquatic larvae live in association with fresh-water sponges, piercing the tissues with their mouthparts.

SITTIDAE. Nuthatches: small sparrow-like birds which nest in natural holes or woodpeckers' borings filled with mud to the right size. They feed first on small insects and later on nuts.

SKELETON. Any structure whose main function is to strengthen and maintain the shape of an animal (see *Endoskeleton* and *Exoskeleton*).

SKELETOGENOUS CELLS. Cells which give rise to skeletal tissues, spicules etc.

SKIN. The outer covering of an animal consisting of an *epidermis* (of ecto-dermal origin) and an underlying *dermis* (of mesodermal origin). The former is sometimes only one cell thick but more usually consists of many layers of stratified epithelium covered by a horny cuticle. The dermis usually contains blood vessels, small muscles, nerve endings and fatty connective tissue. The skin may have embedded in it such structures as hair, feathers, scales, denticles etc. which originate partly from the epidermis and partly from the dermis.

SKULL. The skeleton of the vertebrate head: a composite structure consisting of the *neurocranium* which encloses the brain and sense organs, and the *splanchnocranium* which provides support for the jaws and gill arches where these are present. All these parts may be to a varying degree fused or sutured together to make up the whole skull.

SKULL OF CEPHALOPODA. A cartilaginous covering surrounding and protecting the brain of a cuttle-fish, octopus or similar mollusc.

SLIME TUBE (EARTHWORMS). A mucous envelope secreted by glandular cells in the clitellum or 'saddle' of an earthworm. During copulation this tube invests both worms, holding them tightly together with their ventral surfaces adjacent. At a later stage a similar tube is secreted prior to the formation of a cocoon into which the eggs are laid.

SMALL INTESTINE. That part of the alimentary canal immediately following the stomach in a vertebrate. It consists of the *duodenum*, the *jejunum* and the *ileum* and is chiefly characterized by the presence of a large number of *villi* or finger-like projections on its inner surface.

SMOOTH MUSCLE. See *Involuntary muscle*.

SOFT COMMISSURE. A part of the thalamencephalon in the mammalian brain where the sides of this part, known as the *optic thalami*, are so thickened that they touch one another across the constricted Third Ventricle.

SOLAR PLEXUS (COELIAC PLEXUS). A group of collateral sympathetic ganglia in the mesentery behind the stomach near the point of origin of the anterior mesenteric artery. It is the largest nerve centre in the abdomen and branches from it supply all the principal viscera.

SOLDIER ANTS. Ants with large heads and powerful jaws having the primary duty of defending the colony. Some kinds can also defend themselves by shoot-ing out formic acid from the hind part of the body to a distance of six inches or more. A similar caste of 'soldiers' exists among termites although these are no relation to true ants.

SOLEN-, SOLENO-. Prefix from Greek *Solen*: a channel.

SOLENICHTHYES. Pipe-fish and sea-horses: small fish having the skin covered with bony plates and having a prolonged snout bearing a small ter-minal toothless mouth. The males have a brood pouch formed of two folds of skin on the ventral side of the abdomen. Eggs are retained here till some time after hatching. Sea-horses differ from pipe-fishes in having a prehensile tail and in swimming in an upright position.

SOLENOCYTES. Tubular flame-cells: excretory cells having one or more long flagella hanging into a tubular prolongation from which leads an intra-cellular duct. Waste substances are discharged into the tube and wafted down by the movement of the flagella. Such organs are found in *Amphioxus* and in some annelids. They differ from the flame-cells of Platyhelminthes in being much more elongated (see *Flame-cells*).

SOLENODONTIDAE. West Indian insectivorous mammals without zygo-matic arches and with tritubercular molar teeth in the upper jaw. The male has

236

the penis hanging in an anterior position; the female has two mammae on the buttocks.

SOLENOSTOMIDAE. Small fish having a prolonged snout bearing a small terminal toothless mouth; having wide gill openings, two dorsal fins and broad pelvic fins which form a brood-pouch for carrying the eggs.

SOLLAS' CENTRE. The centre of gravity or morphological centre of a skull from which lines may be drawn to the centre of the foramen magnum and to the junction of the nasal and frontal bones. The angle between these lines is important in the evolution of man as a relative measurement of the size of the fore-brain. In *Homo sapiens* the angle approaches 270°; in fossil man 250–60°; in anthropoid apes about 240° etc.

SOMACTIDS. Radialia: cartilaginous or bony fin-rays of fish.

SOMATIC. Appertaining to the body or the body-wall.

SOMATIC CELLS. Cells other than the reproductive cells.

SOMATIC TISSUES. Tissues of the body-wall.

SOMATIC NERVES. Nerves which convey impulses to and from the body-wall and limbs.

SOMATO-. Prefix from Greek *Soma, somatos*: the body.

SOMATOBLAST. A cell which differentiates into a body-cell.

SOMATOPLEURE. Tissue formed from the fusion of the ectoderm with the somatic or outer layer of mesoderm in the vertebrate embryo, giving rise to parts of the body-wall and to the amnion and chorion.

SOMITES. Segmentally arranged blocks of mesodermal tissue in the vertebrate embryo, giving rise to muscles, vertebrae etc.

SORICIDAE. Shrews: small mouse-like insectivorous mammals with long pointed snouts and with no zygomatic arches. They are found in most parts of the Northern Hemisphere; are usually terrestrial but occasionally aquatic.

SPALACIDAE. Bamboo-rats of Africa and Asia: mole-like rodents with vestigial eyes and short tails.

SPARASSODONTIDAE. Large extinct carnivorous marsupials of Patagonia and elsewhere.

SPATANGOIDEA. Heart-shaped echinoids or sea-urchins with the dorsal parts of the ambulacral grooves having the appearance of flower-petals.

SPATULARIIDAE. Sturgeon-like fish of North American and Chinese rivers, having small isolated scales in the skin and having a long thin flat snout without barbels.

SPECIALIZED CHARACTERS. See under *Secondary Characters*.

SPECIES. A group of animals or plants having a high degree of similarity, able to breed among themselves but not generally to breed with members of another species. Within a species there may be smaller, usually regional groups known as sub-species or varieties and it is sometimes no easy matter to decide whether these should be regarded as separate species or not. There may, for instance, be a series of types extending over neighbouring regions and showing a gradation of differences as one passes from one district to the next. Each may be able to breed with its immediate neighbours but two extreme types living at opposite ends of the chain may not be able to interbreed. In view of these difficulties and in view of the fact that the question of interbreeding has in most cases not been investigated, it is usually more convenient to define a species by its morphological and anatomical features (see also the Appendix on Nomenclature).

SPEMANN'S ORGANIZER. A chemical substance, now known to be a derivative of nucleic acid, which is produced in the early stage of an embryo and induces the specialization of primitive cells to form particular tissues or organs.

SPERM, SPERM CELLS, SPERMATOZOA. Motile male gametes usually having a large nucleus and a tail-like flagellum.

SPERMA-, SPERMO-, SPERMATO-. Prefixes from Greek *Sperma, spermatos*: seed.

SPERMACETI. An oily substance found mixed with fat particularly in the cranial basin of the sperm-whale.

SPERMATHECA (RECEPTACULUM SEMINIS). A sac or receptacle in a female or hermaphrodite animal, in which sperm cells received from another individual are stored until required for fertilizing the eggs.

SPERMATIC ATRIUM. A vestibule or pit into which the *vas deferens* or male genital duct opens, as for instance in some annelids.

SPERMATIC CORD. A cord formed from the spermatic artery, vein and nerve bound together with connective tissue and passing into the testis in some vertebrates.

SPERMATID. An immature spermatozoon in which the flagellum has not yet developed.

SPERMATOCYTE. A cell which, on undergoing meiosis, gives rise to spermatids and afterwards to spermatozoa. Normally a primary spermatocyte with a diploid nucleus forms two secondary spermatocytes with haploid nuclei. These in turn divide by a second meiotic division to form spermatids. One primary spermatocyte therefore forms four spermatids; the latter on modification and development of a flagellum become mature spermatozoa.

SPERMATOGENESIS. The formation of spermatozoa in the testis. Normally this involves the successive formation of spermatogonia, primary spermatocytes, secondary spermatocytes, spermatids and spermatozoa. Meiosis or reduction of chromosomes takes place at the stage when primary spermatocytes give rise to secondary (see *Spermatocyte*).

SPERMATOGONIUM. A germinal cell which gives rise to spermatocytes from which spermatozoa are produced (see *Spermatogenesis* and *Spermatocyte*).

SPERMATOPHORE. A mass of spermatozoa held together by mucilage or similar substance which enables the animal to manipulate or direct it to the receiving parts of the female; produced for instance by many molluscs, crustaceans and other invertebrates.

SPERMATOZOON. The male gamete of an animal, usually consisting of an anterior part containing a large nucleus, a small middle part containing centrosomes and other structures, and a long posterior flagellum which gives the cell the power of locomotion.

SPERM FUNNELS (OF EARTHWORMS ETC.). Ciliated funnels within the seminal vesicles, which collect the sperms as they become detached from the testes and convey them to the vasa efferentia.

SPERMIDUCAL GLANDS. Glands in the neighbourhood of the male aperture, whose function is analogous to that of a prostate gland, *i.e.* to secrete fluid in which the sperm cells are nourished. Glands of this type are found in the earthworm.

SPERM SAC. A sac in which sperm is stored, sometimes a diverticulum from the seminal vesicle or from the *vas deferens*.

SPHECIFORMIA (SPHECOIDEA). Solitary predaceous bees and wasps which sting and paralyse their prey before dragging it into their nests. They may be distinguished from other bees and wasps by the fact that the pronotum does not extend back as far as the wing-bases; an example is the sand-wasp.

SPHEN-, SPHENO-. Prefix from Greek *Sphen, sphenos*: a wedge.

SPHENETHMOID (GIRDLE BONE, RING BONE). A bone forming the front part of the cranium in frogs, toads etc. It is formed by fusion of the two orbitosphenoids which extend to the hinder part of the ethmoid region.

SPHENISCIFORMES. Penguins: flightless marine birds of the Antarctic, whose wings have no quills but are covered with scale-like feathers and used for swimming.

SPHENODONTIDAE. Primitive reptiles of the group Rhynchocephalia, mostly of the Jurassic Period, but including also the present-day *Sphenodon* or Tuatara of New Zealand, noted for the presence of a pineal eye (see *Rhynchocephalia*).

SPHENOID. A wedge-shaped bone in the floor of the cranium in humans and certain other mammals, formed from the fusion of the *presphenoid* and the *basisphenoid*, which in most mammals are distinct and separate.

SPHENO-PALATINE GANGLION. A ganglion innervated by the petrosal nerves (*q.v.*), from which excitor neurones run to the lachrymal glands and the glands of the nose.

SPHENOTIC BONE. One of five bones forming the auditory capsule in bony fish. In higher vertebrates it becomes fused with the others to form the *periotic bone*.

SPHINCTER. A ring of muscles closing an orifice, as for instance the *cardiac sphincter* and the *pyloric sphincter* at the entrance and exit respectively of the stomach.

SPHINGIDAE. Hawk-moths: large stoutly-built moths whose fore-wings are much larger than the hind ones. The proboscis is sometimes extremely long and the antennae end in hooks. The larvae have ten pro-legs and usually bear a horn or spine at the hind end. The Death's Head Hawk-moth, belonging to this family, is the largest British moth, sometimes having a wing-span of over 5 in.

SPHYRAENIDAE. Barracudas: large carnivorous pike-like fish of warm seas, having small cycloid scales.

SPICULES. Calcareous or siliceous bodies, usually needle-like, but often complex and many-pointed, forming an endoskeleton in sponges, corals, protozoa etc.

SPIGOTS. See *Fusulae*.

SPINAL ACCESSORY NERVE. The eleventh cranial nerve in mammals, birds and most reptiles, innervating the dorsal muscles of the shoulders and neck. It arises by several roots in the cervical region of the spinal cord, enters the cranium through the *foramen magnum* and leaves in association with the vagus.

SPINAL CANAL. The canal in which the spinal cord passes through the vertebral column.

SPINAL CORD. The main nerve-trunk of a chordate animal, consisting of a hollow tube running the length of the body in the mid-dorsal position. In vertebrates it is enclosed by the neural arches of the vertebrae.

SPINAL NERVES. Nerves arising from the spinal cord of a vertebrate in the gaps between the vertebrae. Afferent sensory nerves enter the cord in the dorso-lateral position and efferent motor nerves leave it in the ventro-lateral position.

SPINA STERNALIS. An anterior continuation of the sternum between the points of articulation of the coracoid bones in birds.

SPINDLE. A structure formed in a cell during mitosis and meiosis consisting of a number of protoplasmic threads approximately forming an ellipsoid with the chromosomes arranged round its equator. During the anaphase the threads shorten and the chromosomes move towards opposite poles of the cell prior to the formation of two new cells (see *Mitosis* and *Meiosis*).

SPINNERETS. Small tubular appendages from which silk threads are exuded by spiders and some insects. In spiders there may be up to four pairs of spinnerets on the hind-part of the abdomen; the silk comes out in the form of a fluid which hardens in contact with the air. Usually a large number of silk glands open into each spinneret. The size and shape of these glands varies according to whether the silk is to be used for webs, egg-cocoons etc.

The spinnerets of spiders are not homologous with those which produce the cocoons of insects. In caterpillars such as the silkworm there is a median spinneret in the hypopharynx and the silk comes from modified salivary glands.

SPINO-THALAMIC TRACTS. Bundles of nerve fibres within the brain of a vertebrate conveying somatic impulses direct from the spinal cord to the thalamus.

SPIRACLES. (1) *Fish.* A pair of small openings, one on each side of the head, representing the reduced first gill-slits in many cartilaginous fish.

(2) *Insects.* Another name for the stigmata or tracheal apertures along the sides of the body of an insect.

SPIRAL CLEAVAGE. A method of early embryonic development found in Polyclads, Nemerteans, Annelids and Molluscs. The first two divisions of the fertilized egg-cell are in vertical planes at right angles to each other giving a quartet of cells. The next three divisions are in horizontal planes cutting off successive quartets; these cells, however, are not exactly above the four from which they were derived. At each division the new cells are slightly displaced giving a spiral appearance to the young embryo.

SPIRAL VALVE. (1) *Fish.* A spiral structure in the intestines of some fish, formed from the lining of the tube which is partly separated from the wall and twisted. The food passes slowly through the turns of the spiral and so comes into contact with a large surface for absorption.

(2) *Amphibia.* A small twisted flap or septum in the *truncus arteriosus* of a frog or similar amphibian; said to help direct the flow of blood from the heart into the three arterial arches on each side.

SPLANCHNIC MESODERM. The layer of mesodermal tissue surrounding the gut and forming the inner boundary of the splanchnocoel.

SPLANCHNIC NERVES. Communicating nerves running from the anterior and posterior mesenteric ganglia to the spinal nerves in higher vertebrates; part of the sympathetic system.

SPLANCHNO-. Prefix from Greek *Splangchnon*: entrail.

SPLANCHNOCOEL. The perivisceral cavity: the ventral part of the coelom of a vertebrate, lined by *splanchnopleure* on its inner side and *somatopleure* on its outer side.

SPLANCHNOCRANIUM (VISCEROCRANIUM). That part of the skull of a vertebrate formed by the skeletal supports of the gill arches and the jaws; so called from the fact that they lie within the splanchnopleure.

SPLANCHNOPLEURE. A layer of tissue formed embryonically by fusion of endoderm and splanchnic mesoderm, giving rise to the wall of the gut.

SPLEEN. An organ just beneath the stomach of a vertebrate, having the function of making and storing lymphocytes and storing red corpuscles. These are squeezed out into the blood stream when the body needs more in circulation, as for instance in haemorrhage or shock.

SPLENIAL BONE. A membrane bone between the *dentary* and the *angular* bone in the lower jaws of amphibia, reptiles and birds.

SPLENIUM. The posterior part of the *corpus callosum* connecting the two cerebral hemispheres in mammals.

SPLINT BONES. Reduced metacarpals and metatarsals of the second and fourth digits in the horse and similar one-toed ungulates. In the horse the third digit is enlarged and hoofed; the second and fourth are reduced and the first and fifth are absent.

SPONGILLIDAE. Small fresh-water sponges belonging to the class Demospongiae, having a skeleton composed of simple spicules of the monaxon type.

SPONGIN. A horny substance secreted by the cells in the middle layer of a sponge. It may act as a cement, uniting spicules of silica or calcium compounds, or may form a fibrous skeleton without spicules.

SPONGIOPLASM. The protoplasmic network of a cell.

SPOR-, SPORO-. Prefix from Greek *Sporos*: a seed.

SPORANGIUM. An organ which produces and contains spores. The term usually refers to plants but is also used sometimes with reference to protozoa.

SPOROBLAST. A stage which gives rise to sporozoites or other spore-like cells in the life-history of a member of the Sporozoa.

SPOROCYST. A stage in which there is an envelope containing numerous small spores produced by repeated or multiple fission of the parent cell, *e.g.* in members of the Sporozoa.

SPOROGONY. A stage in the life-cycle of certain protozoa at which *sporozoites* are formed by multiple fission of a zygote or *sporont*.

SPORONT. A cell which undergoes *sporogony* or multiple asexual fission to form spores. In those members of the Sporozoa which have a complex life-history the term *sporont* is more strictly applied to the zygote stage which, immediately after syngamy, divides to form *sporozoites*.

SPOROZOA. Protozoa which are nearly all parasitic and have no obvious means of locomotion. Most of them pass through a complicated life-cycle involving alternation of sexual and asexual reproduction and the formation of spores. Some, such as the malaria parasite, live in the blood of the host; others may infest the gut or the muscles. They are often transmitted by blood-sucking insects.

SPOROZOITES. Small mobile individuals produced by multiple fission of a zygote or *sporont* in certain protozoa, particularly of the group Sporozoa.

SPUMELLARIA. See *Peripylaea*.

SPURS. Thick cuticular hairs or spines on the legs of certain insects.

SQUALI (SQUALOIDEA). Sharks with two dorsal fins; the great majority of living sharks (see *Selachii*).

SQUALODONTIDAE. Extinct whales of the Eocene, Miocene and Pliocene Periods, known by their teeth and skulls. The teeth are heterodont, the premolars being conical and one-rooted, while the molars are flattened, serrated and two-rooted.

SQUALORAIIDAE. Extinct cartilaginous fish of the Liassic Period intermediate between sharks and rays.

SQUAM-. Prefix and root word from Latin *Squama*: a scale.

SQUAMATA (LEPIDOSAURIA). A group of reptiles comprising all lizards and snakes as well as the extinct aquatic Dolichosauria and Mosasauria. They are characterized by having a parapsid type of skull (*i.e.* with the lower temporal bar missing) and a movable quadrate bone which enables the upper jaw to be moved relatively to the brain-case. The body is covered with horny scales.

SQUAMOSALS. Paired membrane-bones on each side of the skull. In lower vertebrates they abut on the quadrate bones but in mammals they cover parts of the alisphenoid and the exoccipital bones. On the mammalian skull each squamosal bone has a forward projecting spur which, together with the jugal bone, forms the *zygomatic arch*. On the under side of this spur at its hind-end is a shallow depression into which the lower jaw articulates.

SQUAMOUS EPITHELIUM (PAVEMENT EPITHELIUM). A single layer of flattened plate-like cells which, by being very thin, allow the passage of gases and liquids freely through them. Such cells line the capillaries of the blood-system and the alveoli of the lungs.

STADIUM. The interval between two successive moults of an insect or other arthropod.

STAPES. The stirrup-bone: the innermost of the three auditory ossicles in the mammalian ear, homologous with the *columella auris* and the *hyomandibula* of lower vertebrates. It is a small U-shaped bone attached to the *fenestra ovalis* and having its outer extremity adjacent to the *incus* or anvil (see also *Malleus* and *Incus*).

STAPHYLINIDAE. Rove beetles: a large family of beetles, some carnivorous and some herbivorous, all characterized by the short elytra which leave the abdomen exposed. The larvae are of the campodeiform type with long legs and powerful biting mouth-parts. One of the best known British examples is the 'Devil's Coach-horse'.

STAPHYLINOIDEA. A superfamily which includes the Staphylinidae (Rove-beetles) and the Silphidae (Carrion beetles) and their allies.

STATOBLASTS. Internal buds of Polyzoa from which new colonies develop after the parent polyp has died down.

STATOCYST (OTOCYST). An organ of balance consisting of a fluid-filled sac lined with sensory epithelium and containing one or more granules of calcium carbonate or similar substance. As the animal changes its position these granules, known as *statoliths*, stimulate different sensory cells giving a sense of position to the animal. Simple statocysts of this kind are found in many invertebrates. In vertebrates the semicircular canals of the ear, though more complex, act on the same principle enabling the animal to perceive any rapid acceleration or change of position.

STATOLITH. See *Otolith* and *Statocyst*.

STAUROMEDUSAE. The simplest type of jelly-fish of the class Scyphozoa. Unlike the majority of the class they are sedentary and polyp-like, attaching themselves to sea-weeds by a short stalk and by adhesive organs known as *marginal anchors* on the tentacles. There is no alternation of generations in their life-history and the egg develops into an individual exactly resembling its parent.

STEATORNITHIDAE. Oil-birds or Guarcharos of the mountainous country from Trinidad to Peru: nocturnal owl-like birds which feed on fruit and oily nuts.

STEGOCEPHALIA. The earliest known land vertebrates, found in the Lower Carboniferous, Permian and Triassic Periods. They were lizard-like amphibia in which the dorsal surface of the skull was completely covered by dermal bones, these being also present to a varying degree over the rest of the body.

STEGOSAURIA. Large extinct herbivorous reptiles of the Jurassic Period, attaining a length of 28 ft., but having a minute brain. The legs lifted the body well off the ground and locomotion was plantigrade. The dorsal and tail vertebrae had large neural spines which supported a dermal armour of enormous vertical triangular plates.

STENOPODIUM. The typical slender biramous limb of a crustacean; if flattened it is known as a *phyllopodium*.

STERIORNITHES. A group including a number of large extinct flightless birds from the lower Tertiary strata of South America. The principal genera are *Phororhacos*, *Brontornis*, *Stereornis*, *Patagornis* and *Dryornis*. It is probable, however, that these do not all form a natural group as there is considerable diversity.

STEREOSPONDYLOUS. Vertebrae whose component elements are fused into a single whole (*cf. Temnospondylous*).

STERN-, STERNO-. Prefix from Greek *Sternon*: breast bone.

STERNAL PLATES. Exoskeletal plates forming the ventral covering of each segment of a crustacean or insect.

STERNEBRAE. Segments making up the sternum in some vertebrates.

STERNINAE. Terns or 'sea-swallows': marine birds with long beaks and forked tails. There are about fifty species found in most parts of the world.

STERNITE. See *Sternum*.

STERNORHYNCHA. Plant-bugs of the group Homoptera having a head-flexure so acute that the rostrum appears to arise from between the fore-limbs. The antennae are well developed and do not possess a terminal spine or arista. To this group belong the scale-insects (Coccidae) and the plant-lice (Aphididae).

STERNUM. (1) The breast bone of a terrestrial vertebrate, to which most of the ribs are attached.

(2) The skeletal plate on the ventral side of each segment in an insect or crustacean. It may be a single plate or may be made up of several parts known as *sternites*.

STERRULA. A free-swimming ciliated larva preceding the planula stage in Alcyonaria.

STEWART'S ORGANS. Internal gills of some Echinoids (sea-urchins), formed by pouches of the lantern coelom projecting upwards into the perivisceral cavity (see *Aristotle's Lantern*)

STIGMA. (1) A pigmented eye-spot or photo-receptor, *e.g.* in certain protozoa and other invertebrates.

(2) The spiracle or breathing pore of an insect.

STING. A defensive or offensive organ usually able to pierce the prey and at the same time inject poison.

e.g. (1) Of Bees and Wasps: a modified ovipositor.

(2) Of Coelenterates: a cnidoblast or stinging thread-cell or a battery of such cells.

(3) Of Scorpions: the telson or tail-spine.

(4) Of Sting-rays: the dermal spines of the tail.

In insects the poison usually consists of proteins of low molecular weight and amino-acids. Formic acid is sometimes present but the theory that the main poison is either an acid or an alkali is no longer held.

STIPE (STIPES). (1) Any stem-like structure, as for instance the stalk of a plant or of a sedentary animal such as a Crinoid or a colonial Coelenterate.

(2) Of insects: the second segment of the maxilla, attached at its base to the cardo and bearing at its distal end the palp, the lacinia and the galea.

(3) The eye-stalk of a crab or similar crustacean.

STOLON. A branched structure resembling the runner of a plant: in colonial organisms such as coelenterates, polyzoa and tunicates it forms a stalk on which new polyps or zooids develop.

STOMACH. The principal digestive sac, usually a muscular enlargement of the gut having a lining of glandular cells which secrete enzymes. In vertebrates the stomach is situated between the oesophagus and the duodenum, having at its anterior end the *cardiac sphincter* and at its posterior the *pyloric sphincter*.

STOMO-, STOMATO-. Prefixes from Greek *Stoma, stomatos*: mouth.

STOMATOPODA. See *Hoplocarida*.

STOMIATOIDEA (STOMIATIDAE). Deep sea herring-like fish with luminous spots along the sides of the body.

STOMODAEUM. The fore-gut: that part of the alimentary canal which is lined by ectodermal tissue invaginated or bent inwards through the mouth. Such tissue may consist of stratified epithelium, with or without a cuticle, and differs markedly from the endodermal columnar epithelium lining the *mesenteron* or mid-gut. In mammals the stomodaeum extends as far as the entrance to the stomach.

STONE CANAL. A tube whose wall is usually calcified, leading from the madreporite to the water-vascular ring in star-fish or other echinoderms (see *Hydrocoel*).

STRATIFIED EPITHELIUM. Epithelium consisting of many layers of overlapping cells giving good protection to underlying structures, as for instance in the epidermis of all vertebrates and many invertebrates.

STRATUM CORNEUM. The horny outer layer of the mammalian epidermis consisting of flattened dead cells without nuclei in which the granules of *eleidin*, originally present, become changed into the hard substance *keratin*.

STRATUM GRANULOSUM. That part of the mammalian epidermis below the horny layer and above the Malpighian layer. It consists of stratified epithelium whose cells contain granules of *eleidin* (see also *Stratum corneum*).

STRATUM LUCIDUM. A layer of stratified epithelium of the mammalian epidermis between the granular and the horny layers. It consists of indistinct cells without nuclei and containing transparent hyaline material (*kerato-hyalin*).

STRATUM MALPIGHII. See *Malpighian layer.*

STREPSIPTERA. Minute insects whose larvae parasitize bees and plant-bugs. The females continue as wingless endoparasites throughout life and remain in their original host. The males, however, develop powers of flight with the hind-wings only, the fore-wings being vestigial. They fly to another host for mating but soon die. Insects which are parasitized by Strepsiptera are said to be 'stylopized' (named after the parasite *Stylops*); the result is usually a modification or degeneration of the sexual organs of the host.

STREPTO-. Prefix from Greek *Streptos*: twisted.

STREPTONEURA. Gasteropod molluscs nearly always having a spiral or cone-shaped shell and exhibiting the phenomenon of *torsion*. The main parts of the body are turned through an angle of 180° at an early stage of development. This causes the visceral loop of the nervous system to be twisted into a figure eight and other organs such as the heart and gills to face the reverse way from that found in normal molluscs (see *Torsion*). The group includes such common molluscs as the limpet, the whelk and the winkle.

STREPTOSTYLIC SKULL. A type of skull found in many reptiles (Squamata) and in some birds. The quadrate bone is freely movable relatively to the squamosal bone. This allows the upper jaw to move relatively to the cranium and enables the animal to open its mouth with an extremely wide gape.

STRIATED BORDER (INSECTS). The striated surface of the epithelial cells lining the mid-gut and the Malpighian tubules of insects. These are of two kinds:

(1) The *honey-comb* border made of many little rod-shaped vesicles fused together to form a palisade.

(2) The *brush* type consisting of separate filaments.

There has been much controversy as to the function of these structures which change their appearance during secretory activity.

STRIATED MUSCLE. Striped or voluntary muscle: contractile tissue consisting of bundles of elongated fibres with many nuclei and with conspicuous stripes at right angles to the fibres. Such muscles are capable of rapid contraction under the influence of nervous stimuli and are concerned with voluntary movements such as those of locomotion.

STRIDULATION. A sound emitted by insects and caused by rubbing one hard part of the body against another. In short-horned grass-hoppers the sound is made by rubbing the rough inner surface of the hind femur against a projecting vein on the outside of the tegmen. In long-horned grass-hoppers and in crickets the two tegmina are rubbed together. Temperature has a great influence on the rate of stridulation which becomes accelerated as the temperature rises.

STRIGES (STRIGIDAE). Owls: nocturnal birds of prey which hunt insects, small mammals, birds, reptiles etc. The eyes are directed forwards and are surrounded by a circle of stiff feathers. The strong hooked beak is bent downwards from the base. There are about 150 species in most parts of the world.

STRIGIL (STRIGILIS). A comb-like structure on a bee's leg, used for cleaning the antennae.

STRINGOPIDAE. Kakapos or owl-like parrots of New Zealand living in holes in the ground during the daytime and coming out at night.

STROBILA. The segmented part of a tapeworm formed by strobilation (see below).

STROBILATION. (1) The repeated formation of similar segments or *pro-*

glottides by a process of budding behind the head or scolex of a tapeworm. (2) A similar process in the scyphistoma larva of a jelly-fish.

STROMA. Fibrous connective tissue binding an organ together.

STRONGYLOID LARVA. The second stage in the larval development of a nematode: a stage in which it has the power of locomotion but is contained within a 'cyst' formed of loosely fitting skin. In this stage some parasitic forms leave the soil where they have lived at first and enter the body of their host.

STRUCTURAL COLOURS. Metallic or iridescent colours caused by the physical nature of the cuticle and not by any pigment (as for instance in the wings of some butterflies).

STRUTHIONES (STRUTHIONIDAE). Ostriches: large flightless running birds inhabiting the plains and deserts of Africa and Arabia. They live in companies and are polygamous. Ostriches may be distinguished from other running birds by the fact that they have a naked head and neck and long naked legs with only two toes.

STURNIDAE. Starlings: passerine birds of the Old World having straight or slightly curved strong beaks. They nest in buildings or in any convenient hole, fly in packs and are practically omnivorous.

STYLO-. Prefix from Greek *Stulos*: a pillar.

STYLOHYAL. One of a number of small bones or cartilages forming the anterior horn of the hyoid arch in mammals. The principal hyoid bone is beneath the tongue and is in the form of a flat plate, the *basihyal*, with two anterior and two posterior horns or *cornua* extending from it. The anterior horns extend upwards towards the inner part of the ear-capsule and each consists of a *ceratohyal, epihyal, stylohyal* and *tympanohyal*; there may, however, be fusion or loss of some of these.

STYLOID PROCESS. A somewhat ambiguous name which may be given to the hyoid cornua, to the stylohyal bone, to a process of the temporal bone or to various processes of limb bones: any pillar-shaped process.

STYLOMASTOID FORAMEN. An opening for the seventh cranial nerve to pass between the periotic bone and the hind-part of the tympanic bulla in the mammalian skull.

STYLOMMATOPHORA. Snails with eyes on the tips of the tentacles, *e.g.* the common garden snail, as distinct from *Basommatophora* or water-snails with eyes at the bases of the tentacles.

STYLOPIZATION. See under *Strepsiptera*.

STYLUS (STYLE, STYLET). Any pointed bristle-like appendage. *e.g.* (1) The piercing mouth-parts of insects such as mosquitoes, bugs etc.
(2) Short unjointed abdominal appendages of some insects.

SUB-. Latin preposition and prefix: under.

SUBCHELA. A reversed pincer on the legs of certain shrimps and other crustacea. The last segment of the limb is folded backwards against the preceding one like the blade of a pocket knife.

SUBCLAVIAN ARTERIES AND VEINS. The principal arteries and veins taking blood to and from the anterior limbs of a vertebrate.

SUBCOXA. A segment preceding the coxa on the legs of certain primitive insects. In the majority of insects, however, this segment has become fused with the body-wall and the second segment, the coxa, forms the functional limb-base.

SUBGENITAL PITS (OF COELENTERATES). Ectodermal invaginations forming pits beneath the four gonads on the under surface of a jelly-fish.

SUBGERMINAL CAVITY (EMBRYOLOGY). A fluid-filled space separating the blastoderm (blastodisc) from the underlying yolk in the egg of a bird or reptile during the early embryonic development.

SUBIMAGO. The stage before the imago in the development of the mayfly (Ephemeroptera). It is winged and closely resembles the adult except that it has a thin dull skin covering the whole of the body and wings. When this is cast off the true brilliantly coloured imago is formed.

SUBINTESTINAL VESSEL. A blood vessel, also known as the ventral vessel, running backwards beneath the intestine in earthworms etc.

SUBLINGUA. A fold of skin beneath the tongue in some mammals, *e.g.* Insectivora.

SUBLINGUAL GLANDS. Salivary glands beneath the tongue in mammals.

SUBMAXILLARY GLANDS. Salivary glands behind the angle of the lower jaw in mammals. They are compound racemose glands of the mixed type, producing a serous (watery) and a mucous (sticky) secretion.

SUBMENTUM. The basal segment of the labium of an insect (see also *Mentum* and *Prementum*).

SUBNEURAL. Beneath the nerve cord: *e.g.* subneural blood vessel of an earthworm; subneural gland of a tunicate etc.

SUBOESOPHAGEAL GANGLION. A ganglion or pair of ganglia situated beneath the oesophagus and connecting the ventral nerve cord with the 'nerve-collar' in annelids, arthropods and some molluscs.

SUBOPERCULAR BONE. One of four bones supporting the operculum of a bony fish, the others being the *opercular*, *pre-opercular* and *interopercular* bones. They are extensions of the hyoid arch.

SUBRADIAL TENTACLES. Tentacles on the rim of a jelly-fish (medusa), formed in a late stage of development and occupying the spaces between the original tentacles which are in the *perradial*, *interradial* and *adradial* positions (*q.v.*).

SUBSCAPULAR VEIN. A vein passing beneath the scapula and bringing blood from the pectoral region of the body-wall in some vertebrates.

SUBSTRATE (SUBSTRATUM). A particular substance or group of substances upon which an enzyme acts.

SUBUMBRAL PITS (OF COELENTERATES). Four vertical pits penetrating deeply into the mesenteries in sessile jelly-fish of the group Stauromedusae. They may correspond to the subgenital pits of free-swimming species and probably help to give extra aeration by providing a large surface for diffusion.

SUBUMBRELLAR (SUBUMBRAL). The under or concave surface of a jelly-fish. The term *ventral* is not generally used since this surface does not correspond to the true ventral surface of the bilaterally symmetrical larva.

SUBZONAL MEMBRANE. A layer of cells formed inside the trophoblast of a mammalian embryo, at a later stage forming part of the chorion (see *Trophoblast*).

SUCCUS ENTERICUS. Intestinal juice containing a number of digestive enzymes secreted by Brünner's Glands in the walls of the small intestine. In mammals the enzymes include *erepsin, enterokinase, maltase, invertase* etc. which, together with the pancreatic enzymes, continue the digestive processes begun in the mouth and stomach.

SUCTORIA. Protozoa of the class Ciliophora which usually lose their cilia in the adult stage and possess one or more suctorial tentacles used for catching prey. Most species are ectoparasites attached to their host by a cuticular stalk which may form a cup enclosing the animal. Reproduction is by conjugation or by budding but never by binary fission. The asexual buds are usually produced in pouch-like cavities.

SUDORIFIC GLANDS (SUDORIFEROUS GLANDS). Sweat glands of mammals: coiled tubes with straight ducts leading through the epidermis to the surface. They excrete water and other substances and play an important part in regulating the body temperature. If this rises, as during exercise, more sweat is produced; evaporation takes latent heat from the body and the temperature is automatically brought back to normal.

SUDORIPAROUS GLANDS. See *Sudorific Glands* above.

SUIDAE. Pigs: ungulates with four toes on each foot of which the outer two are short and off the ground; the inner two larger, hoofed and closely adjacent to each other. The animals are omnivorous and have canine teeth which are usually lengthened into tusks. Apart from domestic animals, pigs are confined to the Old World but fossil remains have been found in the Eocene of North America.

SULCUS. (1) Any groove or fissure, particularly those in the cerebral hemispheres.
(2) Of *Protozoa*: a longitudinal groove in which a flagellum lies when at rest, *e.g.* in the Dinoflagellata.

SULCUS LIMITANS. An internal longitudinal groove forming a line of demarcation between the dorsal and ventral regions of a developing vertebrate brain.

SULCUS LUNATUS. The *Simian Fissure*: a groove in the occipital lobe of the brain in lower apes.

SUPER-, SUPRA-. Latin preposition and prefix: above.

SUPERFICIAL OPHTHALMIC NERVES. Two pairs of afferent cranial nerves bringing impulses from the sensory receptors of the snout in a vertebrate. Each pair consists of a branch of the trigeminal and a branch of the facial nerve running closely parallel to one another and entering the brain behind the eyes (see also *Ophthalmic Profundus Nerve*).

SUPERFICIAL PETROSAL NERVE. See *Petrosal Nerves*.

SUPERIOR OBLIQUE MUSCLE. One of the six muscles which move the eye of a vertebrate and are attached to the inside of the orbit. The others are the *inferior oblique* and the four *rectus* muscles.

SUPERIOR PHARYNGEAL BONE. A bone formed by fusion of several of the pharyngobranchial bones in certain fish.

SUPERIOR RECTUS. See *Eye-muscles*.

SUPERIOR VENA CAVA (ANTERIOR VENA CAVA, PRECAVAL VEIN). One of two veins which bring blood back from the head and neck of a vertebrate into the right auricle of the heart.

SUPERLINGUAE. Small paired lobes fused to the sides of the hypopharynx in certain insects.

SUPERPOSITION IMAGE. See *Mosaic Vision*.

SUPINATION. The position of the fore-arm in which the palm of the hand faces upwards and the radius and ulna are parallel. Man and other primates can put their arms into this position but most other mammals have the radius and ulna crossed and the palm facing downwards (see *Pronation*).

SUPRA-ANGULAR. One of the membrane-bones making up the lower jaw of a reptile or bird. In mammals it is vestigial and forms the *Processus Folii*

SUPRA-CLAVICLE. A former name for the *Supra-cleithrum* (see below).

SUPRA-CLEITHRUM. A membrane-bone on each dorsal extremity of the pectoral girdle in bony fish. It is frequently articulated with the post-temporal bone so that the pectoral girdle is connected to the skull.

SUPRA-OCCIPITAL BONE. A cartilage-bone forming the upper part of the occipital or hind region of the vertebrate cranium. With the two lateral *exoccipitals* and the *basi-occipital* it bounds the *foramen magnum*, the large opening through which the spinal cord passes to the brain.

SUPRA-OESOPHAGEAL GANGLIA. Also named the *suprapharyngeal, hyperpharyngeal* or *cerebral* ganglia: a pair of nerve centres situated on the dorsal side of the pharynx or oesophagus in annelids, arthropods, mollusca etc. and connected to the main ventral nerve-cord by a 'collar' formed of two commissures. Sometimes these ganglia are fused with others such as the optic and antennary and may be sufficiently complex to be regarded as a 'brain'.

SUPRARENAL BODIES. Paired endocrine organs in front of the kidneys of a vertebrate; able to secrete the hormone *adrenalin* which has many of the same effects as stimulation of the sympathetic nerves. The name suprarenal was formerly used to denote organs which are now known as *adrenal* bodies. In man and most higher vertebrates these are composite organs having a medulla derived from suprarenal organs and a cortex derived from *inter-renal* glands. In fish and some amphibia, the two types of gland are distinct.

SUPRA-SCAPULA. A bone or cartilage attached to each scapula on its dorsal side. In some vertebrates such as the frog it is a distinct skeletal element but in mammals it is very much reduced.

SUPRA-TEMPORAL ARCADE. A bar of bone separating the inferior from the superior temporal fossa in the diapsid type of skull. It is formed from the *post-orbital* and the *squamosal* bones.

SUPRA-TEMPORAL BONE. A membrane-bone forming part of the dorso-lateral covering of the cranium in some fish, amphibia and reptiles but tending to disappear in higher vertebrates.

SUPRA-TEMPORAL FOSSA (SUPERIOR TEMPORAL FOSSA). A hole or window perforating the membrane-bones of the upper temporal region and leading through into the temporal cavity in certain vertebrate skulls. It is bounded above by the *parietal* bones and below by the *post-orbital* and *squamosal* bones (see *Anapsid, Diapsid, Parapsid* and *Synapsid*).

SUSPENSORIUM. Any part by which an organ is suspended, *e.g.* those parts of a skull by which the jaws are suspended.

SUSPENSORY LIGAMENTS (OF THE EYE). A ring of small ligaments by which the lens of the eye in a vertebrate is attached to the ciliary muscles which control its shape and position and hence help to focus it.

SUTURE. A line of fusion of two adjacent parts such as bones, cuticular plates etc.

SWAMMERDAM'S GLANDS. Glands which secrete calcareous nodules on each side of the vertebral column of a frog or other amphibian. They are connected with the endolymphatic ducts of the ears.

SWIM-BLADDER. A gas-filled sac usually on the dorsal side of the alimentary canal in bony fish. It may be connected by a pneumatic duct to the oesophagus or sometimes to the stomach but in many cases the connection is lost. Its chief function is to regulate the buoyancy of the fish but in lung-fish it has a respiratory function also. The lungs of amphibia and all terrestrial vertebrates are homologous with the swim bladder of a fish and probably evolved while their ancestors still lived in the water. The swim-bladder of a modern fish has apparently developed as a hydrostatic organ by specialization.

SWIMMERET. See *Pleopod.*

SYCON GRADE. A type of sponge in which the flagellated chambers open directly into the central cavity or paragaster without intervening canals.

SYLVIAN FISSURE (FISSURE OF SYLVIUS). A shallow oblique fissure separating the frontal from the temporal lobe in the cerebral hemisphere of a mammal.

SYM-, SYN-. Prefix from Greek *Sun*: with, together.

SYMBIONTS. Organisms living in a state of symbiosis.

SYMBIOSIS. A state in which two organisms (animal or plant) have a close partnership, sometimes one inside the other, to the mutual benefit of both so that neither can be regarded as a parasite. Examples are bacteria in the intestines of many animals; micro-algae (zoochlorellae) in the body of a hydra or similar animal; fungi attached to the roots of trees etc.

SYMBRANCHII (SYMBRANCHIDAE). Eel-like fish without pectoral or pelvic fins and having the gill-openings confluent so that there is only a single slit each side.

SYMPATHETIC SYSTEM. See *Autonomic system.*

SYMPATHIN. A substance chemically similar to adrenalin, produced at the synapses of the sympathetic nerves and believed to be responsible for passing an impulse from one neurone to the next (see *Synapse*).

SYMPHILY. Commensalism: a form of symbiosis in which two organisms share the same food and each benefits from the association.

SYMPHYLA. A group of millepedes (Diplopoda) whose mouth-parts and certain other features show a similarity to those of insects and suggest a common ancestry.

SYMPHYPLEONA. A suborder of Collembola or spring-tails having a shortened body with the thoracic and some of the abdominal segments not clearly demarcated (*cf. Arthropleona*).

SYMPHYSIS. A junction or coalescence of two bones or cartilages, as for instance the two pubic bones.

SYMPHYTA. Sawflies, wood-wasps etc. A suborder of Hymenopterous insects having no narrow constriction between the thorax and the abdomen and having a large ovipositor which acts as a saw or drill for piercing plant tissues. Their larvae resemble the caterpillars of butterflies and moths but differ from them in having up to seven pairs of abdominal prolegs.

SYMPLAST. A syncytium or multinucleate mass of protoplasm arising by repeated division of the original nucleus.

SYMPLECTIC BONE. A cartilage-bone forming part of the hyoid arch in bony fish. It is immediately beneath the hyomandibula and next to the quadrate bone which articulates with the lower jaw.

SYNAPSE. A small gap between the dendrites or terminal arborizations of two consecutive nerve fibres. When a nervous impulse reaches such a gap it produces a chemical substance (*e.g. sympathin* or *acetyl choline*) which then initiates a new impulse in the next nerve fibre. In this way a nervous impulse passes rapidly along a series of neurones.

SYNAPSIDA. Vertebrates whose skulls have a single inferior temporal fossa on each side. In this respect they differ from the *Diapsida* which have two fossae on each side and the *Anapsida* which have none. Synapsida include mammals and certain fossil reptiles known as *Theromorphs* from which mammals probably evolved.

SYNAPTICULA. Bars of cartilage running at right angles to the gill-bars and connecting them with one another in such creatures as *Amphioxus*, Tunicates etc.

SYNAPTIDA. Holothuroids or 'sea-cucumbers' of burrowing habit, with pinnate tentacles round the mouth and with dermal ossicles which are anchor-shaped.

SYNAPTOSAURIA. An alternative name for the Sauropterygia or Plesiosauria, from the fact that their skulls, like those of theromorphs and mammals, have enlarged inferior temporal fossae bordering on the parietal bones.

SYNCARIDA. A small group of primitive fresh-water crustacea belonging to the class Malacostraca but not possessing a carapace; having little specialization of limbs and little difference between the thorax and the abdomen. Some species found in subterranean waters are degenerate and without eyes.

SYNCYTIUM. A tissue containing a number of nuclei embedded in a mass of protoplasm (*e.g.* striated muscle fibres).

SYNDACTYLA. Kangaroos and similar marsupials having the second and third toes bound together in a common mass of tissue.

SYNDACTYLY (SYNDACTYLISM). The uniting of some digits of a hand or foot by a membrane or skin.

SYNDESMO-CHORIALIS. A type of placenta, found for example in sheep, in which the uterine epithelium is the only layer to be 'eroded' (*cf. Haemochorialis* and *Epithelio-chorialis*).

SYNENTOGNATHI. An alternative name for the Scombresocidae, a group of fish including the sword-fish and some flying-fish.

SYNERGISM. The co-ordinated action of groups of muscles to produce a specific movement.

SYNGAMY. Sexual union of male and female gametes.

SYNGNATHIDAE. Pipe-fish and sea-horses: fish whose skin is covered with small bony plates and whose prolonged snout bears a small terminal toothless mouth. They have no pelvic fins and in sea-horses the tail is prehensile. Eggs are retained until after hatching in a brood-pouch beneath the abdomen of the male.

SYNOVIAL CAPSULE. A capsule of fibrous tissue filled with fluid, acting as a cushion between the articulating surfaces of two bones.

SYNOVIAL FLUID. The fluid contained in a synovial capsule.

SYRINX. The vocal organ of a bird at the lower end of the trachea just above its junction with the bronchi. It has a variable and often complex structure with a number of vibrating membranes (see also *Tympanum* and *Pessulus*).

SYRPHIDAE. Hover-flies: a family of two-winged insects which hover over flowers and shrubs and are sometimes mistaken for wasps when in flight on account of the pattern of yellow bands on the abdomen. The small legless larvae feed chiefly on aphides and are to be found frequently among them on rose bushes or fruit trees.

SYSTEMA NATURAE. The work of Linnaeus in which the binomial system of nomenclature, *i.e.* the name of a genus followed by the name of the species, was used for the first time.

SYSTEMIC ARCH. The fourth aortic arch of a vertebrate or of a vertebrate embryo, by which blood flows from the ventral to the dorsal aorta and hence to all the principal parts of the body. Amphibia and most reptiles have a right and a left systemic arch; birds have only the right arch and mammals only the left (see *Aortic arches*).

SYZYGY. The sticking together of gamonts or gametocytes in pairs prior to true sexual union of the gametes. Such an occurrence is common among protozoa, particularly of the class Sporozoa.

T

TABANIDAE. Horse-flies, gad-flies and clegs: a family of two-winged insects, the females of which suck blood by means of mouth-parts combining the piercing method of the mosquito with the filter-feeding method of the blowfly.

TACHINIDAE. A family of two-winged insects comprising three subfamilies, *viz.* the Sarcophaginae or 'flesh-flies', the Calliphorinae or blow-flies and the Tachininae which are all parasitic. The latter are stout bristly insects usually with grey or striped bodies. Their larvae are maggots, often parasitic in other insects, and are important in the biological control of forest pests.

TACHYGENESIS. The omission of certain embryonic stages, as for instance the tadpole stage of a frog or a larval instar in the development of an insect.

TACTILE CORPUSCLE (END BULB). The sensory ending of a nerve surrounded by a bulbous enlargement of the perineurium.

TACTILE RECEPTORS. Sensory receptors able to detect pressure or touch. They may consist of simple nerve endings in the skin, tactile corpuscles, or sensory hairs such as those of insects.

TAENIAE HIPPOCAMPI (OF BRAIN). See *Fornix*.

TAENIOGLOSSA. Gasteropod mollusca of the group Monotocardia having a spiral shell and a single pectinate gill; specially characterized by having a long ribbon-like tongue or radula bearing small horny teeth arranged in rows of seven. An example is the periwinkle.

TAENIOSOMI (LOPHOTIFORMES). Ribbon-fishes: deep-sea fish having a long flat ribbon-like body with dorsal fin extending from the head to the end of the tail.

TAIL. The hindmost part of an animal, especially when it extends beyond the rest of the body. In vertebrates it is defined as that part which extends behind the anus.

TAIL FAN. The typical tail-structure of a lobster, shrimp, prawn etc., formed from the last pair of abdominal appendages, which are flattened and known as *uropods*, together with a flattened median appendage known as the *telson*.

TAIL FIN (CAUDAL FIN). The tail fin of a fish may be one of the following types: *Diphycercal* – primitively symmetrical and pointed as in many fossil and a few living fish; *Gephyrocercal* – similar to diphycercal but a secondary development, as in eels; *Heterocercal* – a turned up tail with unsymmetrical lobes as in a dogfish; *Homocercal* – secondarily symmetrical and forked but derived from the heterocercal type by lengthening of the ventral rays, as in most present-day bony fish.

TAIL FOLD (EMBRYOLOGY). See under *Head Fold*.

TAGMA. A term used to denote a group of similar somites or segments in the body of an arthropod. The various tagmata are distinguished by differences in width, fusion of segments or special features of the limbs. In insects, for instance, the tagmata are the head, thorax and abdomen. In spiders on the other hand there are only two tagmata. The first is the *prosoma* or cephalothorax carrying legs as well as mouth-parts; the second is the *opisthosoma* or abdomen with no obvious appendages.

TALONID. The posterior part or 'heel' of a molar tooth.

TALPIDAE (TALPINAE). Moles: burrowing insectivora with flattened body and pointed snout and with the fore-limbs strongly clawed and modified for digging. The true moles or Talpinae have an extra digit on the fore-foot, but another group, the Myogalinae have webbed feet for swimming and are without this extra digit.

TANAIDACEA. Small marine crustaceans usually inhabiting burrows or tubes made of a mass of fibres which they secrete. They belong to the group Peracarida but, unlike other members of the group, they have a short carapace and stalked eyes.

TAPETUM. A reflecting layer in the choroid causing the eyes of many animals to shine in the dark. It contains crystals of *guanin*.

TAPIRIDAE. Tapirs: ungulates of tropical and subtropical regions having three digits on the hind-foot and four on the front foot. The body is covered with short hair and there is a mobile proboscis like a short trunk.

TARDIGRADA. (1) Sloths. See *Bradypodidae*.
(2) Minute, almost transparent arthropods with four pairs of short legs ending in claws. They are found among moss or other plants on which they feed by piercing the cells and sucking the sap. The tardigrada are difficult to classify; they have been placed among the arachnids but they also appear to have some affinity with *Peripatus*.

TARSI-. Prefix from Greek *Tarsos*: the sole of the foot.

TARSIIDAE (TARSIOIDEA). Arboreal lemur-like mammals of S.E. Asia, having large eyes and ears, a long thin tufted tail, very long tarsal bones and flattened discs on the ends of the fingers and toes.

TARSIPEDINAE. Small long-tailed insectivorous marsupials about the size of a mouse, closely allied to the phalangers and having an extensile tongue.

TARSOMERES. Segments of the tarsus of an insect.

TARSO-METATARSUS. A bone formed by the union of a tarsal with a metatarsal bone, as in the foot of a bird.

TARSUS. (1) Tarsal Bones: the heel-bones of a tetrapod vertebrate. The full number of these is nine, *viz.* a proximal row consisting of the *tibiale*, *intermedium* and *fibulare*; a single *centrale* and five distal tarsals. Frequently some bones are absent, fused together or modified in some way. The fibulare (also called the *calcaneum*) is the principal heel-bone and usually projects back behind the fibula; the *Tendon of Achilles* is attached to it.

(2) Of Insects. The part of an insect's leg immediately distal to the tibia, usually divided into a number of segments (tarsomeres).

TASTE BUDS. Small bulb-like clusters of chemo-sensory cells embedded in the epidermis of the tongue. They are similar in structure to the neuromast cells (*q.v.*) of fish and are connected with branches of the fifth, seventh or ninth cranial nerves according to their position.

TECT-, TECTI-. Prefix from Latin *Tectum*: a roof.

TECTIBRANCHIATA. Gasteropod molluscs such as *Aplysia*, the 'sea-hare', which have their shell reduced and flat and often nearly covered by upgrowing folds of the mantle. They belong to the group Opisthobranchiata which exhibit the phenomenon of torsion (*q.v.*) followed by detorsion.

TECTORIAL MEMBRANE. A ribbon-like membrane covering the Organ of Corti – the system of sensory receptors running the length of the spiral cochlea in the ear of a mammal.

TECTOSPONDYLOUS VERTEBRAE. Vertebrae whose cartilaginous centrum is strengthened by several concentric calcareous rings, as in some Elasmobranch fish.

TECTRICES. Wing-coverts or small feathers covering the bases of the wings of a bird.

TECTUM OPTICUM. The optic lobes: the thickened roof of the mid-brain of a vertebrate.

TECTUM SYNOTICUM. That part of the roof of the chondrocranium which connects the two auditory capsules with one another in the embryonic development of a vertebrate skull.

TEETH. Hard outgrowths on the jaws or in the skin of the mouth and pharynx, used for biting and masticating food. The teeth of vertebrates are essentially similar to the denticles on the skin of a dogfish. They consist of an inner core of dentine or ivory covered by a hard layer of enamel; in the centre is a pulp cavity consisting of nerves, blood vessels etc. Teeth are classified as follows:
Homodont. All alike, as in a fish.
Heterodont. Of diverse kinds: incisors, canines, premolars and molars, as in mammals.
Pleurodont. Attached by their sides to the rim of the jaw or in grooves.
Thecodont. Set in sockets, as in a mammal.
Acrodont. Fused to the edge of the bone, as in lizards.
Monophyodont. Having only one set of teeth.
Diphyodont. Having two sets: milk teeth and permanent teeth.
Polyphyodont. Having many successive sets.
Bunodont. Teeth with distinct cusps, as in a pig.
Lophodont. Teeth having cusps joined by transverse ridges, as in many ungulates.
Herbivorous animals usually have teeth whose pulp cavities are wide open at the base, giving a plentiful blood supply so that the tooth can keep on growing as fast as it wears down. Carnivorous animals on the other hand have hard sharp teeth with a thick covering of enamel and with the pulp cavity almost closed at the base so that the tooth ceases to grow.

TEGMEN. Latin: a covering.
(1) The leathery fore-wing of an insect such as the cockroach or grasshopper.
(2) Of Echinoderms: the flat leathery upper surface of a Crinoid, in the middle of which is the mouth.

TEGMENTUM. (1) A tegmen or covering.

(2) Of mollusca: the upper calcareous layer of each shell-plate in mollusca of the class Amphineura whose shells are in the form of transverse plates.

TEGULA. A small scale-like sclerite overlapping the bases of the wings in certain insects such as Lepidoptera and Hymenoptera.

TEIIDAE. Lizards of the New World sometimes having the limbs reduced or absent; having no osteoderms and having a long forked tongue.

TEL-, TELE-, TELO-. Prefixes from Greek *Tele*: far, distant.

TELENCEPHALON. The most anterior part of the brain of a vertebrate consisting of the olfactory lobes and the cerebral hemispheres.

TELEO-. Prefix from Greek *Teleos*: complete.

TELEOPTILES. The various kinds of feather occurring on an adult bird, as distinct from the *Neossoptiles* or nestling feathers (see *Feather*).

TELEOSAURIDAE. Extinct marine crocodiles of the Liassic and Oolitic Periods, having a long slender snout and having the anterior limbs about half as long as the posterior.

TELEOSTEI. All present-day bony fish except lung-fish and a few other surviving primitive types such as the sturgeon and the gar-pike. Teleostei are characterized by the presence of a swim-bladder whose main function is hydrostatic; by the loss of the spiracle, the spiral valve of the intestine and the conus arteriosus of the heart and also by the fact that the dermal bones of the head have sunk beneath the surface and become incorporated in the skull. The scales of the body are in some ways like the thicker *ganoid* scales of more primitive fish but in the course of evolution they have lost the layer of ganoin and are either very thin or absent altogether.

TELEOSTOMI. A larger group of bony fish than the Teleostei, comprising all except the lung-fish and including therefore the sturgeons, gar-pikes and coelacanths as well as many fossil species and practically all present-day bony fish.

TELOLECITHAL EGGS. Eggs which have the yolk mainly aggregated towards the vegetative pole, that is on the side away from the nucleus.

TELOLEMMA. The sheath or covering of a nerve-ending within a muscle.

TELOMITIC. Having spindle-fibres attached to the ends of the chromosomes.

TELOPHASE. The final stage of meiosis or mitosis when the movements of chromosomes have ceased and the daughter-nuclei are completely formed.

TELOSPORIDIA. Parasitic sporozoa having a single nucleus in the adult stage and producing simple spore-cases containing several sporozoites. The group includes the Coccidiomorpha and the Gregarinidea many of which are pathogenic in man and domestic animals.

TELOTROCH. A pre-anal tuft of cilia on a trochosphere larva (*q.v.*).

TELSON. A single terminal appendage on the last abdominal segment of most arthropods. It may take various forms such as part of the tail-fan of the crayfish or the sting of a scorpion. In insects it is absent except as a rudiment in the embryo.

TELUM. The last abdominal segment of an arthropod.

TEMNOCEPHALEA. An order of flatworms intermediate between the free-swimming Turbellaria and the parasitic flukes or trematodes. Most of them live attached to fresh-water crustacea or fish, using the host as a resting place from which they catch rotifers and other small creatures as food. Members of the order may be easily recognized by the fact that they have anterior tentacles and a large posterior sucker.

TEMNOSPONDYLOUS. Having vertebrae composed of separate unfused units (*cf. Stereospondylous*).

TEMPERATURE OF ANIMALS. The temperature of an animal depends upon a balance being kept between the rate of production of heat and the rate of loss. Heat production depends on the general rate of metabolism and particularly on the rate of oxidation of food substances. Heat loss depends on the total surface area and on the nature of the skin. The larger the animal, the less the surface area in proportion to the volume. A large mammal or bird therefore tends to lose heat less quickly than a small one and needs less heat-producing food in proportion to its size. A small mammal or bird needs relatively a very large quantity of food and a high rate of respiration in order to make up for the rapid loss of heat from the surface. *Homothermous* or 'warm-blooded' animals have a high metabolic rate and are able to regulate their temperature by controlling the loss of heat from the skin. This is done chiefly by varying the rate of evaporation of water from the sweat glands, by the dilatation or constriction of blood vessels in the skin, and by the erection of hairs or feathers in cold conditions. Cold-blooded or *Poikilothermous* animals on the other hand are at the mercy of their environment, thriving in hot climates and frequently hibernating in colder conditions.

TEMPORAL CAVITY. A space in the skull of a vertebrate between the cranium proper (which encloses the brain) and the overlying covering of dermal bones. The space is occupied by the muscles which move the lower jaw and is usually continuous in front with the orbit. The overlying dermal bones may form a complete covering (as in Chelonia), or may be perforated in one or two places (see *Anapsida, Diapsida, Synapsida, Parapsida*).

TEMPORAL FOSSA. A perforation in the covering of dermal bones which overlies the temporal cavity of a vertebrate skull.

TEMPORAL LOBES. The lateral lobes of the mammalian cerebral hemispheres, separated from the frontal lobes by the *Sylvian Fissures*.

TENDON. A tough inelastic cord attaching a muscle to the sheath of a bone so that contraction of the muscle can bring about movement of the bone. Tendons are largely made of white collagen fibres.

TENEBRIONIDAE. Nocturnal beetles which are usually unable to fly owing to the fact that the wing-cases are firmly fused together. Their larvae include the meal worms of bakeries and flour mills.

TENSION RECEPTORS. Sensory structures which respond to stretching or pulling of the cuticle, as for instance at the joints of insects.

TENSOR TYMPANI. A small muscle connecting the malleus to the wall of the tympanic chamber in the ear of a mammal.

TENTACULATA. A group of Ctenophora or 'sea-gooseberries' having tentacles bearing *lasso-cells*. The majority of ctenophora belong to this order, but one small group, the Nuda, are without tentacles.

TENTACULOCYSTS. Small sensory tentacles, each covered by a projecting hood, situated round the rim of a jelly-fish. On or near these are concentrated a number of types of receptor including olfactory cells, light-sensitive pigment spots and statocysts.

TENTHREDINIDAE (TENTHREDINOIDEA). Sawflies: small symphytous Hymenoptera whose ovipositor resembles a tiny saw enabling the female to cut slits or pockets in plants where she lays her eggs. The larvae, which do much damage to the leaves of trees, resemble caterpillars but have more abdominal prolegs.

TENTORIUM. The endoskeleton of the head of an insect consisting of a number of chitinous struts connecting the exoskeleton to a transverse bar in the centre of the head. The whole arrangement forms a rigid framework for attachment of the head-muscles.

TEREBELLIDS. Polychaete worms living in a tube open at both ends and constructed of a mixture of sand, fish-bones etc. cemented together by a parchment-like substance. Unlike the majority of tubicolous worms they are able freely to leave and return to their tubes.

TEREBRA. An ovipositor adapted for boring or stinging as in bees, wasps, sawflies etc.

TEREBRANT. Wood-boring.

TERGUM (TERGITE). A dorsal cuticular plate covering each body-segment of an arthropod.

TERMITARIUM. The nest of a termite colony

TERMITE. A white ant (see *Isoptera*).

TERRICOLA. Turbellarians or free-living flatworms which live on land in contrast to the majority of the group which are aquatic. Some of the tropical species are brightly coloured and much larger than the aquatic varieties.

TEST (OF TUNICATES). A tough translucent coat covering the body of a tunicate; remarkable among animal tissues in being composed of a substance, *tunicin*, which closely resembles cellulose.

TESTICARDINES. Brachiopods having shells with a hinge (*cf. Ecardines*).

TESTIS. The male gonad or organ which produces spermatozoa.

TESTUDINATA. An alternative name for Chelonia or tortoises and turtles.

TESTUDINIDAE. A family of Chelonia having the thoracic vertebrae and ribs immovably united with the plates of the carapace. They include the majority of land and water tortoises but not turtles.

TETRA-. Prefix from Greek *Tetras*: four.

TETRABRANCHIATA. Cephalopod molluscs which have two pairs of ctenidia or gills, as distinct from the Dibranchiata which have only one pair. Ammonites and Nautiloidea belong to the former group; cuttle-fish and octopuses to the latter.

TETRACTINELLIDA. Sponges in which the spicules are of the *tetraxon* type, each with four points directed towards the apices of a tetrahedron.

TETRALOPHODONT. Having four-ridged molar teeth.

TETRAONINAE. Grouse, partridges and quails: pheasant-like birds of moors and mountains in most parts of the world except South America. They nest on the ground in heather and rough grass.

TETRAPHYLLIDEA. Tapeworms with four suckers which are usually stalked outgrowths of the scolex. The larvae are parasitic in small crustacea and the adults in fish, amphibia and reptiles.

TETRAPLOID. Having four times the basic (haploid) number of chromosomes or twice the diploid number.

TETRAPODA. Four-footed animals: amphibians, reptiles, birds and mammals. The group includes all land vertebrates and also those such as birds and whales whose legs have become modified for flying or swimming.

TETRAPTEROUS. Four-winged.

TETRARHYNCHIDEA. Tapeworms having four suckers each with a long spiny process. The adults are parasitic in cartilaginous fish; the larvae in marine invertebrates.

TETRAXON. A four-pointed spicule formed by the fusion of four simple ones in sponges (see *Spicule*).

TETRODONTIDAE. Globe-fishes having erectile spines in the skin and a power of inflating the body so that the fish can float like a prickly ball. The mouth is in the form of a narrow beak.

TETTIGONIIDAE. Long-horned grass-hoppers: elongated insects distinguished from other grass-hoppers and locusts (Acridiidae) by their very long antennae and by their predatory habits. The chief British representative is the Great Green Grass-hopper.

THALAM-. Prefix and root word from Greek *Thalamos*: a receptacle.

THALAMENCEPHALON (DIENCEPHALON). The posterior region of the fore-brain of a vertebrate containing a cavity, the *third ventricle*, roofed over by a vascular membrane known as the *anterior choroid plexus*. Dorsally it bears the stalk of the pineal organ and ventrally the infundibulum of the pituitary body. The sides are thickened to form the *optic thalami* from which the primary optic vesicles develop in the embryo.

THALAMUS. The thickened side of the third ventricle of the vertebrate brain, connected by nerve-fibres to most of the sensory centres of the brain and probably having the function of determining whether an impulse is pleasurable or painful.

THALASSEMYDIDAE. Extinct turtles of the Jurassic Period.

THALATTOSAURIA. Extinct lizard-like reptiles of the Triassic Period, having paddle-like limbs; probably related to the Rhynchocephalia.

THALATTOSUCHIA. Extinct crocodile-like reptiles of the Jurassic Period, secondarily adapted to an aquatic life and having paddle-like limbs.

THALIACEA. Pelagic Tunicates in which the adult has no tail, a degenerate nervous system, an atrium opening posteriorly, and gill-clefts which are not divided by external longitudinal bars.

THEBESIAN VALVE. A rudimentary sinu-auricular valve where the left anterior vena cava enters the right auricle in the mammalian heart.

THEC-, THECO-. Prefix from Greek *Theke*: a sheath, case.

THECA. Any protective covering or case, *e.g.* for a coral polyp; for the proboscis of an insect etc.

THECODONT. Having teeth in sockets, as in mammals and a few reptiles.

THECOIDEA. A fossil group of sessile Echinoderms resembling Crinoids but having no stalk or arms. The mouth was on top and had five food-grooves leading to it.

THECOPHORA. A group comprising the majority of tortoises and turtles in which the thoracic vertebrae and ribs are immovably united to the carapace.

THERI-, THERO-. Prefixes from Greek *Ther*: a wild beast.

THERIODONTA. Extinct reptiles which were probably the ancestors of mammals. The skull had two occipital condyles; the teeth were of the typical mammalian forms, *viz.* incisors, canines, premolars and molars. The dentary bones were large and articulated with the squamosals, leaving the quadrate bones small and loose. The limbs were long and lifted the animal well off the ground.

In classification the Theriodonta form a subgroup of the *Theromorpha* (see below).

THEROMORPHA (ANOMODONTIA). Extinct reptiles of the Permian and Triassic Periods showing many features which were distinctly mammalian in character. The chief of these were: a skull with a zygomatic arch and with teeth in sockets and differentiated into incisors, canines and molars; limbs lifting the body off the ground; pelvic bones fused together. It is not known whether some of these mammals had hair but this is quite possible as, in many respects, they were more mammalian than reptilian.

THEROPODA. Carnivorous dinosaurs dating from the Triassic to the Cretaceous Periods, having a skull set at right angles to the neck and with sharp teeth. The fore-limbs were short and the animal probably walked or jumped on the hind-feet; the digits varied from three to five and had sharp claws. Some of these reptiles were as small as a dog; others as large as an elephant. Many had hollow bones like those of birds.

THEROPSIDA. A name sometimes used to comprise all mammals and mammal-like reptiles including the Theromorpha and the Plesiosauria, all of which have a similar type of skull with a single enlarged inferior temporal fossa on each side.

THIRD EYE. See *Pineal Body*.

THIRD VENTRICLE. The cavity of the thalamencephalon in the brain of a vertebrate.

THORACICA. Barnacles: sedentary hermaphrodite crustaceans with a reduced abdomen and a greatly enlarged thorax covered by five or more calcareous plates. They attach themselves firmly to floating or submerged objects in the sea by means of a cement gland at the anterior end; sometimes this organ of attachment is lengthened to form a stalk. Barnacles are found in large numbers between the tide levels, their covering of plates being an effective means of preventing desiccation.

THORACIC DUCT. The chief collecting duct of the lymphatic system in mammals, leading into the left anterior vena cava.

THORACIC FINS. Pelvic fins which in the course of evolution have shifted forwards until they are below the pectoral fins, as for instance in the perch.

THORACIC MEMBRANE (ANNELIDS). A lateral frill or outgrowth of skin, possibly respiratory in function, found on certain tubicolous polychaete worms such as the serpulids.

THORACOSTEI. Sticklebacks (see *Gasterosteidae*).

THORAX. The part of the body behind the head and in front of the abdomen. In higher vertebrates it contains the heart and lungs and is protected by the ribs; in mammals the diaphragm separates it from the abdomen.

In insects the thorax consists of three segments on which the legs and wings are borne; in crustaceans and other arthropods it is often merged with the head to form the *cephalothorax*.

THROMBIN. An enzyme which brings about the clotting of the blood of a vertebrate by causing the soluble protein *fibrinogen* to change into a mass of solid threads known as *fibrin*. Thrombin is formed by the action of calcium ions on a precursor called *prothrombin*, activated by another enzyme known as *thrombokinase*.

THROMBOCYTES. Blood platelets: small disc-shaped particles which are believed to take some part in the clotting of blood. They have no nuclei and are much smaller than red corpuscles. During the process of clotting these thrombocytes form centres from which filaments of fibrin radiate out.

THROMBOKINASE (THROMBOPLASTIN). A substance released by disintegrating tissues and blood platelets which, in the presence of calcium ions, causes the change of prothrombin into thrombin (*q.v.*) and so initiates the process of blood clotting.

THYLACOLEONTIDAE. Large extinct marsupials from the Pleistocene of Australia, resembling phalangers but probably mainly herbivorous. They had the orbits completely surrounded by bone, a feature unique in marsupials but found in certain higher mammals.

THYMUS. An organ, probably an endocrine gland in the neck or thorax of many vertebrates. Besides secreting hormones it probably has the function of making and storing lymphocytes. In fish it forms a series of paired outgrowths from the gill-clefts but in higher vertebrates it is often a single organ on the ventral side of the heart. In birds it appears to regulate the production of the shell, shell-membranes and albumen of the eggs; in mammals it apparently controls the time of onset of puberty and gradually diminishes in size as the animal grows older.

THYROGLOSSAL DUCT. A passage along which secretions flow from the thyroid gland into the mouth in some vertebrates; in man it is vestigial and closes in the second month of gestation.

THYROHYAL (THYROHYOID) CARTILAGES. The hyoid cornua or horns which project backwards from the hyoid plate in the floor of the mouth; present in most air-breathing vertebrates and possibly homologous with the *ceratohyal* cartilages of a fish.

THYROID CARTILAGES. Paired cartilages which, together with the *cricoid* and *arytenoid* cartilages, strengthen the wall of the larynx in mammals.

THYROID GLAND. An endocrine organ situated on the ventral side of the larynx in mammals and in a corresponding position in lower vertebrates. It produces the hormone *thyroxin*, a complex substance containing iodine, whose action is to speed up metabolism in general and the oxidation processes in particular. In the case of amphibia it hastens the onset of metamorphosis.

THYROXIN. The hormone produced by the thyroid gland: an iodine derivative of the amino-acid *tyrosine* whose general effect is to catalyse oxidation in the tissues. In humans excess of thyroxin causes *exophthalmic goitre* and deficiency causes *cretinism*.

THYSANOPTERA. Thrips: minute insects with short antennae and narrow wings fringed with hairs. They are classed among the Hemimetabola since they do not undergo a complete metamorphosis but there is a resting stage which may be regarded as a kind of primitive pupation. They are a major pest of cereals and fruit trees which they pierce in order to suck the sap.

THYSANURA. Bristle-tails: primitive wingless insects with long antennae, long cerci and a median tail-filament. The mouth-parts are of the biting type and resemble those of the cockroach. Their usual habitat is among stones and dead leaves.

TIBIA. (1) The Shin-bone: the anterior of the two bones between the knee and the ankle.

(2) Of Insects: a segment of the leg between the femur and the tarsus.

TIBIALE (ASTRAGALUS). A small bone adjacent to the tibia in the ankle of a tetrapod vertebrate. It is one of the three which form the proximal row of tarsal bones; the other two are the *intermedium* and the *fibulare*.

TIBIO-FIBULA. A fusion of the tibia and the fibula in many vertebrates.

TICKS. Arthropods of the group Acarina which are ectoparasitic and usually blood-sucking. They differ from mites, which are usually free-living scavengers, in that they have the chelicerae enlarged for piercing and have a sucking channel between these and the hypostome. The stomach has numerous pouches which can be greatly distended when full of blood. As in most blood-suckers the saliva contains an anticoagulant substance (see *Acarina*).

TIEDEMANN'S BODIES. Small glands of uncertain function arranged round the water-vascular ring in a star-fish.

TILLODONTIA. Extinct mammals apparently intermediate between rodents and carnivora from the Cretaceous and Eocene systems of America.

TINAMIFORMES (CRYPTURI, TINAMIDAE). Tinamous: birds of Central and South America which, although powerful flyers, are usually placed with the ostrich, emu and other flightless birds on account of certain primitive features of the skull.

TINE. A slender pointed process, *e.g.* on an antler.

TINEIDAE. A family of small moths which includes the clothes moth and a large number of species whose larvae are injurious to trees. The latter include the Larch Shoot Borer, the Larch Leaf Miner and the Ash Bud Moth.

TIPULIDAE. Crane-flies or 'Daddy-long-legs': slender insects with long fragile legs, a tapering abdomen and two long wings. The larvae are a dirty brown colour and have a tough skin which gives them the name of *leather-jacket*. Most of them are soil-dwellers and do considerable damage to the roots of grass and young plants in general.

TISSUE. A part of an organism consisting mainly of cells of one kind bound together by collagen or other intercellular material.

TISSUE RESPIRATION. Internal respiration involving oxidation and release of energy within the tissues themselves; frequently catalysed by special respiratory enzymes.

TITANOTHERIIDAE. Huge extinct ungulates of the Eocene Period, having three toes on the hind-feet and four on the front feet and having two horn-like processes on the nasal bones.

TONGUE. A variably shaped and often movable muscular organ within the buccal cavity, bearing taste-buds or other chemo-sensory receptors.

TONGUE BARS. Bars of cartilage growing downwards between the primary gill-bars and so dividing each gill-slit into two, as for instance in *Amphioxus*.

TONOFIBRILLAE. (1) Supporting fibrils, as for instance of cilia.
(2) Tension fibrils such as those running vertically through the epidermal cells of insects and connecting the cuticle with the underlying muscles.

TONSILS. Masses of lymphatic tissue near the back of the throat in most lung-breathing vertebrates; embryologically derived from the vestigial first pair of gill pouches.

TOPOTAXIS. Orientation followed by movement towards or away from a stimulus. Many protozoa, for instance, show such a reaction in relation to light, food etc.

TORMOGEN. A cell which secretes the socket of a seta or bristle on the body of an insect (see also *Macrotrichia*).

TORNARIA LARVA. The free-swimming larva of *Balanoglossus* and other Hemichordata, having two bands of cilia and two eye-spots.

TORPEDINIDAE. Cartilaginous fish having a flattened body like that of a skate with electric organs between the pectoral fins and the head (see *Narcobatoidea* and *Electric Organs*).

TORSION. A phenomenon exhibited by embryonic Gasteropod molluscs in which the visceral hump, when it reaches a certain stage of development, suddenly rotates completely round to face the opposite direction. This involves a reversal of direction of the heart, the gills and the anus; the visceral loop of the nervous system becomes twisted into a figure eight. It is possible that the operation is a means of correcting the balance of the animal after its early unsymmetrical development (see also *Detorsion*).

TORTRICIDAE. A family of moths whose caterpillars do considerable damage to trees such as oaks, pines and larches, often completely defoliating them. The Oak-roller moth, *Tortrix viridana*, for instance, causes the leaves of the tree to roll up, shrivel and fall.

TORYMIDAE. Chalcid seed-flies: small apocritous Hymenoptera whose larvae infest and destroy the seeds of conifers.

TOXI-, TOXO-. Prefix from Greek *Toxikon* (*pharmakon*): arrow-poison.

TOXIGLOSSA. Gasteropod molluscs whose bite can inflict a toxic wound owing to the fact that the tongue or radula bears elongated hollow teeth down which poisonous saliva flows.

TOXODONTIA. Small or moderate sized extinct herbivorous mammals from the Eocene of South America, having three toes on each foot and walking with a plantigrade gait.

TOXOTIDAE. Spiny-rayed fish of S.E. Asia and Australia having the habit of throwing a drop of water at an insect near the surface to make it fall in.

TRABECULAE. A pair of cartilaginous bars forming the front part of the floor of the cranium in the vertebrate embryo.

TRACH-, TRACHE-, TRACHO-. Prefixes from late Latin *Trachia*: the windpipe, originally from Greek *Tracheia*: rough.

TRACHEA. (1) The main breathing tube leading to the lungs in air-breathing vertebrates.
(2) Tracheae: the breathing tubes of insects, opening by spiracles situated along the sides of the body.

TRACHEAL GILLS. Respiratory organs of certain aquatic insects consisting of thin outgrowths of the body wall containing numerous tracheoles.

TRACHEOLES. Minute branching tubes leading from the tracheae of insects and penetrating to all parts of the body.

TRACHINIDAE (TRACHININAE.) Elongated carnivorous blenny-like fish of coastal regions, having the pelvic fins in the jugular position and often having spines which can inflict poisonous wounds.

TRACHOMEDUSAE. Jelly-fish having an umbrella-like form with marginal tentacles round the rim; with sense-organs in pits or vesicles, and with gonads situated on the radial canals. They belong to the order Trachylina (see below).

TRACHYLINA. An order of jelly-fish belonging to the class Hydrozoa but differing from others of this group by reason of the fact that the hydroid stage is suppressed and the medusa develops directly from the egg.

TRACTELLUM. A flagellum which pulls a small organism forwards, as for instance in *Euglena* and similar protozoa.

TRAGULIDAE. Chevrotains: small hornless four-toed ungulates of Asia and Africa, having no teeth on the premaxilla and having a three-chambered stomach. They are apparently intermediate between the ruminants and the other artiodactyla.

TRANSPALATINE BONES. Lateral bones on each side of the palatines in many reptiles. They connect the pterygoids with the maxillae and are homologous with the ectopterygoid bones of some fish.

TRANSVERSE LIGAMENT (OF ATLAS). A ligament stretched horizontally across the central cavity of the atlas vertebra, separating the spinal canal from the *Odontoid fossa* into which the axis vertebra fits.

TRANSVERSE PROCESSES. Lateral processes of vertebrae with which the ribs, if present, articulate.

TRAPEZIUM. The first distal carpal bone of the human wrist; sometimes used also for that of other animals.

TRAPEZOID. The second distal carpal bone of the human or mammalian wrist (*cf. Trapezium*).

TREMATODA. Flukes: parasitic Platyhelminthes having a thick spiny skin, an oral and a ventral sucker and a branched alimentary canal without an anus. Like other flatworms they are hermaphrodite with a complex life-cycle involving a series of larval forms in different hosts.

TRI-. Prefix from Greek *Tria*: three.

TRIAENE SPICULE (OF SPONGES). A four-pointed spicule with three short equal points and one point longer than these (see *Spicules*).

TRICHECHIDAE. Walruses: members of the Pinnipedia or aquatic carnivora closely resembling the eared seals or sea-lions but very much larger and having the upper canines modified into tusks.

TRICHIURIDAE. Hair-tails: Tropical fish of the mackerel family, having a laterally compressed body and fine tail.

TRICHO-. Prefix from Greek *Trix, trichos*: a hair.

TRICHOBRANCHIA. The type of gill found in crayfish and similar crustacea, consisting of a central axis with numerous lateral filaments, the whole being formed from an outgrowth of skin from the upper part of the legs or from the sides of the body.

TRICHOCYSTS. Sensitive structures in the ectoplasm of certain protozoa such as *Paramecium*. They consist of small cavities from which a long thin thread can be ejected in response to the stimulation of touch or of certain chemicals. Their function is uncertain.

TRICHODONTIDAE. Spiny-rayed fish of sandy shores in the North Pacific, having slender sharp teeth in bands on the jaws and on the vomer.

TRICHOGEN. An epidermal cell which produces a cuticular hair or scale on the body or wing of an insect.

TRICHOGLOSSIDAE. A group of parrots characterized by the presence of fine horny fibres on the tip of the tongue. They include the lories of S.E. Asia and Polynesia, the Kea and the Kaka of New Zealand and many others.

TRICHOPTERA. Caddis-flies: moth-like insects whose wings are covered with hairs and whose mouth-parts are so reduced that little food can be taken. The aquatic larvae, which are usually herbivorous, live in tubular cases formed of particles of wood, sand, small shells etc. A pair of hooks on the hind-end enables it to cling to the tube and drag it around when walking. Pupation takes place within the tube after the ends have been closed with silk.

TRICLADIDA. Turbellarians (free-living flatworms) in which the mouth is situated mid-ventrally and the gut is divided into three main branches (see *Planarians*).

TRICONODONT. Elongated molar or premolar teeth having three cone-like cusps arranged in a line.

TRICONODONTIDAE. Extinct marsupials of the Jurassic Period having triconodont teeth.

TRICUSPID VALVE. A valve with three cusps or membranous flaps, situated between the right auricle and the right ventricle in the mammalian heart.

TRIGEMINAL NERVE. The fifth cranial nerve of a vertebrate, usually having an ophthalmic, a maxillary and a mandibular branch.

TRIGLIDAE. Gurnards: spiny-rayed fish usually covered with an armour plating of thick bony scales. The pectoral fins are very large and are used for locomotion on the bottom. Three of the rays of these fins are detached from the others and are extremely sensitive to touch. They resemble fingers and are used for raking the sand in search of food. Some gurnards can make a grunting noise apparently from the swim bladder.

TRIGLIIFORMES. Mail-cheeked fish (see *Scleroparei* and *Triglidae* above).

TRIGONODONT. Teeth, usually molars, having three cusps arranged in the form of a triangle.

TRILOBITA. Fossil arthropods of Cambrian and Silurian times, having a flattened oval segmented body with numerous biramous appendages and a pair of antennae.

TRINGINAE. Curlews, redshanks and similar marsh-birds having long narrow beaks and feeding chiefly on insects, worms, crustacea, molluscs and berries. They are good flyers and waders but do not habitually swim. The young are *praecoces*, *i.e.* fully fledged and able to feed themselves as soon as they hatch.

TRIODONTIDAE. Fish of the Indian Ocean having the skin covered with small bony spines, having a dilatable abdomen and having the teeth fused to form a beak.

TRIONYCHOIDEA. Mud-tortoises found in the rivers of Asia, Africa and North America. They are small and carnivorous with a soft flat leathery carapace and with webbed feet each bearing three claws.

TRIPLOBLASTIC. Having three primary embryonic cell-layers, *viz.* the *ectoderm, mesoderm* and *endoderm*. All animals other than protozoa, sponges and coelenterates are of this type.

TRIPYLAEA. See *Phaeodaria*.

TRITOCEREBRUM (METACEREBRUM). That part of the brain of an arthropod formed by the fused ganglia of the third somite.

TRITUBERCULAR TEETH. Trigonid or trigonodont: teeth with three cusps.

TRITUBERCULATA. An extinct family of marsupials of the Jurassic Period having tritubercular teeth. They were probably related to the common ancestors of marsupials and placental mammals.

TRIUNGULIN. TRIUNGULUS. The minute six-legged parasitic larva of certain insects, notably the Strepsiptera and the Cantharidae (oil-beetles).

TROCH-. Prefix from Greek *Trochos*: a wheel.

TROCHAL DISC. The flattened anterior end of a Rotifer on which are usually two rings of cilia sending food particles into the mouth.

TROCHANTER. (1) One of several large processes for attachment of muscles near the head of the femur of a vertebrate.

(2) A short segment of an insect's leg between the coxa and the femur.

TROCHANTIN. A small basal articular sclerite on the trochanter of an insect.

TROCHILIDAE. Humming birds: brilliantly coloured birds of America and the West Indies, closely related to swifts and including some of the smallest birds in the world. The plumage is variegated with a metallic lustre; the beak is long and awl-shaped and the long cleft tongue can be projected in the form of a double tube. They are mostly insectivorous, spending their time flitting from flower to flower in search of food.

TROCHLEA. A pulley-like groove on a bone, as for instance that on the distal end of the humerus over which the articular surface of the ulna moves.

TROCHLEAR NERVE (PATHETIC NERVE). The fourth cranial nerve of a vertebrate, innervating the superior oblique muscle of the eye.

TROCHOPHORE. See *Trochosphere larva* below.

TROCHOSPHERE LARVA. A planktonic larva, more or less round, having a short gut with mouth and anus and a pre-oral ring of cilia known as the *prototroch*. It is the characteristic larva of many invertebrates including polychaete worms and some molluscs.

TROCHUS. The inner ciliated part of the trochal disc of a Rotifer.

TROGLODYTIDAE. Wrens: small passerine birds with short upturned tails, found in many parts of the world but principally in Central America.

TROGONES (TROGONIDAE). Brightly coloured birds of tropical forests, having the first and second toes directed backwards and the third and fourth forwards. The beak is short with bristles in the corners of the wide mouth; they feed on insects and fruit.

TROPH-, TROPHO-. Prefix from Greek *Trophein*: to nourish.

TROPHALLAXIS. The mutual interchange of food and of salivary or other secretions between the adults and the young of certain social insects.

TROPHAMNION. A sheath or membrane surrounding the eggs of some parasitic insects; able to absorb food material from the tissues of the host.

TROPHI. TROPHIC ORGANS. Mouth-parts or any organs connected with feeding.

TROPHOBIONT. See *Trophobiosis* below.

TROPHOBIOSIS. Symbiosis in which the two organisms or trophobionts reciprocally feed one another.

TROPHOBLAST. A layer of cells enclosing the early mammalian embryo and later forming papillae or *trophoblastic villi* by which the embryo becomes implanted in the wall of the uterus of the parent.

TROPHOBLASTIC VILLI. See *Trophoblast*.

TROPHOCHROMATIN. Chromatin of the *meganucleus* which apparently controls the vegetative life of certain protozoa. It periodically degenerates and is replaced from the *idiochromatin* of the micronucleus, a process which apparently revitalizes the whole organism. This phenomenon occurs in most ciliated protozoa which have two nuclei; in other protozoa both kinds of chromatin may be present in a single nucleus.

TROPHOCYTES. Cells containing a reserve of fat or other food substance, as for instance in the body of an insect.

TROPHONEMATA. Trophoblastic villi (see *Trophoblast*).

TROPHONUCLEUS. The larger nucleus or *meganucleus* in a binucleate protozoon; the nucleus which helps to regulate metabolism other than cell-division (see *Trophochromatin*).

TROPHOTAXIS (TROPHOTROPISM). Chemotaxis in which food is the stimulating agent.

TROPHOTHYLAX. A food pocket in the first abdominal segment of an ant larva.

TROPHOZOITE. The adult or feeding stage in a member of the Sporozoa (parasitic spore-forming protozoa).

TROPHOZOOID. A nutritive polyp or zooid in coral and similar colonial coelenterates.

TROPIBASIC SKULL (TROPITRABIC). A skull in which the cranium is formed from trabeculae or cartilaginous bars which are close to one another and fuse in the middle line (*cf. Platybasic*).

TROPITRABIC. See *Tropibasic*.

TRUNCUS ARTERIOSUS. A short thickened ventral aorta from which the main arterial arches branch off on either side, as for instance in a frog.

TRUNK GANGLION (OF TUNICATES). A large ganglion in the trunk of an ascidian tadpole, degenerating in the adult.

TRYGONIDAE. Sting-rays: flattened cartilaginous fish with broad pectoral fins and with a long tail ending in a sharp spine which can cause severe wounds.

TRYPANEIDAE. A family of cyclorrhaphous fruit-flies, the larvae of which are extremely destructive, tunnelling into the pulp of the fruit.

TRYPANOSOMIDAE. Parasitic flagellate protozoa belonging to the order Protomonadina, having one flagellum which is usually united to the body by an undulating membrane; some species however pass through stages in which the membrane or the flagellum may be lost.

Included in the family are some of the most dangerous parasites of man and domestic animals. African sleeping sickness, Oriental sore and Kala-azar are examples of human diseases caused by them.

TRYPSIN. An enzyme or enzyme-complex able to digest proteins in alkaline solution. In mammals it consists of at least two co-enzymes, *viz. trypsinogen* secreted by the pancreas and *enterokinase* secreted by the walls of the duodenum.

TRYPSINOGEN. See *Trypsin*.

TUBE FEET. Locomotory organs of Echinoderms consisting of elongated outgrowths of the body wall, able to be protruded or retracted by alterations of the fluid-pressure in the water-vascular system. In star-fish they are arranged in rows on each side of the *ambulacral grooves* which run radially along each of the five arms on the ventral side.

TUBER CINEREUM. A projection of grey matter from the floor of the brain in the region where the hypothalamus is joined to the stalk of the pituitary gland.

TUBERCULO-SECTORIAL TEETH. Molar teeth, usually of the lower jaw, having three cusps and a posterior 'heel' or talonid.

TUBERCULUM (TUBERCLE). Any small rounded protuberance or swelling.

TUBERCULUM OF RIB. One of the two 'heads' of the rib in birds, mammals and some reptiles. This type of rib is the normal or *dichocephalic* type articulating in two places with the corresponding thoracic vertebra. The tuberculum articulates with the transverse process; the capitulum with the centrum.

TUBERCULUM ACUSTICUM. The *ear-brain*: a bulging part on the dorsal side of the medulla oblongata containing a large number of neurones and afferent fibres from the seventh, eighth, ninth and tenth cranial nerves. These form the so-called *acustico-lateralis* system.

TUBEROSITY. A projection of a bone, to which a muscle or tendon may be attached.

TUBINARES. An alternative name for the *procellarian* group of birds (petrels and albatrosses). The name indicates that they have tubular nostrils.

TUBULE OF KIDNEY. See *Uriniferous tubule*.

TUBULIDENTATA. See *Orycteropodidae*.

TUNICA ALBUGINEA. A capsule of dense white fibrous tissue enclosing the testis, ovary or other organs.

TUNICATA. Urochorda, urochordata, ascidians or sea-squirts: degenerate or highly specialized members of the phylum Chordata having no obvious resemblance to other members of this group. The body is not segmented and there is no bony skeleton. The larva, the so-called 'ascidian tadpole', has a notochord in the tail region only. The adult has a perforated pharynx or branchial chamber with numerous gill-clefts through which water passes into a dorsal atrium. The whole organism is enclosed in a *test* or outer covering of *tunicin*, a material resembling cellulose.

TUNICA VAGINALIS. A layer of pavement epithelium lining the cavity of the scrotal sac and extending round the testes in mammals.

TUNICIN. A substance related to cellulose, forming the *test* or outer skin of a tunicate.

TUPAIIDAE. Insectivorous tree-shrews from S.E. Asia, characterized by having a large cranium, closed orbits and well developed zygomatic arches.

TURACIN. A red pigment containing copper, found in the feathers of certain birds (see *Musophagidae*).

TURACOVERDIN. A green pigment containing iron, found in the feathers of certain birds (see *Musophagidae*).

TURBELLARIA. Free-swimming Platyhelminthes (flatworms) having a ciliated body without suckers but otherwise closely resembling parasitic flukes in their general anatomy.

TURBINAL BONES (TURBINATE BONES, SCROLL BONES). Thin coiled membranous bones in the nasal cavities of mammals, providing a large area for the mucous membrane which not only contains sensory cells but also warms and filters the in-going air.

TURDIDAE. Thrushes, blackbirds, warblers and similar passerine birds having a fairly long, slightly compressed beak with vibrissae at the base. They are omnivorous and are found in most parts of the world.

TURNICIDAE. Small fowl-like birds of Africa and India; solitary, non-migratory and able to run quickly. The males incubate the eggs.

TYLOPODA. An alternative name for the *Camelidae*.

TYMPANAL ORGANS (INSECTS). Auditory organs on the abdomen or legs of certain insects such as grasshoppers and cicadas. In these organs there is a tympanic membrane formed from the cuticle, an underlying air space acting as a resonator and a group of specialized receptors which can detect high frequency vibrations.

TYMPANIC BONE. A bone found only in mammals, forming the base of the auditory capsule and of the external auditory meatus. It is homologous with the *angular bone* of the reptilian lower jaw.

TYMPANIC BULLA. A downward bulge or enlargement of the tympanic bone within which is a space containing the auditory ossicles. It is peculiar to mammals.

TYMPANIC CAVITY. The middle ear: an air-filled cavity between the tympanic membrane and the inner ear in most land vertebrates; connected to the back of the throat by the *Eustachian tube*. In mammals it contains three ossicles (*malleus, incus* and *stapes*), but in lower vertebrates it only contains one, the *columella auris*, by means of which vibrations are transmitted from the tympanic membrane to the inner ear.

TYMPANIC MEMBRANE. The outer membrane of the ear-drum: a thin diaphragm of fibrous tissue covered on the outside by skin and on the inside by a thin layer of pavement epithelium. Sound waves cause it to vibrate and this movement is transmitted across the cavity of the middle ear by the auditory ossicles.

TYMPANOHYALS. A pair of small bones in the mammalian skull, forming the upper parts of the hyoid arch and fused to the cranium just behind the tympanic bones.

TYMPANUM. (1) The ear-drum: the cavity of the middle ear in land vertebrates.

(2) A resonating chamber formed by the enlarged junction of the trachea and the two bronchi in birds. It is traversed by a vertical bony septum or *pessulus* from which hang a number of processes and membranes forming the vocal apparatus or *syrinx*.

TYPHLOPIDAE. Burrowing snakes having a vestigial pelvic girdle and having reduced eyes covered by scales; found in most parts of the world except New Zealand.

TYPHLOSOLE. A dorsal infolding of the intestine of an earthworm along its whole length, giving an increased internal surface for digestion and absorption.

TYPOTHERIA. Small extinct pentadactyl plantigrade mammals showing characters intermediate between ungulates and rodents and having an opposable hallux.

TYRANNIDAE. King-birds or Tyrant-birds of North America: fierce birds which feed largely on flies, beetles or other insects but sometimes eat mice or frogs which they beat to death on a branch. The Crested Tyrant-bird lines its nest with the cast-off skins of snakes.

U

ULNA. The posterior bone of the fore-arm, parallel to the radius and articulating with the humerus at the elbow.

ULNARE. A small carpal bone adjacent to the ulna: one of the three in the proximal row of carpals, the other two being the *intermedium* and the *radiale*.

ULOTRICHOUS. Having curly hair or wool.

UMBILICAL CORD. (1) A vascular cord of complex structure connecting the abdomen of an unborn mammal to the placenta.

(2) A similar cord in birds and reptiles connecting the embryo with the yolk sac.

UMBILICUS. (1) The Navel: a depression in the abdomen left by the separation of the umbilical cord of a mammal after birth.

(2) Of Feathers: a small hole at each end of the hollow stem of a quill-feather. At the proximal end is the *inferior umbilicus*; at the distal end the *superior umbilicus*.

UMBO. The central or oldest part of a bivalve shell. As the shell grows, concentric rings are added round the umbo.

UMBRELLAR SURFACES. The *exumbrellar* and *subumbrellar* (upper and lower) surfaces of a medusa or jelly-fish.

UNCIFORM BONE. A hook-shaped bone of the mammalian carpus formed by fusion of the fourth and fifth distal carpals; more particularly used of the human wrist but sometimes also of other mammals.

UNCINATE PROCESSES. Small processes projecting backwards from the ribs of birds and some reptiles, each overlapping the next posterior rib and so helping to strengthen the wall of the thorax.

UNCINI. Hooked chaetae or bristles by which certain polychaete worms can cling to the insides of the tubes in which they live.

UNDULATING MEMBRANE. A protoplasmic membrane used for locomotion or feeding in certain protozoa. There are two types of membrane similar in appearance but of different origin. In ciliated protozoa, such as *Paramecium*, it is formed from a number of fused cilia in the neighbourhood of the mouth and is used for feeding. In flagellates, such as *Trypanosoma*, it connects the flagellum with the body of the organism and is used for locomotion.

UNGUAL. Possessing nails, hoofs or claws.

UNGUL-. Prefix from Latin *Ungula*: a hoof.

UNGULATA. A large and widely distributed order of herbivorous mammals having hoofs. There are two suborders (more recently ranked as separate orders), *viz. Perissodactyla* or 'odd-toed' including horses, rhinoceros etc. and *Artiodactyla* or 'even-toed' including sheep, cattle and pigs.

UNGULATE. Having hoofs: a member of the ungulata.

UNGULIGRADE. Animals which walk on their hoofs with the rest of the foot off the ground, as for example the horse.

UNI-. Prefix from Latin *Unus*: one.

UNICELLULAR ORGANISMS. Organisms consisting of only one cell. Some of these are definitely classed in the animal kingdom (protozoa); others among plants as simple algae or fungi. Since the distinction often consists merely of the presence or absence of chlorophyll in two organisms which are otherwise similar, the term *Protista* is often used to comprise both types.

UNIPAROUS. Producing only one at a birth.

UNIPOLAR NEURONE. A nerve-cell having two processes or fibres arising from the same point and appearing as one.

UNISETOSE. Possessing one bristle or seta.

UNISEXUAL. An organism bearing male or female organs but not both.

UNIVALVE. A mollusc whose shell is in one piece, *e.g.* a snail.

UNSTRIATED MUSCLE (UNSTRIPED MUSCLE). Involuntary muscle consisting of elongated tapering cells and not of parallel fibres like those of voluntary muscle. It is found in the walls of the alimentary canal and in other parts where slow involuntary movements take place.

UPUPIDAE. Hoopoes: beautifully coloured birds having a long laterally compressed beak, short triangular tongue and long rounded wings. They are found in Europe, Africa and Asia.

URANOSCOPINA. Star-gazers: blenny-like fish with pelvic fins in the jugular position and with eyes on the upper surface of the head.

UREA. $CO(NH_2)_2$, a soluble nitrogenous waste substance excreted by many animals.

UREOTELIC ANIMALS. Animals whose chief excretory product is urea, *e.g.* fish, amphibia and mammals (*cf. Uricotelic*).

URETER. A tube along which urine flows from the kidney to the bladder.

URETHRA. A tube conducting urine from the bladder of a mammal to the exterior.

URICOTELIC ANIMALS. Animals which excrete uric acid instead of urea, *e.g.* most reptiles, birds, insects, snails etc. These animals develop within a shell and would be unable to get rid of soluble urea during the embryonic stage. Uric acid, however, being less soluble can be retained in the organism or in the shell.

URINARY BLADDER. A sac in which urine is stored prior to discharge: an enlargement of the kidney-duct or of the hind-gut or cloaca.

URINE. The waste liquid excreted by an animal, usually containing nitrogenous constituents such as urea, uric acid or ammonium salts.

URINIFEROUS TUBULES. Narrow tubes, often of great length and with several convolutions, conveying urine from the Malpighian bodies of the vertebrate kidney to the ureters (see also *Malpighian Body*).

URINOGENITAL DUCT. A common duct for the passage of excretory and genital products. In frogs and some fish, for example, the *mesonephric* or *Wolffian duct* is a urinogenital duct because the kidney and the testis are both formed from the tubules of the same embryonic organ, the *mesonephros*.

URINOGENITAL SINUS. A chamber or sac into which the urinogenital ducts open.

URO-. Prefix from Greek *Oura*: a tail.

UROCARDIAC OSSICLE. The hindmost ossicle in the anterior chamber of the stomach (proventriculus) of a crayfish or similar crustacean (see *Gastric Mill*).

UROCHORDA. UROCHORDATA. See *Tunicata*.

URODAEUM. A distinct region of the cloaca into which the ureters and genital ducts open in some reptiles and birds.

URODELA (CAUDATA). Newts, salamanders etc. Smooth skinned amphibia with a well developed tail and usually with two pairs of limbs. External gills are present in the larva and may or may not be retained by the adult.

UROPELTIDAE. Small burrowing snakes of Ceylon and South India, having the cranial bones solidly united and having the tail ending in a large shield.

UROPLATIDAE. Lizards from Madagascar resembling geckos except that the nasal bones are fused, the interclavicle small and the clavicles not dilated.

UROPODS. The last pair of abdominal appendages of a crustacean such as a lobster or crayfish; frequently flattened and forming part of the tail-fan.

UROPYGIAL GLAND. The 'preen-gland' on the dorsal side of a bird's tail, secreting oil with which the bird makes its feathers waterproof.

UROPYGIUM. The short tail-stump of a bird.

UROSOME. (1) The hind-part or 'abdomen' of an arthropod.

(2) More specifically, the hind-part of a crustacean such as *Cyclops* in which there is no true abdomen but a certain number of segments at the hind-end are modified to form a 'tail'.

UROSTYLE. A bony rod formed from a number of fused vertebrae at the hind-end of the vertebral column in frogs and toads. Fused to the pelvic girdle it gives greater rigidity to the latter when the frog is jumping.

URSIDAE. Bears: large plantigrade animals with elongated skull, omnivorous in habit although belonging to the order Carnivora. They are found in all regions except Africa and Australasia. Bears were formerly put in the super-family Arctoidea but are now generally included in the Canoidea or dog-like carnivores.

UTERINE MILK. A nutrient secretion from glands in the wall of the uterus, providing the first food for a developing mammalian embryo before the formation of the placenta.

UTERINE VILLI. Structures resembling villi by which a mammalian placenta is embedded in the wall of the uterus.

UTERUS. The womb: a muscular enlargement of the two oviducts forming a paired or a single sac in which the embryo develops in female mammals. The two halves lead into a median vagina which opens to the exterior by the vulva.

UTERUS MASCULINUS. A sac-like structure dorsal to the bladder in the male mammal, homologous with the uterus of the female.

UTRICLE. UTRICULUS. A part of the membranous labyrinth of the ear consisting of a sac into which the three semicircular canals open.

UVEA. A layer of pigmented cells covering the back of the iris in the eye of a vertebrate.

UVULA. A sensitive fleshy process hanging from the back of the soft palate in mammals.

V

VACUOLARIA. Fresh-water flagellate protozoa of the order Chloromonadina having a complex contractile vacuole and numerous chloroplasts. They can carry out photosynthesis and for this reason are sometimes classed as Algae; much of their life is passed in the *palmella* stage in which many individuals are agglutinated together to form a colony.

VACUOLES. (1) Contractile vacuoles of Protozoa. Spaces which periodically enlarge as they fill up with water and then suddenly collapse expelling their contents to the exterior. They may be single or may consist of a number of subsidiary vacuoles leading into the principal one. Their main function is apparently to get rid of an excess of water which enters the organism by osmosis; this theory is supported by the fact that they are rarely present in marine protozoa.

(2) Food vacuoles: spaces in the bodies of protozoa and in certain cells of other organisms where ingested food particles are broken up by enzyme action and absorbed. They are usually temporary and are often formed by pseudopodia which engulf a food particle together with a drop of water, as in *Amoeba*. In more complex protozoa such as *Paramecium* they may be formed at the base of the so-called 'gullet' from which they periodically become detached, move along a complex path through the endoplasm and finally discharge their contents at a temporary 'anus'.

VAGINA. (1) A passage leading from the uterus to the exterior in a female mammal.

(2) A similar dilatation of the oviduct in some invertebrates.

VAGUS. The tenth cranial nerve of a vertebrate, having numerous branches to the various viscera; in fish also to the lateral line organs and to the last four pairs of gills. The visceral branches to the heart, alimentary canal etc. help to regulate unconscious automatic activities. They form part of the parasympathetic system.

VALVE. (1) Any structure which permits the passage of a fluid in one direction only, as in the heart, veins etc.

(2) In Mollusca, Brachiopoda and some Crustacea: one of the units which go to make up the shell.

VALVE OF VIEUSSENS. A thin transverse band of white matter extending between the two anterior peduncles of the cerebellum immediately behind the corpora quadrigemina.

VALVULA CEREBELLI (FORNIX OF GOTTSCHE). A projection of the cerebellum downwards and forwards beneath the roof of the mid-brain in bony fish.

VALVULAE CONNIVENTES. Large folds of mucous membrane in various parts of the alimentary canal, giving an increased area for secretion and absorption.

VALVULA PARADOXA. A small semilunar valve of doubtful function at the entrance to each systemic aorta in amphibia.

VANE (VEXILLUM). The flat distal part of a feather consisting of barbs and barbules linked together by small hooks.

VANNUS. The anal or posterior lobe of an insect's hind-wing: a fan-like expansion separated from the rest of the wing by a furrow.

VARIATION. See under *Mutation* and *Continuous Variation*.

VAS DEFERENS. The main duct by which seminal fluid passes from the testis to the exterior.

VASA EFFERENTIA. Small tubules leading from the testis into the *vas deferens (q.v.)*.

VASCULAR SYSTEM. (1) The blood system, consisting usually of heart, arteries, capillaries and veins.

(2) Water vascular system of Echinoderms: a system of circular and radial canals, usually open to the exterior by a porous plate known as the *madreporite*, supplying water to the tube-feet of a star-fish or similar echinoderm.

VECTOR. An animal which carries parasites or pathogenic organisms to the host, *e.g.* a mosquito.

VEGETATIVE PHASE. The part of a life-cycle in which reproduction is asexual, *e.g.* by fission or by budding.

VEGETATIVE POLE (VEGETAL POLE). The part of an egg which is opposite to the nucleus and contains most of the yolk. Eggs of this type are known as *telolecithal* and are typical of chordates.

VEINS. Vessels which convey blood from the various parts of the body back to the heart.

VELAR. Pertaining to or attached to the velum (*q.v.*).

VELAR TENTACLES. Of *Amphioxus*: ciliated tentacles situated on the velum round the edges of the enterostome.

VELIGER LARVA. An early larval stage of some mollusca: a phase which starts like the trochosphere larva of an annelid but later develops rudimentary organs including a thin shell. Feeding is by means of a ciliated membrane or *velum* which sweeps food particles into the mouth.

VELUM. A name given to various membranous structures.

   e.g. (1) Of Coelenterates: a narrow shelf or membrane running round the inner surface of the rim of a medusa (jelly-fish).

   (2) Of *Amphioxus*: a transverse membrane pierced by the mouth or enterostome at the hind-end of the oral hood.

   (3) Of Mollusca: see under *Veliger Larva* above.

VELUM TRANSVERSUM, A vascular membrane, formed by an in-pushing of the *pia mater*, running transversely across the brain of a vertebrate and separating the telencephalon from the thalamencephalon.

VENAE CAVAE (CAVAL VEINS). The main veins which return blood to the heart in higher vertebrates. There are generally two anterior venae cavae returning blood from the head and fore-limbs, and one posterior vena cava receiving blood from the body and posterior limbs. In mammals and birds these veins lead into the right auricle; in amphibia and reptiles they lead into the *Sinua venosus*.

VENATION (INSECTS). The arrangement of the veins in an insect's wing: often an important means of identifying species. For the purpose of comparison the veins of a generalized wing are named as follows: The *costa*, a thick unbranched vein forming the anterior margin of the wing; the *subcosta*, another unbranched vein immediately below the costa; the *radius, media* and *cubitus* which are usually branched; the *anal* veins, short and unbranched. All these run longitudinally and may be linked by a variable number of *cross-veins* dividing the wing into 'cells'. The latter are very numerous in mayflies (Ephemeroptera), dragonflies (Odonata) and in some primitive fossil insects, but the evolutionary tendency seems to have been towards a reduction in their number.

VENTRAL. The under-side of an animal: that which is normally facing downwards. Usually there is no difficulty in deciding which side is ventral and which dorsal but in some groups, *e.g.* star-fish, the side which is ventral in the embryo or larva may take up some other position in the adult. In such cases it is preferable to use other names such as *oral* and *aboral* rather than ventral and dorsal.

VENTRALIA. *Basiventrals* and *Interventrals*: bony or cartilaginous elements forming the ventral part of a vertebra. In embryos and in some primitive vertebrates they are separate and distinct, but in the adults of most vertebrates they become fused with others known as the *basidorsals* and *interdorsals* to form a complete vertebra.

VENTRAL ROOTS. Nerve-roots coming from the ventral side of the brain or spinal cord and containing motor fibres.

VENTRAL TUBE (OF INSECTS). A tube formed of two fused ventral appendages on the first abdominal segment in insects of the order *Collembola* (springtails). It is probably used for sucking up water.

VENTRICLES. (1) The contractile or muscular chambers of a heart by which blood is pumped into the arteries.

(2) The brain-cavities of a vertebrate, *viz.* the first and second ventricles in the cerebral hemispheres; the third ventricle in the thalamencephalon and the fourth ventricle in the medulla oblongata.

VENTRICULUS. (1) A ventricle (see above).

(2) The gizzard of a bird.

(3) The stomach or mid-gut of an insect etc.

VERMES. Worms: a name given by Linnaeus to any worm-like creature. The group was a purely artificial one and included species which are now placed in many different phyla such as Annelida, Platyhelminthes, Nemertea etc.

VERMIFORM APPENDIX. The narrow terminal part of the caecum in some mammals. Herbivorous animals usually have a larger caecum and appendix than carnivores since it is in these parts that fibres of wood and cellulose are broken down by the action of symbiotic bacteria.

VERMILINGUA. An alternative name for the Myrmecophagidae or South American ant-eaters, indicating that the tongue is long and worm-like.

VERMIS. A worm-like structure forming the median part of the cerebellum in birds and mammals.

VERSON'S GLANDS. The moulting glands of arthropods: epidermal glands which secrete a fluid having the property of liquefying the endocuticle and thus loosening the epicuticle prior to moulting.

VERTEBRA. See *Vertebral Column* below.

VERTEBRAL COLUMN. The 'backbone'. The most characteristic feature of vertebrates, consisting of a series of small bones or cartilages running the length of the body near the dorsal side, enclosing and protecting the spinal cord.

VERTEBRARTERIAL CANALS. Holes for the passage of the vertebral arteries through the cervical vertebrae of mammals, birds and reptiles. Each hole is formed by a short double-headed rib fused to the side of the vertebra so that the canal is actually between the vertebrae and the ribs but appears to go through the vertebrae.

VERTEBRATA (VERTEBRATES). All animals which have a vertebral column, *i.e.* fish, amphibians, reptiles, birds and mammals. The division of the animal kingdom into vertebrates and invertebrates, though retained for convenience, is now considered to be somewhat artificial. The notochord is regarded as a more fundamental feature than the vertebral column and the name *Chordata* is used to include not only all the vertebrates but also all animals which have a notochord at any stage in their development. In this way a number of lower groups (protochordata) are grouped with the vertebrates in the phylum Chordata, since they have many characteristics in common although possessing no vertebral column.

VESICULAR NUCLEUS. The type of nucleus which consists of a nuclear membrane filled with fluid and without the usual fine network of chromatin or plastin. There may be a coarse meshwork and sometimes a karyosome or central mass.

VESICULA SEMINALIS. The seminal vesicle: an organ in the male body in which sperm or semen is stored.

VESPERTILIONIDAE. Small insectivorous bats without nasal appendages and with a long tail contained in the large interfemoral membrane. They are found in almost all parts of the world.

VESPIDAE. The social wasps.

VESPIFORMIA. A group of Hymenopterous insects including both ants and wasps; a larger group than the vespidae.

VESPOIDEA. A superfamily of Hymenoptera including all the social and solitary wasps and hornets.

VESTIBULATA (HYMENOSTOMATA). Ciliated protozoa such as *Paramecium* having a mouth with a gullet or vestibule permanently open and usually with an undulating membrane.

VESTIBULE. Any passage or cavity which leads into another, *e.g.*:
(1) Protozoa: a depression in the ectoplasm leading to the 'mouth', as for instance in *Paramecium*.
(2) Mammals: the cavity leading from the vulva to the vagina in a female mammal.

VESTIGIAL ORGAN. A small or imperfectly developed organ which may have lost or changed its function in the course of evolution, *e.g.* the vermiform appendix in Man.

VEXILLUM. The vane of a feather (*q.v.*).

VIBRACULUM (OF POLYZOA). A specially modified polyp in a colony of Polyzoa, like a long bristle which moves backwards and forwards and prevents parasites or other organisms from settling on the colony.

VIBRISSAE. Whiskers: stiff sensitive hairs on the faces of many mammals.

VIDIAN NERVE. A nerve of the autonomic system in higher vertebrates linking the anterior part of the sympathetic chain with the spheno-palatine ganglion.

VILLI. (1) *Intestinal villi:* numerous small finger-like projections on the inner walls of the duodenum and ileum in many vertebrates, providing a large surface through which digested foods are absorbed.
(2) *Trophoblastic villi:* projections similar to the villi of the intestine, but occurring in the trophoblastic region of the mammalian placenta. They fit into corresponding crypts in the uterine wall, thereby making a large area of contact through which parental blood-plasma can diffuse from the uterus to the placenta.

VIPERIDAE. Vipers: poisonous snakes having movable quadrate bones suspended from the loosely attached squamosals, and having short erectile maxillae bearing a pair of large perforated fangs. Most are viviparous and include genera which may be terrestrial, semi-aquatic, arboreal or burrowing. They are found in most parts of the world.

VIPERINAE. Poisonous vipers and adders of the Old World, differing from those of the New World in not having sensory pits on the sides of the snout (see *Viperidae*).

VISCERAL ARCHES. (1) Partitions or septa between adjacent gill-slits of a fish.
(2) The bony or cartilaginous skeletal arches lying in these partitions.

VISCERAL CLEFTS. Openings leading from the pharynx to the exterior in

Chordates; in fish they form the gill-slits but in land animals the embryonic clefts usually become atrophied in the adult.

VISCERAL HUMP. An enlargement of the body on the dorsal side of a mollusc, usually accommodating the main visceral organs and allowing the ventral part to be used as a foot for locomotion. The visceral hump is covered by the mantle which secretes the shell and in many cases has a spiral shape which greatly increases its size. It is exemplified in the snail.

VISCERAL MASS. The part of an organism where most of the viscera are concentrated.

e.g. (1) The visceral hump of a mollusc (see above).

(2) Of parasitic crustacea: an irregular mass containing the rudiments of internal organs in such degenerate crustacea as *Sacculina* which parasitizes and causes castration in crabs.

VISCERAL POUCHES. See *Gill Pouches.*

VISCERAL SKELETON. The visceral arch skeleton: the bony or cartilaginous elements which make up the *branchial arches* (supporting the gills), the *hyoid arch* (floor of the mouth) and the *mandibular arch* (lower jaw) of a fish. The name may also be used for homologous parts of the skeleton in higher vertebrates.

VISION. See *Eye.*

VISUAL PURPLE (RHODOPSIN). A pigment produced by the retinal rods in the eyes of vertebrates. The breaking down and bleaching of this substance by a photochemical action causes stimulation of the optic nerves (see *Rods*).

VITELLARIUM. A collection of vitelline glands (*q.v.*) usually leading into a common duct.

VITELLINE DUCT. The duct from a vitelline gland.

VITELLINE GLAND. A gland which secretes yolk round the eggs in those animals in which the yolk is not part of the ovum itself: *e.g.* in many invertebrates.

VITELLINE MEMBRANE. A membrane surrounding and secreted by an ovum; sometimes known as the *primary egg membrane*, as distinct from the secondary membrane (secreted by the ovary) and the tertiary membranes (secreted by the oviducts).

VITRELLAE. Cells which secrete the crystalline cone in each ommatidium or optical unit of a compound eye.

VITREOUS HUMOUR. A transparent gelatinous substance filling the space behind the lens in the eye of a vertebrate.

VIVERRIDAE. Civets and mongooses: small carnivorous mammals with long bodies and heads, in many respects intermediate between the cat and the dog families. They are found in most parts of the Old World but not in Australasia nor in any part of America.

VIVIPAROUS. VIVIPARITY. The development of an embryo inside the body of the parent, as in mammals (see also *Ovo-viviparous*).

VOCAL CORDS. A pair of membranes stretching between the thyroid cartilage and the arytenoid cartilages in the larynx of a mammal. The vibration of these membranes produces vocal sounds which can be varied by altering their tension and position.

VOCAL SACS. Resonating sacs on each side of the buccal cavity, *e.g.* in the males of certain frogs.

VOLUNTARY MUSCLE. See *Striated muscle.*

VOLUTE. Spirally twisted, as for instance of a shell.

VOLVOCINA. Flagellate protozoa which closely resemble plants and may be classed either among the Algae or among the Phytomastigina. They usually have a flask-shaped green chromatophore, one or more pyrenoids, starch reserves and a cellulose cuticle. Some species are single; others form globe-shaped colonies.

VOMER. A median membrane-bone above the palatine bones and beneath the nasal cavity in the mammalian skull. It is homologous with the rostral bone of a bird and with the parasphenoid bone in lower vertebrates. The so-called 'vomer' of a frog is now regarded as a prevomer.

VON BAER'S LAW. Any organism approximately recapitulates its evolutionary history during its embryonic development (also known as *Haeckel's Biogenetic Law*).

VULTURIDAE. Vultures: carnivorous birds with hooked beaks and strong claws; the head and upper parts of the neck are naked or only have small down feathers. True vultures are found in Central Europe, Africa and parts of Asia but are absent from China, Malaya, Australasia and Madagascar.

VULVA. The external opening of the vagina in female mammals.

W

WARM-BLOODED. See *Homoiothermic*.

WARNING COLORATION. Bright colours or markings on animals which are poisonous or otherwise dangerous, *e.g.* on many insects and snakes, giving them the advantage of being recognized and avoided by other carnivorous animals.

WATER VASCULAR SYSTEM. See *Hydrocoel*.

WEBBED FEET. Feet adapted for swimming by having a web of skin between the toes. This form of adaptation has taken place independently in a great many birds, in frogs and in such mammals as the duck-billed platypus (*Ornithorhynchus*), the otter and the beaver.

WEBERIAN OSSICLES. A chain of small bones, probably detached portions of some of the anterior vertebrae, connecting the air bladder with the ear capsule in fish of the group Ostariophysi. Included in this group are many common fresh-water fish.

WEBER'S LAW. The perception of a small difference between two sensory stimuli depends on the ratio of that difference to their magnitudes; not on the absolute difference between them. In the perception of light by the human eye, for instance, if the brightness is increased by 1% the difference is just noticeable. For noise the ratio is about $\frac{1}{3}$; for pressure on the skin it varies between $\frac{1}{10}$ and $\frac{1}{30}$.

WEISMANNISM. The doctrine of the continuity of the germ-plasm and the non-transmission of acquired characteristics. According to this theory the reproductive cells are set apart at an early age from the rest of the body-cells and are uninfluenced by them. Since the only connection between one generation and the next is through the germ-cells it follows that characteristics acquired during life cannot be inherited. In the light of recent research on the action of physical and chemical factors which influence chromosomes, the rigid theory as defined by Weismann has had to be considerably modified.

WHALEBONE (BALEEN). Horny plates which hang down from the roof of the mouth on each side of the palate in the toothless whales (Mystacoceti). When feeding, the whale opens its mouth so that the baleen plates hang vertically downwards with their tips just above the floor of the mouth. Water and small organisms enter; the mouth is then closed sufficiently for the lower ends of the baleen plates to rest on the floor of the mouth and water is driven out through the strainer so formed. Small shrimps and other organisms which are retained are licked off and swallowed (see *Balaenoidea*).

WHEEL ORGAN. (1) A ring of cilia by which food is swept into the mouth in Rotifers.

(2) A complex system of ciliated grooves and ridges on the under side of the oral hood in *Amphioxus*.

WHITE CORPUSCLE. See *Leucocyte*.

WHITE FIBROUS TISSUE. Connective tissue containing white inelastic fibres of *collagen*. It is found in a nearly pure state in tendons and mixed with yellow elastic fibres in areolar tissue.

WHITE MATTER. The name commonly given to those parts of the brain and spinal cord which consist predominantly of nerve fibres having medullary sheaths. In contrast to this, the part containing most of the cell-bodies and nuclei forms the 'grey matter'.

WINGS. See *Flight*.

WINSLOW. See *Foramen of Winslow*.

WINTER EGGS. Eggs of Rotifers and other aquatic organisms specially thickened and much-yolked to delay hatching until winter is passed.

WOLFFIAN BODY. An alternative name for the *Mesonephros (q.v.)*.

WOLFFIAN DUCT. The duct from the *mesonephros* or 'middle kidney'. In fish and amphibia this is a urinary duct in the female and a urinogenital duct in the male. In reptiles, birds and mammals the Wolffian duct becomes the *vas deferens* of the male but degenerates in females. This is because the functional kidney in these animals is a *metanephros* or 'hind-kidney' with its own duct, the ureter (see *Mesonephros*).

WOLFFIAN RIDGES. Ridges on an embryo from which the limb-buds arise.

WORKERS. Among social insects such as bees, ants, wasps and termites, the workers are usually sterile females which build the nest, fetch and store food and feed the other members of the colony; they far outnumber all the other castes. In the case of bees the workers develop from fertilized eggs, the larvae being given a diet of nectar and pollen. If given a more complete diet of 'royal jelly' which has been partly predigested and concentrated, the same larvae may become queen bees.

X

XANTHO-. Prefix from Greek *Xanthos*: yellow.

XANTHOPHYLL. A yellow pigment associated with chlorophyll in plants and in the plant-like protozoa (Phytomastigina). It is an oxidation product of the orange coloured pigment carotin and has the empirical formula $C_{40}H_{56}O_2$.

XANTHOPLASTS. Yellow chromatophores or protoplasmic structures containing xanthophyll.

XANTUSIIDAE. Lizards of Central America and Cuba with short tongues, without osteoderms and without movable eyelids.

X-CHROMOSOME. A sex chromosome which is paired in one sex and single in the other. In humans, for instance, the cells of the female body contain two X-chromosomes; those of the male an X and a Y (see *Sex chromosomes*).

XEN-, XENO-. Prefixes from Greek *Xenos*: strange.

XENARTHRA. South American Edentates: sloths, ant-eaters and armadilloes (see *Edentata*).

XENOPELTIDAE. Snakes of S.E. Asia differing from others in having the cranial bones more or less solidly united and in having many teeth.

XENOSAURIDAE. Lizards of Mexico without osteodermal plates and with numerous small cylindrical teeth.

XIPH-. Prefix and root-word from Greek *Xiphos*: a sword.

XIPHISTERNUM. The posterior and usually cartilaginous part of the sternum in many vertebrates.

XIPHODONTIDAE. A small family of extinct ungulates closely resembling deer but having a complete set of teeth. In this respect they differ from true deer which have no upper incisors and have a gap on each side in front of the premolars.

XIPHOSURA. King-crabs: a small group of aquatic arachnids which have remained almost unchanged since Palaeozoic times. The body consists of a large rounded *prosoma* or anterior part separated by a hinge from the *opisthosoma* or hind-part which bears a long tail-spine. The mouth is ventral and is surrounded by *gnathobases* or flattened jaw-like appendages on the legs. Respiration is by means of gill-books.

XIPHYDRIIDAE. Alder and willow wood-wasps: a family of hymenopterous insects allied to the larger wood-wasps (Siricidae). In the British species the head and thorax are black flecked with white and there is a distinct neck; the abdomen is black with the middle segments red. Eggs are laid in slits in the bark and the larvae tunnel into the wood.

XYLOCOPIDAE. Carpenter-bees: large bees, usually dark violet or black in colour, which burrow into timber. They are widely distributed in all parts of the world but are not native to Britain.

Y

Y-CHROMOSOME. A chromosome which is unpaired and is present only in one sex. In humans, for instance, there are two types of sex chromosome known as X and Y. Egg cells always contain an X-chromosome but sperm cells may contain either an X or a Y. When an egg cell is fertilized by an X-sperm cell, a female offspring will be produced; when fertilized by a Y-sperm cell a male will be produced (see *Sex chromosome*).

YELLOW ELASTIC TISSUE. Connective tissue containing a network of yellow fibres composed of *elastin*. It is found nearly pure in ligaments and mixed with white fibres in areolar tissue.

YELLOW CELLS. See *Chloragogens*.

YELLOW SPOT (MACULA LUTEA). That part of the retina which is most sensitive to colour and which contains numerous cones and very few if any rods. It is present in the eyes of humans and higher apes and is in the centre of the retina immediately opposite the lens. Clear perception of colour therefore only takes place when the eyes look straight at an object and the image falls on this part.

YOLK. The food store of an egg consisting largely of proteins and fats. Usually it is produced within the egg-cell and is contained by the vitelline membrane; sometimes, however, as in the Platyhelminthes, it is secreted by separate *vitelline glands*.

YOLK PLUG. A mass of yolk which protrudes through and partly blocks the blastopore in the developing embryo of such animals as the frog.

YOLK SAC. A sac containing yolk communicating directly with the gut in the embryos of birds, reptiles and some fish. In mammalian embryos there is a similar sac which, however, is small and contains no yolk. It absorbs nutritive secretions from the uterus but ceases to be functional when the placenta is formed.

Z

Z-DISC. A thin membrane, also called *Krause's membrane*, traversing the fibrils of striped muscle at right angles and dividing each fibril into *sarcomeres*.

ZEIDAE. Spiny-rayed marine fish of temperate regions, covered with minute scales; having an air-bladder and having on each side three complete gills and a hemibranch. The 'John Dory' (*Zeus faber*) is a well-known example.

ZEORHOMBI. Spiny-rayed fish with a laterally compressed body, including the *Zeidae* (see above) and the *Pleuronectidae* or flat fish. The latter include the plaice and the sole, both of which are extremely compressed and have a twisted head with both eyes on the same side.

ZEUGLODONTA (ARCHAEOCETI). Primitive whales of the Eocene Period attaining a length of over 60 ft. but differing in many respects from present-day whales. The teeth were heterodont; the neck was elongated with separate vertebrae instead of being compressed and shortened. The skin probably had an armour of dermal plates. It is possible that they were related to the Pinnipedia (seals and sea-lions).

ZEUGOPODIUM. The fore-arm or shank.

ZIPHIINAE. Bottle-nosed whales: a type of whale about thirty feet long characterized by an elongated snout with a few teeth in the lower jaw only. The upper part of the head forms an asymmetrical cranial basin containing spermaceti.

ZOAEA LARVA. The larva of crabs and similar crustacea characterized by a large abdomen and relatively small thorax on which are only two pairs of small appendages. On the abdomen the last pair of limbs usually appears before the foremost ones.

ZOANTHARIA. An order of Coelenterates comprising sea-anemones and the majority of corals. They differ from Alcyonarian polyps in having a large number of tentacles (usually a multiple of six); in having a variable number of mesenteric filaments in the body-cavity and in having two ciliated grooves (siphonoglyphs) in the stomodaeum.

ZONA PELLUCIDA (ZONA RADIATA). A striated membrane which normally surrounds the ovum of a mammal. It is secreted by the follicle cells of the ovary and therefore ranks as a secondary membrane (see *Egg membranes*).

ZONA RADIATA. See *Zona pellucida* above.

ZONURIDAE. Lizards of South Africa and Madagascar having dermal ossifications on the trunk and tail; having no abdominal ribs, and in some cases having a snake-like body with reduced limbs.

ZOO-. Prefix from Greek *Zoon*: an animal.

ZOOBIOTIC. Living parasitically on or in an animal.

ZOOCHLORELLAE. Unicellular green algae living as symbionts within the tissues of other organisms such as sponges, coelenterates, worms etc. By carrying out photosynthesis they produce carbohydrates which their host can utilize, whilst they in turn can feed on the nitrogenous waste products of the host.

ZOOCYTIUM. A gelatinous substance embedding masses of ciliated protozoa.

ZOOECIUM. The stiffened body-wall which forms a cup-shaped capsule enclosing the tentacled 'polypide' in the Polyzoa.

ZOOERYTHRIN. A red pigment in feathers.

ZOOFULVIN. A yellow pigment in feathers.

ZOOIDS. Individual polyps of a colony, as for instance of coelenterates. Although basically similar they may show considerable variation in detail and in function. They may, for example, be *gasterozooids* for feeding, *gonozooids* for reproduction, *dactylozooids* for stinging and catching prey etc.

ZOOMASTIGINA. Flagellated protozoa usually with more than two flagella; not containing chlorophyll and therefore unable to carry out photosynthesis. Many species are parasitic.

ZOOMELANIN. A black pigment in feathers.

ZOOPHYTE. An animal with plant-like appearance (*e.g.* a sea-anemone).

ZOOPLANKTON. Animal life (mostly microscopic) living on or near the surface of water.

ZOOTHECIUM. See *Zoocytium.*

ZOOXANTHELLAE. Unicellular yellow algae containing xanthophyll and living as symbionts in the tissues of higher organisms such as sponges, coelenterates, worms etc. (see *Zoochlorellae*).

ZORAPTERA. A small group of insects closely related to booklice and belonging to the order Psocoptera. They live in decaying wood or sometimes in termite colonies. They comprise both winged and wingless individuals but little is known of their life.

ZYG-, ZYGO-. Prefix from Greek *Zugon*: a yoke. Usually signifies anything joined together.

ZYGANTRUM. A depression on the posterior surface of the neural arch of a snake's vertebra into which fits a wedge-shaped process, the *zygosphene* of the next vertebra.

ZYGAPOPHYSES. Facets by which vertebrae articulate with one another: usually a posterior pair (post-zygapophyses) resting on the anterior pair (pre-zygapophyses) of the succeeding vertebra. In this way a certain amount of movement is possible without dislocation of the vertebrae.

ZYGOCARDIAC OSSICLES. Toothed ossicles on each side of the proventriculus or 'stomach' of a crustacean. They form an important part of the *gastric mill* which grinds and crushes food by a rhythmic to and fro movement of a median tooth between the two rows of lateral teeth.

ZYGODACTYLOUS. Of birds: having the first and fourth toes directed backwards and the second and third forwards (*cf. Heterodactylous*).

ZYGOMATIC ARCH (ZYGOMA). An arch forming the cheek-bone of a mammal consisting of a backward process of the jugal bone joined to a forward process of the squamosal bone.

ZYGOPTERA. A group of dragonflies with slender bodies and narrow-based wings which are held vertically above the abdomen when at rest. The nymphs of this group have three elongated caudal gills at the hind-end of the abdomen.

ZYGOSPHENE. See *Zygantrum.*

ZYGOTE. A fertilized ovum: a cell formed by the union of the male and female gamete.

ZYGOTENE. A stage in meiosis at which the chromosomes come together in homologous pairs. The two of a pair are usually of the same shape and size but one is of paternal and the other of maternal origin.

ZYGOTE NUCLEUS. The nucleus formed by the fusion of male and female nuclei when two gametes combine to form a zygote.

SUPPLEMENT

A

ACANTHODII. Spiny sharks: small Elasmobranch fish ranging from the Silurian to the Permian periods, characterized by the presence of a broad spine in front of each fin.

ACARI. An alternative name for *Acarina*: ticks and mites.

ACHILLES' TENDON. The tendon of the heel: a strong tendon connecting the Gastrocnemius and Solaeus muscles to the calcaneum in tetrapods.

ACOELA. Small primitive Turbellarians living in the sea. Although related to the fresh-water Planarians they are much smaller and resemble the ciliated planula larvae of some coelenterates.

ACROTERGITE. An intersegmental dorsal plate sometimes found as a narrow band attached to the front of a notum in certain insects. It is formed by secondary segmentation of the thoracic tergites.

ACROTHORACICA. Minute naked barnacles which bore into and inhabit mollusc shells and corals.

ACTIN. A protein whose presence forms light bands running transversely across the fibrils of striated muscle. These are the so-called I-Bands (formerly known as J-Discs). When the muscle is relaxed it is believed that the actin molecule is attached to an ATP molecule whose energy is released when the muscle is stimulated. (See also *Myosin*.)

ACTINOPODA. Sarcodina or amoeboid Protozoa possessing axopodia, *i.e.* long needle-like pseudopodia having a central axis.

ACULEI. An alternative name for *Microtrichia*, the minute hairs on the wings of certain insects.

ADAPTIVE CONVERGENCE. A superficial similarity between different species of animals due to similarity of habits, or sometimes due to the fact that both show protective resemblance to some other species, *e.g.* several flies, beetles, spiders etc. show a protective resemblance to ants.

ADAPTIVE ORIENTATION. A habit, shown by many insects as well as some birds and other animals, of standing or resting in such an attitude as to make the best possible use of their protective coloration.

ADAPTIVE RADIATION. The tendency for a group of animals, the members of which are generally similar, to produce local races which are suited to different environments, *e.g.* coloured for camouflage in different habitats.

ADECTICOUS PUPAE. Pupae in which the mandibles are reduced and are not used for escaping from the cocoon or cell.

ADELOGNATHA. A group of weevils (Curculionidae) having a short rostrum and temporary mandibles which are later cast off. The larvae usually live in soil, feeding on roots.

ADELOSPONDYLI. See *Microsauria*.

ADENINE. A purine base, *i.e.* a nitrogenous compound with a double ring structure: one of the four bases which are combined with sugars and phosphate groups in the DNA and the RNA molecule. The importance of these bases lies in the fact that their various possible arrangements form the so-called 'Genetic Code'.

ADENOTROPHIC VIVIPARITY. A phenomenon in certain insects (*e.g.* in *Glossina*) whereby the larvae after hatching are retained in the 'uterus' of the parent, develop by feeding on uterine secretions, moult twice and are finally deposited as mature larvae almost ready to pupate.

ADNEXA. A general name for foetal membranes, placenta or other extra-embryonic structures.

ADP. Adenosine diphosphate. See under ATP.

ADRENOTROPHIC HORMONE. A hormone produced by the Anterior Pituitary Lobe which stimulates activity of the Adrenal Glands.

AEDEAGUS. The male intromittent organ or 'penis' in certain insects.

ALCIFORMES. Razorbills, guillemots, auks and puffins. See *Alcidae*.

ALCIOPIDAE. A family of free-swimming Polychaetes characterized by having remarkably well developed eyes with lens and cornea capable of forming an image and with an efficient mechanism for accommodation.

ALLEN'S LAW. Mammals of cold regions show a tendency to have shorter extremities (feet, tail and ears) than similar animals of warmer regions. In this way they are better able to conserve heat. North American rabbits and hares are a classical instance.

ALLOCRYPTIC COLORATION. Adventitious concealing coloration: the use of any available material by an animal in order to disguise itself. Some crustaceans for instance decorate themselves with seaweed; some insects coat themselves in chalk, sand or lichen.

ALLOEOCOELA. Small Platyhelminthes (flatworms), mostly marine, having a simple or branched enteron and having three or four pairs of longitudinal nerve cords. They are intermediate between Acoela and Tricladida and were formerly classed with the Rhabdocoela

ALLOMETRIC GROWTH. Differential rates of growth whereby the sizes of certain parts of the body are a constant exponential function of the size of the whole animal. In male stag-beetles, for example, as the beetle grows larger the mandibles become relatively much larger in proportion to the body.

ALLOSEMATIC PROTECTION. A method of protection in which an animal regularly associates itself with another which is poisonous, distasteful or dangerous, *e.g.* a hermit crab in association with a sea anemone.

AMBLYPYGI. Tropical arachnids of nocturnal habit, having a flattened body, long legs and a pair of powerful pincers (pedipalps). They walk sideways with a crab-like gait and feed on insects.

ANAPSIDA. Reptiles in which the skull has a complete covering of dermal bones with no temporal fossae, although the bones may be emarginated from the hind-end. They include the fossil Cotylosauria as well as tortoises and turtles.

ANASPIDA. Ostracoderm fish of the Silurian and Devonian periods having a laterally compressed fusiform body, a heterocercal tail and long thin rectangular scales arranged in parallel rows on the body.

ANASPIDACEA. Mountain Shrimps: small crustaceans without a carapace, found in mountain lakes at high altitudes. See *Syncarida*.

ANCODONTA. An infra-order of Artiodactyla (even-toed ungulates) including the hippopotami and a number of closely related animals from the Eocene period onwards.

ANECDYSIS. A long passive period between two moults of an arthropod, during which there appears to be no preparation for the next moult.

ANEPIMERON. The upper part of the epimeron of an insect in those cases in which it is divided by a transverse suture. The lower part is called the *katepimeron*.

ANEPISTERNUM. The upper part of the episternum of an insect, the lower part being called the *katepisternum*.

ANISOZYGOPTERA. Fossil dragon-flies of the Mesozoic period showing features intermediate between the Zygoptera and the Anisoptera.

ANKYLOSAURIA. Heavily armoured turtle-like reptiles of the Cretaceous period.

ANOPHELINI. A group of mosquitoes many of which are vectors of malaria and other diseases. They differ from the Culicine mosquitoes chiefly in the fact that the body is not covered with overlapping scales but only fine hairs.

ANOPLA. Nemertean worms whose proboscis is not armed with a stylet. They include the two orders *Palaeonemertini* and *Heteronemertini*.

ANTHURIDEA. Marine and fresh-water crustaceans of the order Isopoda having the first pair of legs large and subchelate and the first pair of pleopods forming an operculum covering other pairs.

ANTIBODY. A substance, usually a soluble protein, produced in the body of an animal as a protection against invasion by parasites or other foreign bodies such as bacteria and viruses. The foreign body contains substances called *Antigens* which cause the production of specific antibodies. The interaction of the two usually causes the death of the invading parasites.

ANTIGEN. See *Antibody*.

ANTIPATHARIA. Black Corals: members of the class Anthozoa having a branched chitinous and often spiny skeleton secreted by the ectoderm.

ANTISQUAMA. An additional small lobe near the base of the wing between the Squama and the Alula on certain insects.

APATETIC COLORATION. Coloration which protects an animal by misleading its enemies. In some cases the outline of the animal is broken up (see *Disruptive Coloration*); in others the markings suggest the presence of an eye or other organ where in fact there is none.

APNEUSTIC. A term used to describe insects which are without spiracles. There is a closed tracheal system into which air can diffuse either through the whole body surface or through specialized extensions known as tracheal gills. Such an arrangement is common among aquatic insect larvae.

APODA. (1) Legless Amphibia. See *Gymnophiona*.
(2) Burrowing sea-cucumbers (Holothuroids) without tube-feet.
(3) Parasitic barnacles having a degenerate maggot-like body without mantle or trunk appendages.

APODIFORMES. A group of birds with short legs, long wings and deep-keeled sternum. They include swifts and humming birds which were formerly classed as Cypseli, a group of the Coraciiformes.

APOSTOMATIDA. Marine parasitic and commensal Protozoa having the body covered with cilia arranged in a spiral.

APOSYMBIOTIC. An organism which normally lives in symbiosis with another but has been separated from its partner.

ARANEI. An alternative name for *Araneida*: spiders.

ARCHAEOCYATHA. A group of reef-forming organisms of the Cambrian period, possibly related to the calcareous sponges.

ARCHAEOCYTES. Primitive animal cells able to develop into various other types. Sometimes, as for instance in sponges, they take the form of large amoebocytes with blunt pseudopodia and large nuclei.

ARCHAEOGASTROPODA. An alternative name for *Aspidobranchiata* or *Diotocardia*: Gastropod mulluscs having two auricles, two kidneys and two fern-like gills. The shell is either spiral or secondarily symmetrical. An example of the latter is the limpet. See also *Streptoneura*.

ARCHOSAURIA (ARCHAEOSAURIA). A large and somewhat vague group of reptiles all having a diapsid skull. They include the various kinds of Dinosaur, Pterosaurs and crocodiles.

ARCIFERAL GIRDLE. A pectoral girdle in which there is a mid-ventral overlapping of some of the bones, *e.g.* the epicoracoids of certain frogs.

ARCULUS. An arc-shaped vein forming a cross-connection between other veins in the wings of certain insects, *e.g.* that connecting the radial and median veins in dragonflies.

ARIXENIINA. Ectoparasitic earwigs (Dermaptera) having eyes, mandibles and forceps greatly reduced. They are found in the breast-pouches of certain bats.

ARTHROPODIN. A water-soluble protein found in association with chitin in the exoskeletons of insects. Chemically it is similar to the silk protein *sericin*.

ASCHELMINTHES (NEMATHELMINTHES). Worm-like pseudocoelomate animals without internal segmentation, usually aquatic but occasionally parasitic, having no respiratory or circulatory system and having the sexes separate. In recent classification this phylum comprises six classes which were formerly regarded as separate phyla: *viz. Rotifera, Gastrotricha, Kinorhyncha, Nematoda Nematomorpha* and *Priapulida*.

ASCOSPERMOPHORA. An order of millepedes having about thirty segments whose terga each bear lateral carinae and three pairs of bristles. The last segment has a pair of silk glands opening on its dorsal side.

ASELLOTA. Marine and fresh-water crustacea of the order Isopoda having all the pleopods used as gills and having all the abdominal segments fused to form a large plate.

ASTRAPOTHERIA. An order of fossil ungulates of which the principal representative was *Astrapotherium*, a large South American animal of the Oligocene and Miocene periods, having a large head, canine tusks and possibly a short proboscis.

ATP. Adenosine triphosphate. A compound formed from the purine base *adenine* combined with the sugar *ribose* and three phosphate groups. It is built up by the mitochondria of cells and forms their chief energy store. It is a comparatively unstable substance and readily breaks down to adenosine *di*phosphate ADP releasing a large amount of energy in the process.

AURICULARIA LARVA. The typical larval form of a Holothuroid, having a band of cilia arranged in a complex pattern down each side and folded back at each end where the body is extended to form a pre-oral and a post-anal lobe.

AUTECOLOGY (AUTO-ECOLOGY). The study of a particular species in relation to its environment (*cf. Synecology.*)

AXOCOEL. That part of the coelom which forms the *hydroporic canal* by which the water-vascular system opens to the outside in starfish and other Echinoderms. It is derived from the embryonic *protocoel.*

B

BALANOMORPHA. Acorn barnacles: sessile members of the order Thoracica having no peduncle and having the wall of the carapace surrounded by paired terga and scuta. (See *Thoracica.*)

BASILAR MEMBRANE. A membrane forming the base of the *Organ of Corti* and supporting the sensory cells in the cochlea of the mammalian ear. It separates the *Scala tympani* from the *Scala media.*

BASILOSAURIDAE. Ancestral whales of the Eocene period having a serpentine body which sometimes attained a length of 55 ft. (See also *Zeuglodonta.*)

BASIPHIL. An alternative spelling of *Basophil (q.v.)*

BATHYNELLACEA. Small blind cave-dwelling crustacea without a carapace, belonging to the group *Syncarida.*

BDELLOIDEA. Bdelloid Rotifers: those which move by a looping action like that of a leech.

BDELLONEMERTINI. Nemertean worms living commensally in the mantle cavities of certain molluscs. They are characterized by the absence of eyes and other special sense organs.

BENTHOHYPONEUSTON. Organisms which live on the sea-bed but rise to the surface at night.

BERGMANN'S LAW. A general tendency for mammals living in cold climates to be larger than similar species inhabiting warmer climates. A large animal has a relatively smaller surface area in proportion to its volume and is therefore able to conserve heat better than a smaller animal of the same shape. The law only applies in cases of animals which are similar in other respects. A variety of factors such as the thickness of fur and fat can upset or even reverse the tendency.

BILATERIA. A collective name for all the phyla of animals which are bilaterally symmetrical or which have developed a secondary radial symmetry. They include the whole animal kingdom other than Protozoa, Porifera, Coelenterata and Ctenophora.

BIPINNARIA LARVA. The planktonic larva of certain types of Echinoderm, having two or more ciliated lateral projections or 'arms'.

BLATTARIA. Cockroaches: a suborder of insects formerly included in the Orthoptera under the name of *Blattidae (q.v.)* but now placed in the order *Dictyoptera.*

BRACHIOLARIA LARVA. A late larval stage of certain Echinoderms having several ciliated 'arms' at one end and three adhesive 'arms' at the other. Between the bases of these arms is a glandular sucker by which the organism attaches itself to a suitable substrate.

BRACHYRHYNCHA. A group containing the majority of crabs characterized by having a more or less round carapace and a very small rostrum. They include marine, fresh-water and land crabs as well as a few highly specialized types such as the Fiddler Crab *Uca*, the Ghost Crab *Ocypode* and the Coral Gall Crab *Hapalocarcinus*.

BRONTOTHERIOIDEA (BRONTOTHERIIDAE, TITANOTHERIIDAE). A group of large primitive horse-like mammals living from the Eocene to the Oligocene periods. They had a relatively small brain and poorly developed teeth. The feet had hoofed toes: four on the manus and three on the pes.

BRUSH BORDER. See *Striated Border*.

C

CAENOLESTOIDEA. Opossum-rats. See *Epanorthidae*.

CALANOIDA. Minute free-living planktonic copepods which exist in enormous numbers in the sea and form an important part of the food of many fish. (See *Copepoda*).

CALIGOIDA. Marine copepods commonly living as ectoparasites on the gills of fish; often called 'fish-lice'. See *Copepoda*.

CAPRELLIDEA. Crustacea of the order Amphipoda having vestigial appendages and fused abdominal segments. They include the 'skeleton-shrimps' *Caprellidae* and the whale-lice *Cyamidae*.

CAPRIMULGIFORMES. See *Caprimulgidae*.

CARDIAC OSSICLE. A large median ossicle in the proventriculus or 'stomach' of certain crustaceans. With other ossicles it forms the gastric mill.

CARIDEA. A group of Decapod crustaceans which includes the majority of free-swimming shrimps. In these the third pair of legs are without chelae and the gills are phyllobranchiate.

CAROTID BODY. (1) Of Amphibia: The Carotid Labyrinth (*q.v.*)

(2) Of Mammals: A minute structure close to the carotid sinus, consisting of a network of tiny blood vessels innervated by branches of the glossopharyngeal nerve. Its function appears to be connected with control of the rate of respiration.

CAROTID SINUS. A slight enlargement of the common carotid artery where it bifurcates into the internal and external carotids.

CEPHALOCARIDA. Minute shrimp-like crustaceans discovered as recently as 1955 in the sand off Long Island. They have a horse-shoe shaped head without eyes and with two pairs of short antennae. The rest of the body consists of 19 uniform segments with primitive appendages on the first nine of these. Little is known of their anatomy or their habits.

CEPHALODISCIDA. Small worm-like Hemichordata living as colonies of separate individuals within a common tube secreted by the parent zooid. They belong to the class *Pterobranchia* and have four to nine pairs of branched ciliated arms.

CERATOMORPHA. Tapirs and rhinoceroses: members of the order Perissodactyla or odd-toed ungulates having a pachydermatous skin, reduced dentition, toes never reduced below three but the middle one always the largest.

CERIANTHARIA. Anemone-like animals usually classed with the Zoantharia but differing from others of this group in having numerous septa and numerous tentacles arranged in two whorls.

CERVIX. (1) The neck or any neck-like structure.

(2) *Cervix Uteri.* The neck of the uterus opening into the vagina of mammals.

CESTODARIA. An alternative name for *Monozoa*: small non-strobilating tapeworms found in the bodies of fish.

CESTOIDEA. A class of endoparasites including the true tapeworms or *Cestoda* and the non-strobilating forms or *Cestodaria*. In the older system of classification these were known as *Merozoa* and *Monozoa* respectively (*q.v.*).

CHAETOTAXY. The identification and classification of insects by the structure and arrangement of the principal hairs or setae on its body. Sometimes, as for instance in some mosquito larvae, this may be the only method of distinguishing between two species.

CHALICOTHERIOIDEA. Three-toed mammals of the Miocene and Pliocene periods having affinities with the Perissodactyla but having incompletely developed hoofs.

CHEILOSTOMATA. Ectoprocta (formerly classed as *Polyzoa*) in the form of colonies of box-like zooids adjacent but having separate walls. The orifice of each zooecium can be closed with an operculum when the animal withdraws.

CHELICERATA. One of the two main groups of Arthropoda comprising all those bearing chelicerae, pedipalps and four pairs of legs. Unlike insects with their head, thorax and abdomen, the chelicerata have the body divided into two main regions, *viz.* an anterior cephalothorax or *prosoma* and a posterior abdomen or *opisthosoma.*

The group includes spiders and harvestmen, ticks, mites and scorpions as well as the sea-spiders (*Pycnogonida*), king-crabs (*Xiphosura*) and the extinct giant sea-scorpions (*Eurypterida*). All these were formerly classed as Arachnida but this name now has a more restricted use.

CHILOGNATHA. A group containing the majority of millepedes: those which have a peculiar mouth appendage known as a *gnathochilarium* (*q.v.*) derived from the first maxillae.

CIBARIUM. A small pouch at the base of the mouth in certain insects. It may be for food storage or may be modified into a sucking pump.

CINGULATA. Armadilloes and Glyptodonts: members of the order Edentata having the body covered with bony scutes to form a carapace.

CLADOCOPA. A small order of marine Ostracoda without trunk appendages. (See *Ostracoda.*)

COENOTHECALIA. An order of Alcyonarian corals including *Heliopora*, the 'blue coral' characterized by large pits or thecae in which the polyps are lodged.

COLEOIDEA. A subclass of Cephalopod molluscs comprising *Octopoda* and *Decapoda, i.e.* octopuses, squids, cuttlefish etc.

COLIIFORMES. Mouse-birds. See *Colii.*

COLOBOGNATHA, Tropical millepedes having 30-70 segments and having the mouth-parts reduced and modified for sucking. Most of the segments bear a pair of defensive repugnatorial glands which secrete a poisonous or caustic fluid.

CORPORA CARDIACA. An alternative name for the pharyngeal ganglia of insects. These are immediately behind the brain and connected to it by a pair of nerves. They apparently secrete a number of hormones.

CREMASTER MUSCLE. A thin muscle running along the spermatic cord in vertebrates.

CRITHIDIAL STAGE. A stage of the development of *Trypanosoma*, the parasitic flagellate which causes sleeping sickness. In this stage, which is passed in the Tse-tse fly *Glossina*, the parasite has its flagellum attached in the middle to form an undulating membrane for half its length. It is so called on account of its similarity to the genus *Crithidia*.

CRYPTIC COLORATION. Any coloration which helps the concealment and thus the protection of an animal. In some cases the colours may cause an animal to match or blend with its surroundings; in others the outline of the animal is broken up by patches of contrasting colour. See also *Obliterative shading*.

CRYPTIC RESEMBLANCE. See *Cryptic coloration*.

CTENOSTOMATA. Ectoprocta (formerly classed as *Polyzoa*) in the form of compact colonies of zooids with a membranous or chitinous zooecium.

CYCLOPOIDA. One of the largest groups of Copepods: small crustaceans having a median eye, long antennae and a biramous telson. The common fresh-water *Cyclops* is typical (see *Copepoda*).

CYSTIDEA. See *Cystoidea*.

CYTOSINE. A pyrimidine base consisting of a single ring-structure containing two nitrogen and four carbon atoms: one of the four bases which are combined with sugars and phosphate groups in the DNA and the RNA molecules. The importance of these bases lies in the fact that their various possible arrangements form the so-called 'Genetic Code'.

D

DANCE OF BEES. A behaviour pattern of bees which has long been known but which has recently been accurately observed and interpreted by Von Frisch as a 'language' by which in-coming worker bees are able to communicate information regarding direction and distance of food supplies. There are two types of movement known respectively as a 'round dance' (performed when food is close) and a 'wagtail dance' (when food is further away). In the latter case the movement resembles a figure of eight in which the two loops are separated by a straight run. The orientation of this and its frequency give other bees the required information.

DECTICOUS PUPA. An exarate pupa having powerful mandibles which help the insect to escape from its cell or cocoon. See *Pharate*.

DENDROID. Branching or tree-like, as for instance of certain Graptolites.

DERMAL DENTICLES. Tooth-like structures, often minute, embedded in the skin of Elasmobranch fish.

DEUTEROSTOMA (DEUTEROSTOMIA, DEUTEROSTOMATA). A group of phyla which appear to form a natural evolutionary line in which there are many common features as well as a similarity in embryonic development. They include the Chaetognatha, Echinoderms, Pogonophora, Protochordata and all higher Chordates. In all these the early embryo develops by radial indeterminate cleavage; the mesoderm develops from pouches in the archenteron and the mouth arises at some distance from the blastopore. In the larval stage there are typically three regions known as the *protosoma*, *mesosoma* and *metasoma* each containing a separate coelomic compartment.

DIASTOLE. Relaxation of the heart or of any chamber of it: commonly used to refer to the ventricles of the human heart (*cf. Systole*).

DICTYOPTERA. An order of insects comprising the cockroaches (Blattaria) and the Mantids (Mantodia). These were formerly included with the grasshoppers and locusts in the order Orthoptera, but there are many points of difference.

DICYEMIDA. Minute worm-like parasites inhabiting the nephridia of certain molluscs. They are classed with the Mesozoa and are possibly a degenerate form of flatworm. (See *Mesozoa*.)

DIECDYSIS. The period in the moulting cycle of an arthropod between metecdysis and proecdysis (*q.v.*).

DINOCERATA. Large pentadactyl hoofed mammals of the Eocene period, nearly as large as an elephant and having three pairs of bony horn-like prominences on the skull. (See *Amblypoda*.)

DIPLOCARDIAC. Having the right and left sides of the heart distinct and separate.

DISPLAY. An action, movement or other behaviour pattern by which an animal draws attention to itself. It may be a sexual display in which usually the male shows itself off in front of the female. Frequently, however, displays are of a warning nature when an animal is attacked. Examples of the latter are the sudden increase of size or inflation of the porcupine-fish; the sudden presentation of a warning colour in certain snakes; the rattle of the rattle-snake etc.

DISRUPTIVE COLORATION. A form of animal camouflage in which patches of bright colour or of light and dark serve to deceive the eyes of enemies by breaking and masking the true outline of the animal.

DIURESIS. Increased production of urine, caused for instance by various drugs or other chemicals.

DIXIDAE. Two winged insects (Diptera) which resemble mosquitoes but have a smaller proboscis, long slender antennae and have no scales on the wings. Their larvae inhabit shady pools and could be mistaken for those of *Anopheles*.

DNA. Desoxyribonucleic acid: the principal substance contained in chromosomes and forming the basis of the hereditary material of genes. A DNA molecule is a long double helix, each half of which is a chain of sugar molecules and phosphate groups. The helices are joined by pairs of bases *Adenine* to *Thymine* and *Guanine* to *Cytosine*, these bases being attached to the sugar molecules of either chain. This forms a ladder-like structure whose sides consist of sugar and phosphate groups and whose rungs are formed by pairs of bases. The sequence of these pairs of bases forms a 'genetic code' determining the types of proteins and enzymes to be built up by the cell. The whole DNA molecule is capable of replicating itself as the two helices separate at the bases and each half then builds up a partner identical with that which has been detached. (See also under *Genetic Code*.)

DOCODONTA. Primitive mammals of the Jurassic period having multituberculate teeth: known only from a few fragmentary fossils and believed to be ancestors of the Monotremes.

DOLIOLARIA. A late planktonic larval stage of a Holothuroid (sea-cucumber): a minute barrel-shaped organism having three to five flagellated girdles.

DROMIACEA. Primitive crabs in which the last pair of legs are dorsal and modified for holding objects over the body. *Hypoconcha*, for instance covers itself with one half of a bivalve shell, its body being modified to fit into this.

Another genus *Dromia* uses its pincers to cut out a piece of sponge which it fits upon its body like a cap.

DYER'S LAW. A generalization relating to the rates of linear growth of various parts of insects. In most Lepidoptera, for example, the width of the head increases in a geometric progression with a ratio of 1.4 in successive instars. Although there are many exceptions the principle holds good with various measurements in many different insects.

E

ECHINODERA. See *Kinorhyncha*.

ECOLOGY. The study of plants and animals in relation to their environment.

ECOSYSTEM. A system consisting of a living community of plants and animals together with their physical environment of air, water, soil etc., the two continually interacting by means of food chains and chemical cycles so that the same elements are used over and over again.

EDRIOASTEROIDEA. An alternative name for *Thecoidea*: a group of fossil echinoderms resembling Crinoids but with no stalk or arms.

EMBRITHOPODA. Extinct ungulates of the Oligocene period closely allied to the Amblypoda and characterized by having two immense horns growing from the nasal bones.

ENDOCHORION. The inner layer of the 'shell' or chorion of an insect's egg.

ENOPLA. Nemertean worms whose proboscis is armed with barbs or stylets.

ENTEROCOELA. A group of phyla containing all those animals which have an enterocoel, *i.e.* a coelom which, in the embryonic stage, has been in communication with the archenteron. These include all the Chordata and Protochordata, Echinodermata, Pogonophora and Chaetognatha.

ENTOPROCTA. See *Endoprocta*.

EOSUCHIA. Extinct crocodile-like reptiles from the Permian to the Eocene periods.

EOTHERIA. Ancestral Sirenia of the Eocene period, having the pelvis well developed but without hind limbs.

EPICARIDEA. Degenerate crustaceans of the order Isopoda living as parasites on other crustaceans. The females are sometimes without segmentation or appendages.

EPITOKY. The formation of a reproductive individual with specialized secondary sex characters so that it differs markedly from the non-sexual form, *e.g.* in certain Polychaetes.

EPIZOITE (EPIZOIC). An animal which lives on another for the purpose of anchorage, protection or dispersal. This cannot be regarded as a parasitic or a symbiotic relationship as there is no particular advantage or disadvantage to the host.

EPIZOOTIC. A term corresponding to epidemic of humans: a disease affecting a large number of animals at the same time.

ERRANTIA. Free-swimming Polychaete worms usually having a large number of similar segments bearing parapodia and chaetae. (See *Polychaeta*.)

EUHYPONEUSTON. Organisms which normally live very close to the surface of the water (in the upper 5 cm.).

EUMALACOSTRACA. All Malacostraca except the small order *Nebaliacea* which differ from others in having seven abdominal segments. Eumalacostraca comprise about three quarters of the world's crustaceans. See also under *Malacostraca* and *Caridoid Facies*.

EUMELANIN. Black melanin as distinct from the brown variety or *phaeomelanin*.

EUMETAZOA. All multicellular animals other than Mesozoa and Parazoa (*q.v.*).

EUNICIDAE. Errant Polychaete worms having an eversible pharynx with a pair of chitinous lower jaws and up to five pairs of similar upper jaws.

EURYTHERMOUS. Able to survive by adapting themselves to a wide range of temperature.

EUTARDIGRADA. Fresh-water Tardigrada having no lateral cirri (see *Tardigrada*).

EXOCHORION. The thick outer layer of the 'shell' or chorion of an insect's egg.

F

FAUNA. All the animals in a particular locality or habitat.

FELOIDEA. See *Aeluroidea*.

FERAE. A name formerly given to a miscellaneous collection of carnivorous mammals, but now used to denote a superorder in which the Carnivora are the sole living representatives.

FERAL. Living in a wild state: usually referring to animals which have been domesticated or whose ancestors have been domesticated.

FERUNGULATA. A cohort or group of mammalian orders comprising the Ungulates, Carnivores and Tubulidentata.

FIBROIN. A tough elastic protein which forms the inner core of the silk threads of insects, spiders etc.

FLABELLIFERA. Crustacea of the order Isopoda having large coxal plates and a large tail-fan. Some are wood-borers and some are ectoparasites of fish.

FLASH COLOURS. Brightly coloured parts which are concealed by an animal at rest but suddenly come into view when it moves, *e.g.* the brilliant hind-wings of some moths. It is thought that the sudden appearance of such colours helps to confuse would-be predators.

FORFICULINA. Earwigs: free-living Dermaptera (*q.v.*) usually with membranous semicircular hind-wings covered by short leathery tegmina; having typical biting mouth-parts and large forcipulate cerci.

G

GALATHEIDEA. Members of the *Anomura*, apparently intermediate between crabs and lobsters, having a well developed tail-fan beneath the thorax. They include the Plated Lobster *Galathea* and the Porcelain Crab *Porcellana*.

GAMMARIDEA. Crustacea of the order Amphipoda, mostly marine or freshwater forms but including also a few semiterrestrial and parasitic genera. They are characterized by a laterally compressed body without a carapace; well developed coxal plates on the thoracic legs and palps on the maxillipedes.

GASTRODERMIS. An alternative name for the Endoderm: the innermost of the primary germ layers in a diploblastic or a triploblastic animal.

GENERAL PROCRYPSIS. Having a colour which generally helps to conceal an animal in the variety of habitats in which it lives, *e.g.* the colour of a mouse or a rabbit: not as definite as the special cryptic colour patterns of some animals.

GENETIC CODE. This is the stored information, also referred to as a *gene*, which is carried in chromosomes and passed on from cell to cell. It enables each generation of cells to build up the particular kinds of proteins characteristic of those cells and the particular enzymes necessary to govern their activity. The principal substances involved are DNA (desoxyribonucleic acid) and RNA (ribonucleic acid). Briefly the sequence of events is as follows: A DNA molecule contains a long chain of bases *adenine, cytosine, guanine* and *thymine* in a particular sequence. These act as a template for the production of molecules of *Messenger RNA* which in turn gives rise to *Transfer RNA* containing the four bases adenine, cytosine, guanine and *urocil*. It follows therefore that the original sequence of bases in the DNA molecule will ultimately determine the order of the bases in each molecule of Transfer RNA. This order constitutes the *genetic code*. The long line of bases is like a sentence which can be divided into a large number of 3-letter words or triplets (groups of three bases). Each triplet, by reason of its structure and shape, is capable of picking up a particular aminoacid molecule and transferring it to its right position on the template formed by the messenger RNA. Since a protein is merely a long chain of amino-acids arranged in a particular order, it follows that the genetic code will determine the types of protein to be produced in the cells. Examples of code triplets which can attach themselves to particular amino-acids are: CCG (alanine), UGG (glycine), GGU (tryptophane), AUU (tyrosine) etc. An RNA molecule containing a sequence of bases CCG – UGG – GGU – AUU should therefore theoretically be capable of building up a protein molecule containing alanine, glycine, tryptophane and tyrosine in that order. In actual fact there are only about twenty common amino-acids, but many thousands of molecules of these are necessary to build up one large protein molecule. (See also under DNA and RNA).

GEOPHILOMORPHA. Slender burrowing centipedes sometimes having up to 180 pairs of short legs, short antennae and no eyes.

GLYCERIDAE. Burrowing Polychaete worms having a long proboscis with at least four jaws.

GNATHIIDEA. Crustaceans of the order Isopoda having the abdomen much narrower than the thorax. The larvae are ectoparasites on marine fish.

GONADOTROPHIC HORMONES. Hormones secreted by the anterior lobe of the pituitary body which stimulate the ripening and activity of the gonads.

GORDIACEA. An alternative name for *Nematomorpha*: long slender aquatic worms similar in many ways to the Nematoda, usually starting life as parasites and later living a free existence.

GORDIOIDEA. Fresh water Nematomorphs usually starting life as parasites in insects and later becoming free-swimming.

GÖTTE'S LARVA. A free-swimming larval stage of certain marine Turbellaria, similar to *Müller's larva (q.v.)* but differing from it in having only four ciliated lobes.

GUANIN. A purine base whose molecule $C_5H_5ON_5$ has a double ring structure. It is found in the excretory products of many animals but its main importance is due to the fact that it is one of the four bases combined with sugars and phosphate groups in the DNA and RNA molecules. The arrangement of these bases forms the so-called *genetic code* (*q.v.*).

GYMNOLAEMATA. Marine Ectoprocta consisting of polymorphic colonies of zooids usually with a circular lophophore and a thick zooecium forming an exoskeleton.

GYMNOPLEURA. Primitive crabs with elongated carapace, subchelate first legs and the remaining legs flattened for burrowing.

GYN-, GYNO-. Prefix from Greek *Gyne*: woman, female.

GYNANDROMORPH. An individual which shows both male and female features but is not a true hermaphrodite. In some butterflies, for instance, the wings on one side show the female pattern and those on the other the male.

GYNATRIUM. A female genital pouch or vestibule.

GYRODACTYLOID. A free-swimming larval stage of some ectoparasitic flukes (Monogenea). In *Polystoma*, for instance, this larva has five rings of cilia, four eye-spots and a ring of hooks at the hind end. It lives freely for about two days before entering the gill chamber of a tadpole and finally maturing in the cloaca of an adult frog.

H

HAEMOCOELOUS VIVIPARITY. A phenomenon which occurs in certain insects which have no oviducts. Eggs escape into the haemocoel, develop into larvae and finally escape through secondary openings in the body wall.

HARPACTICOIDA. Minute Copepods which crawl and burrow in the sea-bottom or live between sand grains. The body is roughly cylindrical and the antennae reduced. Like all copepods they have a single median eye. See *Copepoda*.

HELIOFLAGELLATA. Sarcodina or amoeboid protozoa possessing both axopodia and flagella.

HEMIMERINA. Wingless earwigs (Dermaptera) ectoparasitic on rats in tropical Africa.

HEMIPNEUSTIC. A term used to denote insects in which one or more pairs of spiracles are non-functional.

HERPETOLOGY. A study of the zoology and biology of reptiles; often extended to include Amphibia.

HESIONIDAE. Aquatic Oligochaeta having two air-sacs opening into the anterior part of the intestine. These probably have a hydrostatic function.

HETEROTARDIGRADA. Marine Tardigrada having a pair of lateral cirri on each side. (See *Tardigrada*.)

HEXACORALLIA. Corals in which the polyps have tentacles arranged in multiples of six. They include the majority of large reef-forming corals. (See *Zoantharia*.)

HIPPIDEA. Mole-crabs: burrowing crustaceans belonging to the group Anomura intermediate between crabs and lobsters. The body is more or less cylindrical and the abdomen is flexed beneath the thorax. They live on open beaches and, as each wave recedes, they burrow rapidly backwards with their abdomens, leaving only the antennae visible.

HIPPOMORPHA. Horse-like mammals: members of the Perissodactyla or 'odd-toed ungulates' having the central toe greatly enlarged. They include present-day horses and a number of fossil types, but exclude the tapirs and rhinoceroses.

HIS' BUNDLE. A bundle of muscle and nerve fibres connecting the auricles and ventricles of the heart and transmitting impulses from the pacemaker.

HISTAMINE. A derivative of the amino-acid Histidine produced by damaged cells of vertebrates. When released it has the effect of dilating capillaries and lowering blood-pressure. It is also believed to be one of the active substances in the stings of bees, wasps etc.

HOLOPNEUSTIC. A term used to describe insects which have ten pairs of functional spiracles, usually eight pairs on the abdomen and two pairs on the thorax. This is the maximum number present in any living adult insect with the exception of a few primitive Diplura.

HOMOIOSTASIS. The phenomenon by which an animal is self-regulating and maintains its body in any particular physiological state, *e.g.* temperature regulation in mammals, osmo-regulation in aquatic animals etc.

HONEYCOMB. Wax cells made by the honey-bee for storing honey and as brood-cells. They are made from wax secreted by four pairs of glands situated in the membrane below the 3rd, 4th, 5th and 6th abdominal segments. This wax is produced as small white scales which the bee manipulates with its feet and mandibles to make regular hexagonal prisms of uniform size. The geometrically perfect shape ensures the most economical use of the wax.

HONEYCOMB BORDER. See *Striated Border*.

HOPLONEMERTINI. Nemertean worms having a specialized proboscis armed with a large barb or stylet. They include members living in salt and fresh water.

HOST. An organism on which or in which a parasite lives. (See *Primary Host* and *Secondary Host*.)

HYDROPYLE. A specialized region of the shell of an insect's egg through which water can enter.

HYPERIIDEA. Marine crustacea of the order Amphipoda having very large eyes, reduced coxae and no palps on the maxillipedes.

HYPONEUSTON. Surface plankton: organisms living immediately below the surface of water.

HYPONYCHIUM. A layer of epidermis lying under a nail.

HYPOTHALAMUS. The thickened floor of the third ventricle in the vertebrate brain. In mammals it contains the nerve centres which control the temperature of the body.

I

I-DISC. Formerly called the J-disc: a light coloured band running transversely across the fibres of striated muscle. A region occupied by filaments of the protein *Actin* between two successive *Myosin* filaments.

IMPENNES. Penguins: birds having no true feathers and whose wings are well adapted for swimming but incapable of flight. Usually classed as *Sphenisciformes*.

INDUSTRIAL MELANISM. The existence of black varieties of moths or other insects in industrial regions. The best known example is the Peppered Moth *Pachys betularia*. The exact cause of this mutation is obscure but, once the new variety is established, it tends to increase in smoke-blackened regions where its dark colour helps to camouflage it. In some industrial parts of Britain the black varieties of peppered moth have almost entirely replaced the normal members of the species.

INSTINCT. An innate disposition which enables an animal to perceive an object or situation and to respond to it by a specific pattern of behaviour. Such behaviour-patterns are often extremely complex and little understood. The main distinction between instinct and intelligence is that in the former the behaviour-pattern is relatively immutable whereas the latter enables an animal to 'learn', *i.e.* to modify its actions with every changing situation.

IRREGULARIA. An alternative name for *Exocyclica*: irregular heart-shaped Echinoderms.

ISOGENIC. (1) Animals of the same sex from the same litter.
  (2) Embryonic tissues having a similar origin.
  (3) Syn. *Homozygous*.

ISOPNEUSTIC. A term used to denote insects in which all the spiracles are similar, functional and regularly arranged. Most present-day insects have some degree of specialization or reduction of spiracles and the isopneustic arrangement is considered to be primitive.

J

JULIFORMIA. Millepedes with 40 or more cylindrical segments each bearing poison glands. (See *Opisthospermophora*.)

K

KATEPIMERON. The lower part of the epimeron of an insect in those cases in which it is divided by a transverse suture. The upper part is called the *Anepimeron*.

KATEPISTERNUM. The lower part of the episternum of an insect, the upper part being called the *Anepisternum*.

KINETODESMA. A fine fibril connecting the basal granules of a row of cilia in some protozoa.

KINETOSOME. The basal granule of a cilium.

KINETY. A row of basal granules of cilia connected by a protoplasmic fibril or *kinetodesma*.

KINORHYNCHA (ECHINODERA). Microscopic spiny worm-like creatures living in the sea. The cuticle is divided into 13 segments and the anterior end can be invaginated into the posterior segments.

KREBS' CYCLE. The citric acid cycle: a series of chemical actions by which oxidation takes place and energy is released in cells. The net result is the conversion of pyruvic acid (a breakdown product of sugar) to carbon dioxide and water. The stages of the cycle are: *citric acid, cis-aconitic acid, isocitric acid, oxalosuccinic acid, α-ketoglutaric acid, succinic acid, fumaric acid, malic acid, oxalo-acetic acid* and so to citric acid again. Each stage is catalysed by a specific enzyme and the process repeats itself endlessly until all the substrate is oxidized.

KRONISM.   The killing and eating of eggs or young by the parent.

KUPFFER CELLS.   Stellate phagocytic cells in the sinuses of the liver. They form part of the reticulo-endothelial system and apparently have as their main function the ingestion of dead red corpuscles.

L

LAMARCKISM.   The evolutionary theory propounded by Jean Lamarck (1744–1829) but now largely discredited. Its main principles were: (1) That frequent use of an organ caused it to become larger and better developed, whereas its disuse caused it to become small or degenerate.

(2) Such changes acquired during life could be inherited and eventually lead to the formation of new species. (See also *Neo-Lamarckism.*)

LEPADOMORPHA.   Goose-barnacles: members of the order Thoracica having a long flexible stalk or peduncle by which they attach themselves to floating or submerged objects.

LEPOSPONDYLI.   Fossil amphibia of the Carboniferous and Permian periods, closely related to present day salamanders. They were characterized by having vertebrae in which the centrum was formed by direct ossification round the notochord, the latter often being persistent in the adult.

LEPTOMONAD.   A stage in the development of the parasitic flagellate *Leishmania* which causes the disease known as 'Oriental Sore'. In the leptomonad stage the parasite has a flagellum growing from one end. It is named after the genus *Leptomonas*, also known as *Herpetomonas*, which it resembles. It passes this stage of its life in the sand-fly *Phlebotomus* which acts as a vector.

LERNAEOPODOIDA.   Minute copepods living as ectoparasites on the gills of marine and fresh-water fish. Thoracic appendages are reduced and the second maxillae are modified for attachment to the host. (See *Copepoda*.)

LIMACOMORPHA.   Small blind, tropical millepedes having about 20 segments and having the last pair of legs modified as clasping gonopods.

LITHOBIOMORPHA.   Centipedes having fifteen pairs of legs, tergal plates alternately large and small, long antennae and compound eyes. They are to be found mostly in tropical and subtropical regions.

LIME SACS.   An alternative name for *Swammerdam's Glands* (*q.v.*).

LORICA.   A protective external case either secreted by an organism or composed of foreign material cemented together.

LUNGFISH.   See *Dipnoi*.

M

MADREPORARIA.   An order of Zoantharians comprising all the true corals, *i.e.* the larger reef-building types. (See *Corals* and *Zoantharia*.)

MANCA STAGE. A larval stage of certain crustacea in which the carapace is incompletely developed.

MANDIBULATA. One of the two main groups of Arthropods, the other being the *Chelicerata*. Mandibulata include crustaceans, insects and myriopoda, all of which are characterized by the possession of antennae, maxillae and mandibles.

MELANIC MUTATION. A mutation in which an animal shows an abnormally black colour in comparison with others of the species, *e.g.* the Peppered Moth *Pachys betularia*. (See also *Industrial Melanism*.)

MELANISM. Excessive blackness of skin, hair etc. due to presence of *melanin*.

MELANOPHORE. A black pigment-cell such as those found beneath the epidermis in many animals. Variation in the amount or distribution of melanin in these cells may cause a change of colour in the skin, as for instance in frogs.

MERISTIC VARIATION. A variation involving the number or arrangement of the parts of an organism.

MEROSTOMATA. A class of equatic chelicerate arthropods comprising the King Crabs (Xiphosura) and the extinct giant water-scorpions (Eurypterida). Both are characterized by a spike-like telson at the hind end of the body.

MESAXONIA. Mammals in which the axis of symmetry of the foot passes up the middle digit. They thus include the Perissodactyla and some related fossil types. The middle toe may be the only one (*e.g.* in present day horses), or may be the largest of three, as in some ancestral horses and in the hind-foot of the rhinoceros.

MESOCOEL. The coelomic cavity in the *mesosoma* or middle segment of a deuterostome larva or embryo. (See *Deuterostoma*.)

MESOGASTROPODA. An alternative name for the *Pectinibranchiata* also known as *Monotocardia*: a group of Gasteropod molluscs whose heart has only one auricle and whose single gill is in the form of a comb. They include the whelk, the periwinkle and many others. (See also *Streptoneura*.)

MESOSOMA. The middle of three regions into which the body of a Deuterostome larva is commonly divided. It is situated between the *Protosoma* and the *Metasoma* and contains a separate coelomic cavity, the mesocoel.

MESOZOA. Minute ciliated worm-like animals commonly found as parasites in certain molluscs. Their life-cycle usually involves alternation of sexual and asexual phases and during the former, the whole organism consists merely of a few reproductive cells enclosed within a thin layer of somatoderm. (See also *Nematogen*.)

MESSENGER RNA. A form of ribonucleic acid whose molecules carry the message or 'genetic code' during the growth and reproduction of cells. Each molecule is a chain of ribose molecules joined to phosphate groups and having specific bases as side branches. There are four of these, namely *Guanine*, *Cytosine*, *Adenine* and *Urocil*. It will be noted that RNA differs from DNA in having the base *Urocil* instead of *thymine*. The sequence of these bases determines the positions to be taken up by molecules of *Transfer RNA* which carry the amino-acids and build them up in the correct positions to form specific protein molecules. (See also under headings DNA, RNA and *Genetic Code*.)

METACOEL. A separate coelomic cavity contained in the *metasoma* or hind region of a Deuterostome larva.

METASOMA. The hindmost of three regions into which the body of a Deuterostome larva is commonly divided, the other two being the *protosoma* and the *mesosoma*. It contains its own separate coelomic cavity, the metacoel.

METECDYSIS. The period after a moult in arthropods when the new cuticle is hardening and the animal is returning to its normal state.

METRATERM. The terminal portion of the oviduct or 'uterus' in Trematoda such as the liver fluke.

MIACOIDEA. Small primitive mammals of the Eocene period: probably ancestors of present-day Carnivora.

MILLEPORINA. Stinging corals: see *Hydrocorallinae*.

MODIOLUS. The central axis of the spiral cochlea of the ear.

MONOGASTRIC. Having a single stomach or a single gastric chamber. A term used chiefly by veterinarians to distinguish animals with one stomach from those such as sheep and cattle which are *polygastric*.

MONOPLACOPHORA. A group of primitive mulluscs, hitherto known only as Cambrian fossils, but recently found living in the Pacific Ocean. They have a single dorsal shell and a body showing some degree of metameric segmentation. They thus form a possible evolutionary link between Mollusca and Annelida.

MONSTRILLOIDA. Marine Copepods whose larvae are parasitic in polychaete worms. See *Copepoda*.

MORGAN. A hypothetical unit, named after T. H. Morgan, for measuring the distance apart of genes in a chromosome. One morgan represents a cross-over value of 100% (when the genes are at opposite ends of a chromosome). A cross-over value of 10% is a *decimorgan*; 1% a *centimorgan* etc. (See *Crossing Over*.)

MUTICA. A group or cohort comprising all present-day toothed and toothless whales as well as the extinct *Archaeoceti* of Eocene and Miocene times (see *Cetacea*).

MYIASIS. Infestation by parasitic dipterous larvae.

MYODOCOPA. Marine Ostracoda in which the two halves of the shell have notches at the anterior end making a hole through which the antennae can protrude when the valves are closed. (See *Ostracoda*.)

MYOGLOBIN. A protein contained in muscle and acting as a temporary storehouse for oxygen. Its structure, recently elucidated by Kendrew, has been shown to consist of a compact polypeptide chain of about 150 amino-acids together with a single *haem* group containing iron. The whole chain is bent and folded in a most complex fashion. Haemoglobin, the red substance of blood, consists of molecules which are each made up of four myoglobin-type molecules.

MYOSIN. An important protein contained in muscle fibres. The well known striped appearance of voluntary muscle is believed to be caused by the interlocking of relatively thick filaments of myosin with thinner filaments of another protein *Actin*. Myosin forms the dark 'A-bands' and actin the lighter 'I-bands'. The principal function of this actomyosin complex appears to be a catalytic action which converts ATP to ADP and releases energy for muscle contraction.

MYSTACOCARIDA. Minute crustacea about 4 mm. in length recently discovered in the sand off Long Island. They resemble copepods but are in many ways more primitive.

MYZOSTOMARIA. Flattened annelids, sometimes classed as Polychaetes, ectoparasitic on Echinoderms.

N

NATANTIA. Shrimps: those Decapod crustaceans which are well adapted for swimming as distinct from the *Reptantia* (crabs and lobsters) which are better adapted for crawling. Natantia have slender legs, a laterally compressed body and a well developed abdomen.

NEBALIACEA. Marine crustaceans of the subclass *Leptostraca* having a large carapace which is not fused with the sides of the thorax and having a pointed rostrum articulating with the front of the head. See *Leptostraca*.

NECTONEMATOIDEA. Marine Nematomorphs usually starting life as parasites in Crustaceans and later becoming free-swimming.

NEMATHELMINTHES. See *Aschelminthes*.

NEMATOGEN. The parasitic sexual phase of a Mesozoan (*q.v.*), consisting usually of a single sexual cell surrounded by a layer of somatoderm.

NEOGASTROPODA (STENOGLOSSA). Gasteropod molluscs having a single comb-like gill and a bipectinate osphradium, a sensory organ resembling a gill. Many are carnivorous and some are poisonous. See also *Streptoneura*.

NEO-LAMARCKISM. A partial revival of the theory of inheritance of acquired characteristics originally propounded by J. B. Lamarck. Its chief exponent at the present day is the Russian geneticist Lysenko.

NEONYCHIUM. (1) A soft protective pad covering each claw or nail in the embryonic stage of certain mammals.
(2) A similar but more horny covering for the claws of birds before hatching.

NEPHROPSIDEA. Decapod crustacea with cylindrical carapace, well developed rostrum and large chelipeds. They include many varieties of lobster and crayfish.

NEREIDAE. Ragworms: free-swimming polychaete worms having a pair of eversible horny jaws, four pairs of oral tentacles, two pairs of simple eyes and a row of regular parapodia running along each side for the whole length of the body.

NEUROHAEMAL ORGAN. A structure consisting of a group of nerve endings in close relationship to a group of blood vessels. Their probable function is the discharge of neuro-secretory substances into the blood.

NEUROHORMONE. A neurosecretion: a substance of physiological importance secreted at the ends of nerve fibres, *e.g. Acetylcholine*.

NEURULA. A stage in the embryo of a chordate at which the neural folds have covered the neural plate and the latter has begun to roll up to form the spinal cord.

NEURULATION. The process of formation of the neural cord in a chordate embryo. (See *Neural Folds* and *Neural Plate*.)

NICHE. The ecological position which best suits a particular plant or animal: a habitat in which it is in equilibrium with its neighbours.

NOTODELPHYOIDA. Small copepods which live commensally or parasitically in certain molluscs, tunicates etc. (See *Copepoda*.)

NOTOUNGULATA. Extinct ungulates of the Eocene and later periods including the *Toxodontia* (*q.v.*) and some forms which more nearly resembled rodents.

NUCLEIC ACIDS. Complex substances contained in chromosomes and forming the basic hereditary material of cells. Their molecules, which have the ability to replicate themselves during cell-division, consist of long chains of sugars, phosphate groups and bases. The order of sequence of the latter constitutes the so-called *genetic code* (*q.v.*). See also under headings DNA and RNA.

O

OBLITERATIVE SHADING (COUNTERSHADING). A form of camouflage exhibited by most animals in which the upper or dorsal surface is darker than the ventral and there is a gradual shading off from one to the other. When the animal is illuminated from above, the gradation of colouring often exactly counterbalances the fact that the ventral part of an animal is in shadow. In this way the shape and solid appearance of the animal becomes almost invisible and gives a high degree of protection from predators.

OCTOCORALLIA. Corals in which each polyp has eight tentacles and eight septa. (See *Alcyonaria*.)

ODOBENIDAE (TRICHECHIDAE). Walruses: aquatic carnivores characterized by the absence of incisors but having large upper canines in the form of tusks.

ODONTOSTOMATIDA. Ciliophora with few cilia and having the body enclosed in a wedge-shaped capsule or *lorica*.

OLIGOPNEUSTIC. A term used to denote insects with a reduced number of spiracles. See also *Propneustic, Metapneustic* and *Amphipneustic.*

ONISCOIDEA. Slaters and Woodlice: amphibious and terrestrial crustaceans of the order Isopoda having the coxal plates fused with the body; having distinct abdominal segments and terga forming transverse plates enabling the animal to roll up into a ball.

ONISCOMORPHA. Pill millepedes: a group found largely in the tropics and characterized by having ventrally flattened segments enabling the body to be rolled up into a ball.

OPILIONES. An alternative name for *Phalangida* or 'Harvestmen': long-legged arachnids resembling spiders but having no silk-glands. (See *Phalangida*.)

OPISTHONEPHROS. An embryonic kidney consisting of tubules of the *mesonephros* and *metanephros* (*q.v.*).

OPISTHOPORA. Earthworms and certain fresh-water and amphibious worms all characterized by having the male gonopores situated some distance behind the testes. They include the common earthworms (*Lumbricidae*) and the Giant Australian Earthworms (*Megascolecidae*).

OPISTHOSPERMOPHORA. One of the largest orders of Millepedes found in most parts of the world and characterized by having forty or more cylindrical segments each bearing two pairs of legs.

OPPOSABLE THUMB. An arrangement found only in the Primates whereby the thumb or pollex can be brought into a position facing the other digits. This is of great help in arboreal animals as it facilitates the grasping of tree-branches. In many apes and monkeys a similar arrangement is found in the hallux or first toe.

ORTHONECTIDA. Minute worm-like parasites found in the tissues of a number of invertebrates. They are classed with the Mesozoa and undergo a complex life-cycle with alternation of generations. In the asexual phase the organism is a multinucleate amoeboid plasmodium which reproduces by fission giving rise to free-swimming sexual forms. (See *Mesozoa*.)

OXYRHYNCHA. Spider Crabs: marine types with the body roughly triangular and the carapace narrowed in front to form a pseudorostrum.

OXYSTOMATA. Box-Crabs etc. Sluggish bottom-dwellers with box-like carapace. In some species the chelipeds are flattened and can be folded tightly across the face leaving two small apertures through which water is inhaled into the branchial chamber.

P

PAENUNGULATA. Also called *Subungulata:* a superorder of mammals dating from Cretaceous times and forming a miscellaneous group including Proboscidea (elephants), Sirenia (dugongs) and Hyracoidea (conies). All these, though not true ungulates, appear to be closely related to them.

PAGURIDEA. Hermit-crabs, stone-crabs and some land-crabs. Members of the *Anomura*, intermediate between crabs and lobsters and having an asymmetrical soft abdomen. In the hermit-crabs the abdomen is modified to fit into the spiral shells of whelks or similar molluscs.

PALPIGRADI. Minute arachnids found in the soil or beneath stones in caves. They differ from spiders in having a thin pale integument, long pedipalps used as legs, and a jointed tail-filament.

PANTODONTA. See *Amblypoda*.

PANTOTHERIA. Ancestral mammals of Jurassic times: probably the original Therian stock giving rise to both Marsupials and Eutherians.

PARASEMATIC CHARACTERS. Deflective marks. Features which help to protect an animal by deflecting the attacks of enemies from vital parts of the body, *e.g.* eye-like patterns on the wings of insects, on the bodies of snakes etc.

PARATENIC HOST. An intermediate host of a parasite, acting as a vector between two other hosts and not essential for the life-cycle.

PARAXONIA. An alternative name for *Artiodactyla* or even-toed ungulates in which the first digit is missing and the axis of symmetry of the foot is between the third and fourth digits.

PAUROPODA. Minute soft-bodies myriopoda resembling millepedes but having only nine pairs of legs. They are usually less than 2 mm. in length and inhabit decaying forest litter.

PECILOKONT. A flagellum or other similar or homologous structure.

PELECANIFORMES. An order of birds comprising a large number of fish-eating types including cormorants, gannets, pelicans etc. They usually nest in colonies near water and are well adapted for diving and swimming. The feet are webbed and have four toes. The mouth is usually large and, in the case of the pelican, has a large storage pouch.

PELTATE TENTACLES. See *Shield-shaped tentacles*.

PELVIS. (1) See *Pelvic Girdle*.

(2) An expanded part of the ureter into which the uriniferous tubules of the kidney open.

PENAEIDEA. Free-swimming decapod shrimps in which the third legs are chelate and the gills are branched.

PENTACULA. A hypothetical ancestor of Echinoderms: a bilateral animal with five pairs of hollow tentacles around the mouth.

PERAMELOIDEA. Bandicoots: burrowing insectivorous marsupials differing from most others in having an allantoic placenta.

PERIPATIDAE. Typical members of the *Onychophora*: primitive worm-like arthropods with soft skin and numerous clawed appendages. They apparently form an evolutionary link between Annelids and Anthropods. All present day forms are terrestrial and inhabit tropical forests. (See *Onychophora*.)

PERIPNEUSTIC. Having spiracles arranged along the sides of the body: the normal arrangement in insects.

PHAEOMELANIN. A brown form of melanin. (*cf. Eumelanin*.)

PHANEROGNATHA. A group of weevils (Curculionidae) characterized by the absence of mandibles but having conspicuous maxillae not hidden by the mentum. (*cf. Adelognatha*.)

PHARATE INSTAR. A stage in the development of an insect during which the cuticle has become separated from the hypodermis but has not yet been ruptured or cast off. A pharate pupa, for instance, is within but separated from the last larval cuticle; a pharate adult is within but free from the pupal cell or cocoon.

PHARYNGOBDELLIDAE. Terrestrial and aquatic leeches having a non-protrusible pharynx with no teeth but with one or two stylets.

PHOCAENIDAE. Porpoises: small whales belonging to the group *Odontoceti*, having uniform teeth, a single external nostril or blow-hole and a pair of broad flippers.

PHORONIDA. See *Phoronidea*.

PHREATOICIDEA. Fresh-water isopods of Australia and South Africa having a laterally compressed body, abdominal segments not coalesced and uropods in the form of stylets.

PHYLACTOLAEMATA. Fresh-water Ectoprocta (formerly classed with the *Polyzoa*) consisting of colonies of uniform zooids, each with a horse-shoe shaped lophophore and a non-calcified zooecium.

PHYLLOCARIDA. Small marine crustaceans having flattened biramous appendages and a large carapace which is not fused with the sides of the thorax. (See *Leptostraca*.)

PHYLLODOCIDAE. Free swimming polychaete worms having uniramous parapodia and flattened leaf-like oral cirri.

PHYTOSAURIA. Aquatic thecodont reptiles of the Triassic period, superficially resembling crocodiles by convergent evolution.

PILOSA. Sloths and anteaters: the hairy members of the Edentata as distinct from the armadilloes whose body is covered with bony scutes. (See *Edentata*.)

PLACODERMI. Primitive fish of the Devonian period, thought to be ancestors of the Chondrichthyes. They are among the earliest true fish but differed from most modern types in having bony plates in the skin and in having functional gill clefts in front of the hyoid arch.

PLANKTOHYPONEUSTON. Organisms which accumulate near the surface of water at night but live at lower levels during the day.

PLANKTOSPHAEROIDEA. Hemichordata which up to now are only known as planktonic larvae in the form of transparent ciliated spheres.

PLASMODROMA. Protozoa of the classes Mastigophora, Sarcodina and Sporozoa, typically having nuclei of only one kind and moving, if at all, by means of pseudopodia or flagella. The group therefore includes all protozoa other than the Ciliophora which, owing to their complexity, are regarded by some authorities as not being very closely related to the rest.

PLATYCOPA. A small order of marine Ostracoda having biramous second antennae and a single pair of trunk appendages. (See *Ostracoda*.)

PLESIOPORA. Aquatic Oligochaeta distinguished by having the male gonopores in the segment immediately following that containing the testes. They include such well known fresh-water worms as *Tubifex* and *Stylaria*.

PLOIMA. Rotifers in which locomotion is performed only by the ciliated disc, the tail being reduced, retractile or absent.

POGONOPHORA. A small phylum of aquatic invertebrates, the first of which was discovered off Indonesia in 1900. They are sessile, worm-like and live in chitinous tubes on the sea-bottom. The most characteristic feature is the 'beard', a bunch of up to 200 ciliated tentacles at the anterior end. There is no digestive tract and the method of nutrition is still somewhat of a mystery.

POLYDESMOIDEA. See *Proterospermophora*.

POLYGASTRIC. Having several gastric compartments as for instance in Ruminants. (*cf. Monogastric*.)

POWDER DOWN. Incipient feathers which fail to develop but partly disintegrate into powder.

PRIAPULIDA. Worm-like creatures, all marine, formerly included in the Annelida. They have a superficial segmentation, a terminal mouth and anus, and usually a number of warty appendages at the hind-end.

PRIMARY HOST. The host in which a parasite spends most of its adult life.

PROCRYPTIC COLORATION. See *Cryptic Coloration*.

PROECDYSIS. The period of preparation for a moult in an arthropod.

PROPNEUSTIC. A term used to denote insects whose only functional spiracles are those in the prothorax. Such a condition is rare but is found in the pupae of certain Diptera.

PROSOPORA. Aquatic or parasitic Oligochaeta distinguished by having the male gonopores in the same segment as the testes.

PROSPHASE. The whole period during which a cell prepares for mitosis.

PROTARRHENOTOKY. Producing all male offspring before producing any females.

PROTEROSPERMOPHORA (POLYDESMOIDEA). Flat-backed millepedes in which the segments are flattened dorsoventrally and have lateral carinae. Poison glands are present on alternate segments.

PROTOARTHROPODA. Primitive Arthropoda having affinity with the Annelida. They include the three phyla *Onychophora, Tardigrada* and *Pentastomida*.

PROTOCOEL. A separate coelomic cavity contained in the *Protosoma* or anterior region of a Deuterostome larva.

PROTOSOMA. The anterior of three regions into which the body of a Deuterostome larva is commonly divided. It contains a separate coelomic compartment, the *protocoel*. Such an arrangement is found, for instance, in the Hemichordata.

PROTOSTOMATA (PROTOSTOMIA). A group of phyla which appear to form a natural evolutionary line in which there are many common features as well as a similarity in embryonic development. They include flatworms, annelids, arthropods and molluscs as well as a number of smaller phyla.

PROTOTHELYTOKY. Producing all female offspring before producing any males.

PROTOUNGULATA. A miscellaneous group of ancestral ungulates together with a few present-day related types such as the African *aardvark*. They include the *Condylarthra, Tubulidentata, Notoungulata, Litopterna* and *Astrapotheria*, most of which lived from the Eocene period onwards but a few in late Cretaceous times. (See also *Orycteropodidae*.)

PROTRACTOR LENTIS MUSCLE. A muscle present in the eyes of Amphibia by which the lens can be moved forward for focusing near objects.

PSELAPHOGNATHA. Small millepedes with long leg-like maxillae and a soft integument bearing tufts of bristles.

PSEUDAPOSEMATIC CHARACTERS. False warning colours. Batesian mimicry in which a harmless animal bears a resemblance to a poisonous or dangerous one and is thus given a degree of protection since predators will tend to avoid both.

PSEUDEPISEMATIC CHARACTERS. Features which enable an animal to lure its enemies to destruction, *e.g.* the lights on some fish which attract small crustacea and molluscs on which the fish feeds; the 'fishing line' of the Angler Fish etc.

PSEUDOCARDIA. See *Pseudo-heart*.

PSEUDOCOELOMATA. Animals whose body-cavity is developed from the blastocoel and is therefore not a true coelom. They include the three phyla *Entoprocta, Aschelminthes* and *Acanthocephala*.

PSEUDOPARENCHYMA. In addition to its botanical meaning, this word is sometimes used to denote *Mesenchyma*, a syncytial network of cells whose spaces contain fluid and wandering amoeboid cells. Such tissue is found in certain primitive phyla such as the Platyhelminthes and in the embryos of higher animals.

PSEUDOPLACENTAL VIVIPARITY. A phenomenon found in certain insects in which the egg has very little yolk and the embryo is nourished through placenta-like structures in contact with maternal tissue.

PSEUDOSCORPIONIDA. False scorpions: tiny arachnids a few millimetres in length which superficially resemble scorpions and inhabit leaf mould.

PSITTACIFORMES. Parrots. See *Psittaci*.

PYROTHERIA. Large elephant-like mammals of the Eocene and Oligocene periods having a well developed trunk, two pairs of incisor tusks in the upper jaws and one pair in the lower.

R

RADIATA. A group consisting of the three phyla *Porifera, Coelenterata* and *Ctenophora* in all of which the radial symmetry is believed to be primary or primitive.

RALLIFORMES. See *Rallidae* and *Gruiformes*.

RAPHIDIOIDEA. Snake flies. See *Raphidiidae*.

REGULARIA. An alternative name for *Endocyclica* or globular Echinoderms.

RELICT FAUNA. Animals which have survived from a previous age long after most of the other animals occupying the same habitat have become extinct. Such animals are usually found in small communities and have gradually adapted themselves to changing conditions.

REPTANTIA. Decapod crustaceans in which the first pair of legs usually have large pincers and the others are strong and well adapted for crawling. The pleopods, when present, are usually small and not well suited for swimming. The group includes crabs, lobsters, prawns, crayfish etc. (See *Brachyura* and *Macrura*.)

RETICULOCYTE. An immature erythrocyte having reticular cytoplasm.

RETRACTOR. A muscle which pulls any organ or part of an organ backwards.

RETRACTOR BULBI. A muscle which pulls the eyeball in, depressing the roof of the mouth and helping in the action of swallowing. It is present in frogs and other amphibia and in some reptiles.

RETRACTOR LENTIS. A muscle present in the eyes of bony fish. It connects the lens to the back of the eye, so that when it contracts the lens is moved back enabling the fish to focus more distant objects.

RHABDOPLEURIDA. Small tube-dwelling Hemichordata living as branching colonies of zooids which remain attached to the parent. They belong to the class *Pterobranchia*.

RHYNCHOCOELA. An alternative name for the proboscis worms (*Nemertina*, *Nemertea* or *Nemertini*). The name refers to the tubular cavity or sheath which encloses the eversible proboscis.

RIBOSOMES. Minute particles lining the cytoplasmic membranes of cells and consisting largely of proteins and ribonucleic acid (RNA). Their function appears to be the building up of proteins with the help of 'Messenger RNA' which carries the *genetic code* (*q.v.*).

RICINULEI. Small spiders found in damp caves etc. in Africa and Central America, characterized by having a thick, heavy, sculptured cuticle and a curious hood-like organ which can be lowered over the mouthparts.

RNA. Ribonucleic acid. An important constituent of all cells on account of the part it plays in heredity and particularly in the *genetic code*. Its molecular structure, like that of DNA, consists of a chain of sugars, phosphate groups and bases. During the growth and reproduction of cells each DNA molecule in a gene acts as a 'template' for the building up of an RNA molecule; the latter in turn acts as a template on which a protein molecule is built up. In addition to template or 'messenger RNA' there is also another type of RNA known as 'transfer RNA' whose main function appears to be the gathering of specific amino-acid molecules and transferring them to their appropriate places on the template in order to build up particular proteins. (See also *DNA*.)

S

SALENTIA. Jumping, tailless amphibia: frogs and toads. An alternative name for *Anura*.

SARCOPTERYGII. Fish having thick fleshy fins and usually having internal nostrils. They include the *Crossopterygii* (*q.v.*) and the lungfish.

SCHIZOCOELA. A group of phyla comprising all those animals whose coelom arises by splitting of the mesoderm. They include about 17 phyla, all invertebrates.

SCHWANN CELLS. Cells which wrap themselves round nerve fibres forming the *neurolemma* or outer sheath. In medullated nerve fibres they also give rise to the fatty or myelin sheath.

SCOLOPENDROMORPHA. Large tropical centipedes, some up to one foot in length, having about 23 segments, four small ocelli on each side of the head and long antennae of about 20 segments.

SCUTIGEROMORPHA. Centipedes with 15 pairs of very long legs, long slender antennae and compound eyes.

SCYLLARIDEA. Spiny lobsters: common forms of edible lobster caught off the Gulf and Pacific coasts of America and elsewhere. The body is covered with spines and the first four pairs of legs are without chelae.

SECONDARY HOST. The host in which a parasite lives for a short time (usually in its larval stage) before being transferred to spend most of its life in the primary or definitive host.

SEDENTARIA. Polychaete worms which either burrow or dwell in tubes and rarely, if ever, swim freely. See *Polychaeta*.

SEISONACEA. Marine Rotifers, very elongated and having a poorly developed ciliary organ.

SERICIN. A water-soluble gelatinous protein forming an outer layer surrounding an inner core of fibroin in the silk threads of insects, spiders etc.

SEYMOURIAMORPHA (COTYLOSAURIA). Extinct vertebrates generally regarded as reptiles but having some of the features of Amphibia. The skull was of the *Anapsid* type, *i.e.* with a complete covering of dermal bones having no temporal fossae.

SIBLINGS. Offspring of the same parents: brothers and sisters.

SIPUNCULIDA. See *Sipunculoidea*.

SOLPUGIDA. Sun-spiders: large arachnids of tropical and semitropical deserts, having chelicerae in the form of powerful, vertically articulating pincers. They have voracious appetites and savagely tear apart any small animals which they may seize.

SOMATODERM. The outer layer of cells, usually less than 24, which surround the reproductive cell in Mesozoa.

SPIROTRICHA. Ciliated Protozoa having well developed buccal cilia but having few or none over the rest of the body.

SPONGOCOEL. An alternative name for the *paragaster*: the chief cavity in a sponge.

SQUAMAE. Small wing-lobes close to the *alulae* in certain insects such as the house-fly.

STENOGAMY. The ability to mate in a confined space.

STENOGLOSSA. An alternative name for *Neogastropoda* (*q.v.*). An order of Gasteropod molluscs having a single comb-like gill and a bipectinate *osphradium*, a sensory organ resembling a gill. Many are carnivorous and some are poisonous. See also *Streptoneura*.

STENOLAEMATA. See *Stenostomata*.

STENOPODIDEA. Free-swimming decapod shrimps in which the first three pairs of legs have chelae and the gills are filamentous.

STENOSTOMATA (STENOLAEMATA). Ectoprocta (formerly classed as *Polyzoa*) consisting of colonies of tubular zooids each having a calcified zooecium or exoskeleton fused with those of neighbouring zooids. The group is sometimes known as *Cyclostomata* but, as this name is also used for a group of Chordates, it is best for it not to be used for the Stenostomata.

STENOTHERMOUS. Adapted to a very narrow temperature range.

SUBUNGULATA. See *Paenungulata.*

SUIFORMES. Pig-like, non-ruminating even-toed ungulates. They include pigs, peccaries, hippopotami and a number of fossil types.

SUINA. Pigs and peccaries: members of the *Artiodactyla* (even-toed ungulates) having short stout limbs with four digits; having tusk-like canines which are triangular in section, the remaining teeth being bunodont. True pigs inhabit the old World; peccaries the New.

SYLLIDAE. Small free-swimming Polychaete worms with long delicate bodies and uniramous parapodia.

SYNAPOSEMATIC COLORATION. Having warning coloration common to different species of animal. See *Batesian Mimicry* and *Müllerian Mimicry.*

SYNECOLOGY. Ecology in which the plants and animals of a particular locality are regarded as a single community or ecological unit. (*cf. Autoecology.*)

SYNSACRUM. A number of fused sacral vertebrae supporting the pelvic girdle of a bird.

SYSTOLE. Muscular contraction of the heart or of any chamber of it: commonly used to refer to the ventricles of the human heart. (*cf. Diastole.*)

T

TAENIODONTA. Fossil mammals of the Eocene period having many features in common with the ground sloths: probably fore-runners of the Edentata.

TENRECOIDEA. See *Centetidae.*

TESTACIDA. Fresh-water amoebae enclosed in a single-chambered shell: sometimes known as *Monothalamia.*

TEUTHOIDEA. See *Myopsida*: Squids.

THALASSINIDEA. Marine burrowing shrimps in which the first pair of legs form unsymmetrical chelipeds.

THERAPSIDA. An order of reptiles in which the teeth, limbs and many other skeletal features resembled those of mammals. They lived from Permian to Jurassic times and can be regarded as the ancestors of mammals.

THERIA. A general name for all mammals other than Monotremes.

THEROCEPHALIA. Reptiles of the Permian period having an affinity with the ancestors of mammals but probably more primitive than the Triassic *Cynodonts.*

THIGMOTRICHIDA. A small group of marine and fresh-water Ciliophora which live in association with bivalve molluscs and attach themselves by a tuft of cilia at the anterior end.

THYMINE. A pyrimidine base characterized by a single-ring structure containing two nitrogen and four carbon atoms: one of the four bases which are combined with sugars and phosphate groups in nucleic acids.

THYROTROPHIC HORMONE. A hormone produced by the anterior lobe of the pituitary body which stimulates and maintains the activity of the thyroid gland.

TINTINNIDA. Free-swimming, usually marine protozoa belonging to the class Ciliophora, having a chitinous capsule or *lorica* into which they are anchored by an aboral process.

TONUS (TONE, TONIC CONTRACTION). The condition of a muscle which remains in a continuous state of partial contraction enabling an animal to maintain its posture.

TRAGUS. A small process resembling a second pinna inside the main pinna of the ear, *e.g.* of certain bats.

TRANSFER RNA. A form of ribonucleic acid whose molecules consist of short double helices capable of picking up specific amino-acid molecules and transferring them to their appropriate places during the building up of protein molecules. (See also *RNA*, *DNA* and *Genetic Code*.)

TRICHOSTOMIDA. Protozoa with uniform ciliation over the body but with no buccal cilia.

TRILOBITOMORPHA. See *Trilobita*.

TRITORAL. Adapted for grinding, as for instance of the surface of a tooth.

TROGONIFORMES. See *Trogones*.

TURTLE. See *Chelonidae*, *Chelydidae* and *Chelydridae*.

U

UNGUICULATA. A cohort or group of mammalian orders comprising all those which have distinct claws or nails. These are the *Insectivora*, *Dermoptera*, *Chiroptera*, *Edentata*, *Pholidota* and *Primates*.

UROPYGI. Whip-scorpions: nocturnal Arachnids of tropical and subtropical regions, ranging in size from a few millimetres to about three inches. They appear to be intermediate between spiders and scorpions, having a pair of powerful pincers in front and a whip-like flagellum at the rear of the abdomen. When irritated they elevate this and squirt a stream of poisonous fluid at their adversary.

V

VALVIFERA. Marine crustaceans of the order Isopoda in which many of the abdominal segments are fused together and the thoracic limbs bear large coxal plates.

VANADOCYTES. Green blood-cells containing the element vanadium, found in certain Ascidians.

VASO-CONSTRICTOR. Causing constriction of blood vessels, *e.g.* by the action of certain nerves or hormones.

VASO-DILATOR. Causing the enlargement of blood vessels, *e.g.* by nerve or muscle action.

VECTOR TISSUE. Specialized tissue in the bodies of certain leeches connecting particular regions of the skin to the ovisacs. When copulating the sperm, in the form of a gelatinous spermatophore, is deposited on the skin. It perforates this and makes its way along the vector tissue to the eggs.

VEGETATIVE NUCLEUS. An alternative name for *Meganucleus* (*q.v.*).

VERRUCOMORPHA. Box-like barnacles on short stalks and having a movable lid or operculum formed from one tergum and one scutum.

Z

ZOOSIS. A disease produced by animal parasites.

Corrigenda

CORPUS ALBICANS. In addition to the meaning which we have given, this word is also sometimes used synonymously with *Corpus mammillare*.

EXOSKELETON. We referred only to the exoskeletons of invertebrates. In fact, however, some vertebrates also have an exoskeleton (*e.g.* the tortoise.)

GLOW WORM. This is defined as the wingless female of the beetle *Lampyris noctiluca*. In some species, however, the males are also luminous.

GRAPTOLITES. These are now classed with the Protochordata and not with Coelenterates.

LIPOTERNA. This should read *Litopterna*.

HELMINTHOLOGY. This should read 'The study of *parasitic* worms.'

HEMICHORDATA. We stated that these had certain vertebrate features. It would be more correct to say *chordate* features, but there is some doubt as to whether the notochord is in fact homologous with that of other chordates. Some authorities have given Hemichordata the rank of a separate phylum.

ONYCHOPHORA. We stated that this group contains only a single genus. Our latest information is that there are two families containing nine genera and about 65 species in all.

OSSIFICATION. We defined this as the transformation of cartilage into bone. This is of course not necessarily the case, as ossification can take place without the previous presence of cartilage. We have treated the matter further under the heading *Membrane Bones*.

POROCYTES. We referred to the porocytes of sponges. Any cell with an intracellular passage may be called a porocyte, *e.g.* those in nephridia.

ZOOPLANKTON. We stated that this lives on or near the surface of water. In fact zooplankton is found down to the greatest depths. *Phytoplankton* (plant-plankton) is confined to the surface layers where light penetrates.

APPENDIX ON CLASSIFICATION AND NOMENCLATURE

SYSTEMATIC classification of the animals of the world dates from the publication in 1758 of the tenth edition of the SYSTEMA NATURAE of Linnaeus. In this work the Binomial System of nomenclature, previously applied to plants, was applied to the animal kingdom for the first time. According to this system every animal or plant has two names, the first being the name of the genus and the second the name of the species. Thus within the genus *Mustela* we have a number of species, *viz. Mustela erminea* (stoat), *Mustela nivalis* (weasel), *Mustela putorius* (polecat) and *Mustela furo* (ferret). There has been much argument as to the exact definition of a species but it may roughly be defined as a compact group whose members can breed with one another but cannot breed with other species to produce fertile offspring. Thus the horse, the ass and the zebra are distinct species all belonging to the genus *Equus*. The horse and the zebra will not breed together, neither will the ass and the zebra. The horse and the ass will breed but produce a sterile offspring, the mule. Since in practice it is rarely known whether two kinds of animal will in fact breed together, the definition of a species is usually based on morphological or anatomical features. This causes a certain amount of confusion because it is often difficult to say whether certain differences truly indicate distinct species or whether they merely show variations and strains within the species. Since verbal descriptions and diagrams may be insufficient or ambiguous it is customary whenever possible to keep a 'type specimen' in a museum or other suitable place so that any doubt as to the identity of a specimen can be settled by reference to the original type.

Well over a million known species of animals have been described and these are classified for the sake of convenience into a hierarchy of groups. In the original Linnaean classification this grouping was based on morphological features and much of the system is now regarded as rather artificial. Many of the Linnaean groups are still used but there has been a considerable amount of regrouping in the light of modern knowledge on genetics and evolution. An ideal classification should be a natural one based as far as possible on known evolutionary relationships and should include both fossil and present-day types.

In Linnaeus' original classification the following names of groups

312

were employed: *Empire, Kingdom, Class, Order, Genus, Species, Variety*. The first was soon dropped and two additional names added, *viz. Phylum* between Kingdom and Class; and *Family* between Order and Genus. By international agreement the following names are now obligatory in *formal* classification:

> Kingdom.
> Phylum.
> Class.
> Order.
> Family.
> Genus.
> Species.

If a phylum contains a very large number and diversity of species, additional group names are frequently used for convenience. The name *Branch* may be placed between Subphylum and Class; the name *Cohort* between Class and Order, and the name *Tribe* between Family and Genus. In addition to these the prefixes *Super-, Sub-* and *Infra-* may be added to any group name, *e.g.* Superfamily, Subphylum, Infra-order etc. In practice it is rarely necessary to employ all these. The terms *Division* and *Grade* are occasionally used to subdivide other groups but their use is not recognized as part of the *formal* classification.

As an example of the use of the above system, a full classification of the lion *Felis leo* is shown below. Some of the terms, however, are rarely if ever used.

KINGDOM.	Animalia.
SUBKINGDOM.	Metazoa.
PHYLUM.	Chordata.
SUBPHYLUM.	Craniata.
BRANCH.	Gnathostomata.
CLASS.	Mammalia.
SUBCLASS.	Theria.
INFRA-CLASS.	Eutheria.
COHORT.	Ferungulata.
SUPERORDER.	Ferae.
ORDER.	Carnivora.
SUBORDER.	Fissipedia (now discontinued).
SUPERFAMILY.	Feloidea (formerly Aeluroidea).
FAMILY.	Felidae.
SUBFAMILY.	Felinae.
TRIBE.	Felini.
SUBTRIBE.	Felina.
GENUS.	Felis.
SPECIES.	Leo.

It should be noted, however, that in recent years the lion has been renamed *Panthera leo*. The cat and the lynx are retained in the genus *Felis*; the lion, leopard and tiger are placed in the genus *Panthera*. This involves corresponding changes in nomenclature of all groups up to the level of Subfamily. Not all zoologists have accepted these changes and the name *Felis leo* is still often used.

In the naming of the various groups, particularly of the larger ones, there has been much confusion and it is only recently that a certain amount of order has been attained by the efforts of an International Commission. This was first appointed by the Zoological Congress at Leyden in 1895 and in due course published a code of recommendations which was formally ratified in 1901 at the congress held in Berlin.

The Code originally consisted of 41 articles and 20 recommendations dealing with the formation, spelling and use of names denoting families, genera and species. Since then there have been a number of additions and alterations. The present Code dates from the XVth International Congress of Zoology held in London in 1958. This contains 87 articles and a large number of recommendations designed to amplify these. Some of the principal requirements are quoted below.

Article 11

'The scientific names of animals must be either Latin or latinized, or must be so constructed that they can be treated as Latin words.'

'A genus-group name must be a noun in the nominative singular or be treated as such, *e.g. Canis, Felis* etc.'

'A species-group name may be a simple or compound word and must be treated as:

(*a*) An adjective agreeing in gender with the generic name, *e.g. Felis marmorata*.

(*b*) A noun in the nominative singular in apposition to the generic name, *e.g. Felis leo*.

(*c*) A noun in the genitive, *e.g. rosae, cuvieri* etc.'

Article 28

'Names of the family- and genus-groups must be printed with a capital initial letter and names of the species-group with a lower case initial letter.'

'The author of a scientific name is that person who first published a description of a particular species. If it is desired to cite the author's name it should follow the scientific name without the interposing of a comma or other punctuation; the name Linnaeus is commonly abbreviated to L.'

Examples. *Fasciola hepatica* L., *Anopheles maculipennis Meigen.*
With regard to the naming of groups higher than Genus, the Code
lays down the following rules:

Article 29

'A family-group name is formed by the addition of -IDAE to the
stem of the name of the type genus in the case of a family and -INAE
in the case of a subfamily, *e.g.* Canidae, Caninae.'

One difficulty which arises here is in deciding what is the stem of
a generic name. If the word is a true Latin substantive or a latinized
Greek substantive, the root is obtained by removing the ending from
the genitive singular, thus:

Genus.	*Homo* (= man) (Latin).
Genitive.	*Hominis.*
Root.	*Homin-*
Family.	*Hominidae.*

Genus.	*Dinornis* (Ornis = bird) (Greek).
Genitive.	*Ornithos.*
Root.	*Ornith-*
Family.	*Dinornithidae.*

In cases where the generic name is not a true Latin or Greek word
there may be differences of opinion as to what constitutes the stem.
The International Commission meeting in Paris in 1948 accepted the
following resolution:

'The stem need not be that used in the grammatical sense with
reference to classical Latin, but in a practical sense applicable to
Scientific Latin, having regard to simplicity and euphony.'

The International Commission has not laid down any definite
rules for the naming of categories higher than the family, but in
1916 Van Duzee suggested that the ending -OIDEA be added to the
stem to designate a superfamily and the ending -INI to designate a
tribe. These suggestions were ratified in the 1958 Code.

One difficulty about using -OIDEA for a superfamily is that certain
well established names with this ending have been long used to
indicate other groups. The names SEPIOIDEA, NAUTILOIDEA and
ECHINOIDEA, for instance, denote a suborder, an order and a class
respectively. Sometimes also the ending -OIDEA has been added to
the stems of Greek words for common animals and these have been
in use for so long that there is reluctance to discontinue them. Thus
CYNOIDEA, AELUROIDEA and ANTHROPOIDEA come from the Greek
words for dog, cat and man respectively. According to the more
modern recommendation these should be CANOIDEA, FELOIDEA and
HOMINOIDEA linking them with the Latin roots of the type genera.

Larger group names sometimes take the form of adjectives or participles with the Latin ending -*a* agreeing with the neuter plural *Animalia* (CHORDATA, ACOELOMATA, CRUSTACEA, THALIACEA). They may be plural nouns or adjectives with such endings as -*fera* (Latin) and -*phora* (Greek) both signifying 'bearer' (PORIFERA, ROTIFERA, CILIOPHORA, CTENOPHORA). Most orders of insects end in -*ptera* signifying 'wings' (HYMENOPTERA, COLEOPTERA, LEPIDOPTERA, DIPTERA etc.). Some orders of invertebrates end in -*poda* signifying 'feet' (ISOPODA, DECAPODA, CEPHALOPODA etc.).

Many of the group names of fish have the masculine plural endings -*i* or -*es* agreeing with the Latin word *pisces*, e.g. ELASMOBRANCHII, JUGULARES, GADIFORMES. Similarly the names used for birds may have a feminine plural ending agreeing with the Latin *Aves*, thus RATITAE-, CARINATAE. The higher categories of fish and birds may end in the Greek words -*ichthyes* (fish) and -*ornithes* (birds), thus OSTEICHTHYES (bony fish), CHONDRICHTHYES (cartilaginous fish), ARCHAEORNITHES (ancient birds), NEORNITHES (modern birds).

The Latin ending -*formes* signifying 'having the form of' has long been used for most of the orders of birds and some of fishes, *e.g.* PASSERIFORMES (sparrow-like), ANSERIFORMES (goose-like) and GADIFORMES (cod-like). The Greek ending -*morpha* has the same meaning as the Latin -*formes*, thus THEROMORPHA (animal-like).

We see therefore that there is very great diversity and until the International Commission establishes rules for the higher categories it is recommended that the first name proposed for a group in an unambiguous manner should be accepted regardless of its ending.

Transliteration of Greek Words

Endings

When changing a Greek word to a corresponding Latin or latinized form, the ending of a word is usually changed to the corresponding Latin ending in the nominative of the same gender. The Latin plural is also used. Thus:

Masculine ending. Greek -ος, Latin -us. e.g. θαλλος = Thallus (a young shoot); Pl. Thalli.

Feminine ending. Greek -η, Latin -a. e.g. θηκη = Theca (case, box); Pl. Thecae.

Neuter ending. Greek -ον, Latin -um. e.g. Τυμπανον = Tympanum (drum); Pl. Tympana.

Vowels

For the most part these become the corresponding Latin vowels. Thus:

$$a = a.$$
$$\omega \text{ or } o = o.$$
$$\iota = i.$$
$$\epsilon = e.$$

η = e except at the end of a word (see above).

It should be noted, however, that the Greek letter υ is always changed to y, thus:

δακτυλος (finger) = dactyl.
πτερυζ (wing) = pteryx.
φυλλον (leaf) = phyll.

Dipthongs

αι = ae, χαιτη (hair): Chaeta.

ει = i, χειρ (hand): Chiroptera (bats).

οι = oe, or -e, οικος (home): Dioecious, Ecology.

ου = u, βουνος (mound, hillock): Bunodont.

ευ and αυ usually remain unchanged as eu and au, γλαυκος (sea-green); Latin Glaucus, English glaucous.

Consonants

The majority of consonants remain unchanged, but θ = th, ϕ = ph, χ = ch, ψ = ps. κ usually becomes c, κοιλομα (hollow): coelom. ρ becomes rh at the beginning of a word, but r in the middle of a word, ροδον (rose): rhodo-; μικρος (small): micro-.

Double Consonants

γ before a consonant usually becomes an n, thus $\gamma\zeta = nx$, $\gamma\gamma = ng$, $\phi\alpha\lambda\alpha\gamma\zeta$, $\phi\alpha\lambda\alpha\gamma\gamma\epsilon\varsigma = phalanx$, *phalanges*.

$\gamma\chi = nch$, $\beta\rho\alpha\gamma\chi\iota\alpha$ (gills): *branchia*.

References

For further information on general aspects of classification and nomenclature the reader is referred to the following works:

International Code of Zoological Nomenclature, published in English and French by the International Trust for Zoological Nomenclature. (London, 1961.)

SCHENK AND MCMASTERS. *Procedure in Taxonomy*. (Stanford University Press, California, 1948.)

MAYR, GORTON LINSLEY AND USINGER. *Methods and Principles of Systematic Zoology*. (McGraw-Hill Inc. New York, 1953.)

JULIAN S. HUXLEY (Ed.) *The New Systematics*. (Oxford, 1940.)

T. H. SAVORY. *Latin and Greek for Biologists*. (University of London Press, 1946.)

G. G. SIMPSON. *The Principles of Classification and a Classification of Mammals*. (Bull. Amer. Mus. Nat. Hist. No. 85. 1945.)

LORD ROTHSCHILD. *A Classification of Living Animals*. (Longmans, 1961.)

N. W. PIRIE. *Preliminary List of New Words in Biology and Related Subjects*. (J. Inst. Biol. Vol. 11, No. 3. 1964.)

N. W. PIRIE and P. W. TALBOYS. *Second List of New Words in Biology*. (Ibid. Vol. 12, No. 4. 1965.)

N. W. PIRIE and P. W. TALBOYS. *Third List of New Words in Biology*. (Ibid. Vol. 13, No. 4. 1966.)